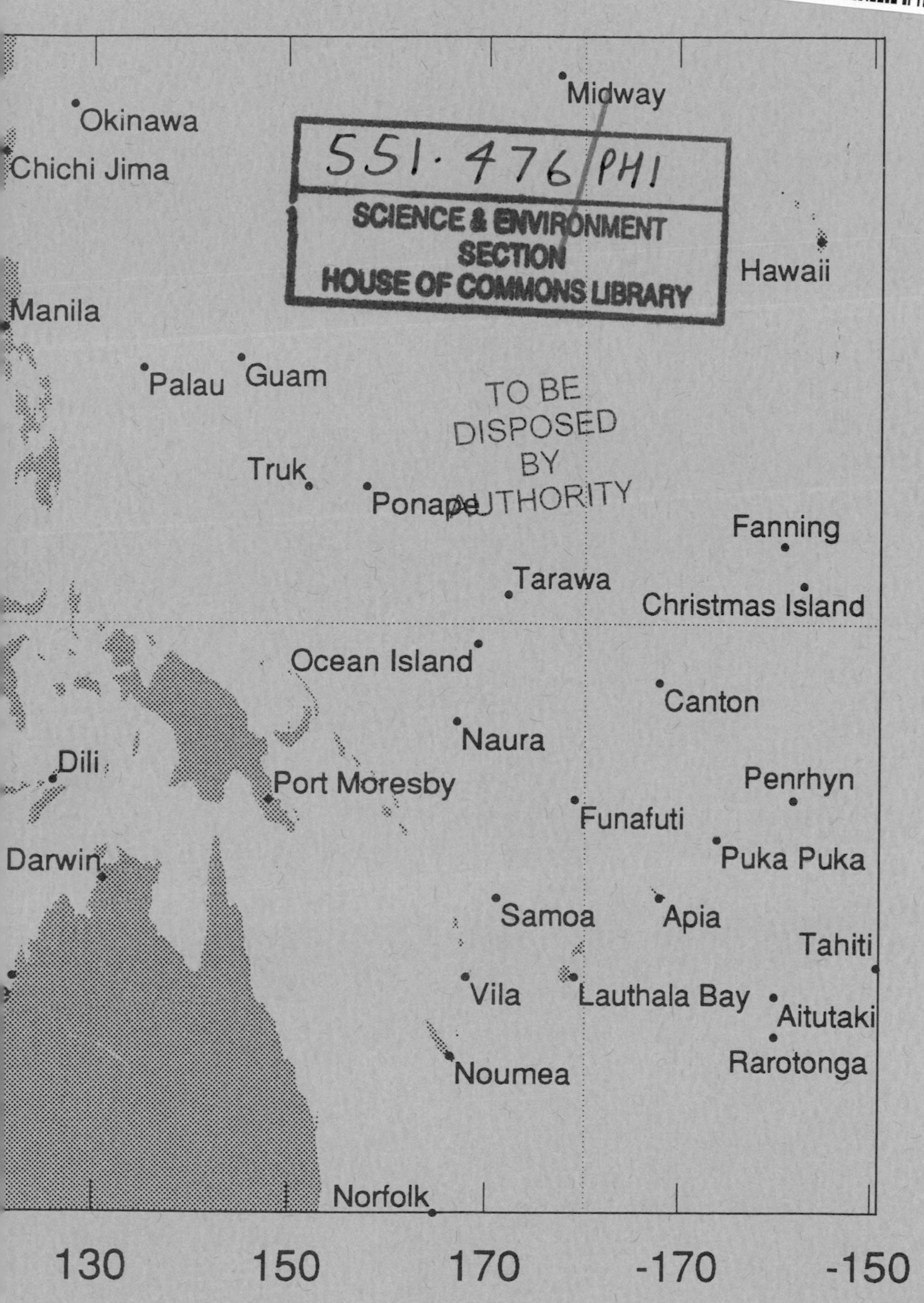
Midway
Okinawa
Chichi Jima
Hawaii
Manila
Palau
Guam
Truk
Ponape
Fanning
Tarawa
Christmas Island
Ocean Island
Canton
Naura
Dili
Port Moresby
Penrhyn
Funafuti
Darwin
Puka Puka
Samoa
Apia
Tahiti
Vila
Lauthala Bay
Aitutaki
Rarotonga
Noumea
Norfolk
130
150
170
-170
-150

El Niño, La Niña, and the Southern Oscillation

This is Volume 46 in the
INTERNATIONAL GEOPHYSICS SERIES
A series of monographs and textbooks
Edited by RENATA DMOWSKA and JAMES R. HOLTON

A complete list of the books in this series appears at the end of this volume.

El Niño, La Niña, and the Southern Oscillation

S. George Philander
Geophysical Fluid Dynamics Laboratory / NOAA
Princeton University
Princeton, New Jersey

ACADEMIC PRESS, INC.
Harcourt Brace Jovanovich, Publishers
San Diego New York Berkeley Boston
London Sydney Tokyo Toronto

Cover photograph courtesy of Central Weather Bureau Taiwan, Republic of China; provided by L. Miller.

Academic Press, Inc.
San Diego, California 92101

United Kingdom Edition published by
Academic Press Limited
24–28 Oval Road, London NW1 7DX

Library of Congress Cataloging-in-Publication Data

Philander, S. G. H. (S. George H.)
El Niño, La Niña, and the southern oscillation / by S. G. H. Philander.
p. cm. — (International geophysics series)
Bibliography: p.
Includes index.
ISBN 0-12-553235-0 (alk. paper)
1. El Niño Current. 2. Southern oscillation. 3. Global warming.
I. Title. II. Title: El Niño and the southern oscillation.
III. Series.
GC296.8.E4P48 1989
551.47'6—dc20 89-31225
CIP

Printed in the United States of America
89 90 91 92 9 8 7 6 5 4 3 2 1

Contents

Preface

Northern America suffered an exceptionally hot and dry summer in 1988. India occasionally experiences a disastrous failure of the monsoons. In Africa, severe droughts can persist over large areas for extended periods. These are all examples of climate variability that is caused by complex interactions between the atmosphere and the water, ice, and land surfaces beneath it. Efforts to understand and simulate these interactions, especially those between the ocean and atmosphere, have thus far focused on a phenomenon known as the Southern Oscillation, an irregular interannual fluctuation between warm El Niño and cold La Niña states. This oscillation has its largest signature in and over the tropical Pacific and Indian Oceans, but it affects oceanic and atmospheric conditions globally. The Southern Oscillation attracted enormous public attention in 1983 when its warm El Niño phase attained an exceptionally large amplitude and was associated with devastating droughts over the western tropical Pacific, torrential floods over the eastern tropical Pacific, and damaging weather patterns over various parts of the world. That event caught oceanographers and meteorologists completely by surprise. When a group of experts met in Princeton, New Jersey in October 1982 to discuss plans for a program to study El Niño, no one was aware that the most severe episode of the past century was occurring at that time. Although the interactions between the ocean and atmosphere that cause the Southern Oscillation were reasonably well understood by 1982, little had been done to put that knowledge to practical use. Matters were very different by the time of the next El Niño in 1987. By then the National Meteorological Center in Washington, D.C. had started to issue a monthly bulletin that describes, in detail, current oceanic and atmospheric conditions related to the Southern Oscillation. It was possible to follow the erratic development of El Niño of 1987 as it occurred. Such information is now routinely available because the tropical Pacific is being monitored with a variety of instruments. The data arrive, by satellite, at various centers shortly after the measurements are made. The oceanographic data, which complement the meteorological data available on the Global Telecommunication System, are far too sparse to paint a coherent picture of the currents and density field of the tropical Pacific. A realistic General Circulation Model of the ocean is therefore being used to integrate the measurements and to provide maps of oceanic conditions, the equivalent of weather maps.

Not only was it possible to follow the development of El Niño of 1987 month by month, but coupled ocean–atmosphere models, developed since 1982, succeeded in predicting the event a few months in advance. The models can be used to determine whether or not a warm event is likely to develop during the coming months, but at this stage they are too crude to predict how an event will evolve or what amplitude it will attain. Sophisticated models capable of realistic simulations are under development and will shortly be available for routine predictions. It is hoped that success in explaining and predicting the Southern Oscillation will lead to similar success with other phenomena that contribute to the broad spectrum of climate variability.

This book summarizes what is currently known about El Niño, La Niña, and the Southern Oscillation, and about the interactions between the oceans and atmosphere that cause this phenomenon. It also identifies some of the many remaining problems in this area of active research. The introduction, a brief history of studies of this topic, describes how, for approximately 100 years, the meteorological investigations of the Southern Oscillation and the oceanographic investigations of El Niño proceeded independently until Professor J. Bjerknes of the University of California, Los Angeles pointed out that these are two aspects of the same phenomenon. The first two chapters describe measurements of the atmospheric and oceanic variability associated with the Southern Oscillation. The next three chapters discuss theories that explain the measurements. Progress since Bjerknes' studies has been marked by fruitful interactions between observationalists and theoreticians. To retain this aspect, the chapters on measurements include qualitative explanations and the chapters on theories describe some of the measurements that motivated the oceanographic and meteorological models. The last chapter concerns interactions between the ocean and atmosphere and explains the basis for long-term predictions.

I have assumed that my readers, like the Reverend Sydney Smith—"give me a short book on an inescapable subject"—desire brevity. To this end, the book has numerous figures, footnotes that guide the reader to papers with more detailed information, and an extensive list of references.

Progress in our studies of the oceans and atmosphere is dependent on close collaboration among various groups of scientists: oceanographers, meteorologists, observationalists, and theoreticians. To coordinate the numerous individual efforts it is necessary to organize large programs. They are usually known by mundane or unpronounceable acronyms such as GATE, INDEX, and FGGE. A participant in one of these programs is not as romantic a figure as the traditional oceanographer, an individualist who explores the unknown from ships with such evocative names as *Challenger*, *Meteor*, and *Atlantis*, but the rewards are richer. The frequent meetings to prepare plans and to synthesize results afford numerous opportunities to form close professional and personal ties across institutional and national boundaries. This book is dedicated to all the participants in the various large programs whose unselfish cooperation contributes to the considerable progress we are making in understanding the oceans and atmosphere. A special, warm word of thanks to my friends and colleagues at the Geophysical Fluid Dynamics Laboratory (Princeton), where they have created a nearly perfect environment for research; to Don Hansen, David Halpern, Ants Leetmaa, Stan Hayes, and Gene Rasmusson, who make EPOCS an enjoyable and rewarding experience; to Eli Katz, Robert Weisberg, Jacques Merle, Philippe Hisard, Christian Colin, Silvia Garzoli, and Phil Richardson, whose enthusiasm for measuring the ocean motivates us all; to Dennis Moore, who introduced many of us to this topic and who, together with Jay McCreary, Jim O'Brien, and David Anderson, made sure that meetings of the INDEX Theoretical Panel were lively and stimulating forums for the discussion of many of the ideas described here; and to Mark Cane and Ed Sarachik, who always question and who sometimes have answers.

I am indebted to Stan Hayes and David Neelin for reading the manuscript and offering many helpful suggestions for improvements. The following journals, publishers, and organizations granted me permission to reproduce previously published figures: the American Meteorological Society, the American Geophysical Union, Elsevier Science Publishers, Gordon and Breach Science Publishers, Pergamon Journals Limited, Wiley Interscience, *Science*, *Nature*, *Tellus*, the Royal Meteorological Society, and the *Journal of Marine Research*. The writing of this book has been an entirely private undertaking at home, and the preparation of the manuscript and of the original figures has been at my own expense.

George Philander

Introduction

In the year 1891, Señor Dr. Luis Carranza, President of the Lima Geographical Society, contributed a small article to the Bulletin of that Society, calling attention to the fact that a counter-current flowing from north to south had been observed between the ports of Paita and Pacasmayo.

The Paita sailors, who frequently navigate along the coast in small craft, either to the north or the south of that port, name this counter-current the current of "El Niño" (the Child Jesus) because it has been observed to appear immediately after Christmas.

As this counter-current has been noticed on different occasions, and its appearance along the Peruvian coast has been concurrent with rains in latitudes where it seldom if ever rains to any great extent, I wish, on the present occasion, to call the attention of the distinguished geographers here assembled to this phenomenon, which exercises, undoubtedly, a very great influence over the climatic conditions of that part of the world.[1]

"That part of the world" is usually a barren desert adjacent to a cold ocean that teems with fish and other forms of marine life.[2] The warm, southward El Niño current moderates the low sea surface temperatures during the early months of the calendar year. Every few years the current is more intense than normal, penetrates unusually far south, is exceptionally warm, and is accompanied by very heavy rains.[3] This was the case in 1891 when

it was then seen that, whereas nearly every summer here and there there is a trace of the current along the coast, in that year it was so visible, and its effects were so palpable by the fact that large dead alligators and trunks of trees were borne down to Pacasmayo from the north, and that the whole temperature of that portion of Peru suffered such a change owing to the hot current that bathed the coast.[4]

Such years were known as *años de abundancia* (years of abundance) when

the sea is full of wonders, the land even more so. First of all the desert becomes a garden. . . . The soil is soaked by the heavy downpour, and within a few weeks

> the whole country is covered by abundant pasture. The natural increase of flocks is practically doubled and cotton can be grown in places where in other years vegetation seems impossible.[5]

Other wonders that an unusually intense El Niño current sometimes brings to the barren shores of Peru include long yellow and black water snakes, bananas, and coconuts. On such occasions, however, the birds and marine life that are usually abundant disappear temporarily.

Not until the 1960s did oceanographers realize that the unusually warm surface water off the coast of Peru during *años de abundancia* extends thousands of kilometers offshore and is but one aspect of unusual conditions throughout the upper tropical Pacific Ocean. The tide-gauge data analyzed by Professor Wyrtki of the University of Hawaii provided one of the first indications that the interannual appearance of exceptionally warm surface waters off the coast of Peru is a consequence of changes in the circulation of the entire ocean basin in response to changes in the surface winds that drive the ocean. To explain El Niño it is necessary to explain how the ocean adjusts to the changes in the surface winds. Originally, studies of the response of the ocean to variable surface winds concentrated on the Indian Ocean rather than the Pacific. When Sir James Lighthill of Cambridge University first calculated, in the 1960s, that the time it takes the ocean to adjust to a change in the winds should decrease rapidly with decreasing latitude, attention focused on the Somali Current. Its seasonal reversal, in response to the reversal of the monsoon winds, was viewed as a prime example of rapid oceanic adjustment in low latitudes and it motivated several measurement programs. This happened before it was fully appreciated that El Niño is also an example of the rapid response of the equatorial oceans to changing winds. Over the past two decades, studies of the seasonal and interannual variability of each of the three tropical oceans have contributed to considerable progress in our understanding of the dynamics of the oceans in low latitudes. This has led to convincing explanations for El Niño and to the development of models capable of simulating this phenomenon realistically.

From an oceanographic point of view El Niño is caused by changes in the surface winds over the tropical Pacific Ocean. But what causes the interannual wind fluctuations? Efforts to describe these wind variations, and more generally efforts to document interannual variations in the circulation of the tropical and global atmosphere, started towards the end of the nineteenth century. Sir Gilbert Walker initiated this research, which was originally motivated not by El Niño but by the occasional disastrous failures of the monsoons. Walker became the Director-General of Observatories in India in 1904, shortly after the famine of 1899 when the monsoons failed. He set out to predict the interannual variations of the monsoons, an activity begun by his predecessors after the catastrophic drought and famine of 1877. Walker was probably unaware that both 1877 and 1899 were *anõs de abundancia* along the coast of Peru—he was probably unaware of El Niño—but he knew of evidence that interannual pressure fluctuations over the Indian Ocean and eastern tropical Pacific are out of phase: "When pressure is high in the Pacific Ocean it tends to be low in the Indian Ocean from Africa to Australia." This irregular fluctuation, which he named the Southern Oscillation,[6] reinforced Walker's view that the monsoons are part of a global phenomenon and he set out to document its full scope in the hope that it held the key to predictions of the monsoons. To accomplish this, Walker organized a team of statisticians and clerks to compute correlations between numerous variables from stations around the world. The variables included surface pressure, temperature, rainfall, sunspots, and even the magnitude of the annual flood of the Nile. Papers published between 1923 and 1937 synthesized the results and established that the Southern Oscillation is correlated with major changes in the rainfall patterns and wind fields over the tropical Pacific and Indian Oceans, and with temperature fluctuations in southeastern Africa, southwestern Canada, and southeastern United States of America. However, attempts to translate these findings into predictors of the monsoons failed. Walker's contemporaries expressed doubts about these statistical relations inferred from relatively short records, with no a priori hypotheses or theoretical support. Many years later the analyses of much longer records would resoundingly vindicate Walker, but because of the skepticism of his colleagues and because early investigators were unable to identify the physical processes implied by the persistence, over several seasons, of anomalous conditions associated with the Southern Oscillation, interest in this phenomenon waned during the decades following Walker's publications.

The sea surface temperature data available to Walker were inadequate to determine whether the ocean is involved in the Southern Oscillation. A better data set, and an act of nature that coincided with the last two years of Walker's life, finally set investigators on the trail that linked the Southern Oscillation to oceanographic variations in the tropical Pacific. In 1957 and 1958, during the International Geophysical Year, both atmospheric and oceanographic conditions in the tropical Pacific were remarkably anomalous. The coast of Peru experienced an exceptionally strong *año de abundancia*, the strongest since 1941–1942. It was noted that the warm surface waters were not confined to the coast of South America but extended far westward to the date line. This coincided with weak trade winds and heavy rainfall in the central equatorial Pacific, a normally arid region. Walker and others had previously documented a relation between weak winds and heavy rains in this region but the relation with unusually warm surface waters had not been noted before.

Professor Bjerknes of the University of California, Los Angeles proposed that the coincidence of unusual oceanographic and meteorological conditions was not unique to 1957–1958 but occurs interannually. This is indeed the case. At present the term *El Niño* describes not a local seasonal current off the coast of Peru but the infrequent *años de abundancia* and the associated changes in the circulations of the tropical Pacific and the global atmosphere. El Niño is that phase of the Southern Oscillation when the trade winds are weak and when pressure is low over the eastern and high over the western tropical Pacific. Not only has our use of the term changed, but our view of El Niño has become pejorative. El Niño is now associated primarily with ecological[7] and economic[8] disasters that coincide with devastating droughts over the western tropical Pacific, torrential floods over the eastern tropical Pacific, and unusual weather patterns over various parts of the world. El Niño episodes are separated by generally benign periods when oceanic and atmospheric conditions are complementary to those during El Niño. The term *La Niña*[9] is apposite for this other phase of the Southern Oscillation when sea surface temperatures in the central and eastern tropical Pacific are unusually low and when the trade winds are very intense.

In 1969 Bjerknes proposed a physical relation between the interannual oceanographic and meteorological variations in the tropical Pacific. He explained how dry air sinks over the cold water of the eastern tropical Pacific and flows westward along the equator as part of the trade winds. The air is warmed and moistened as it moves over the progressively warmer water until it reaches the western tropical Pacific, where it rises in towering rain clouds. Return flow in the upper troposphere closes this Walker Circulation, a term Bjerknes introduced. He proposed that the sea surface temperature gradients—the cold water off Peru and the warm water in the western tropical Pacific—are necessary for the atmospheric pressure gradients that drive the Walker Circulation. A warming of the eastern tropical Pacific Ocean weakens the Walker Circulation and causes the convective zone of heavy rainfall to move eastward, from the western into the central and eastern tropical Pacific. In other words, the Southern Oscillation is caused by the interannual sea surface temperature variations of the tropical Pacific. Recent calculations with realistic General Circulation Models of the atmosphere confirm that this is indeed the case.

Interannual sea surface temperature variations in the tropical Pacific cause the Southern Oscillation, but from an oceanographic point of view the sea surface temperature changes are caused by the surface wind fluctuations associated with the Southern Oscillation. From these circular arguments Bjerknes inferred that interactions between the ocean and atmosphere are at the heart of the Southern Oscillation. He described how an initial change in the ocean could affect the atmosphere in such a manner that the altered meteorological conditions in turn induce oceanic changes that reinforce the initial changes. For example, a slight relaxation of the trade winds, which drive the warm surface waters westward and expose cold water to the surface in the east, can cause a modest warming of the central and eastern tropical Pacific. This in turn can cause a further relaxation of the winds and further warming so that El Niño gradually develops. These arguments can also be reversed to explain the evolution of La Niña. Bjerknes proposed "a never-ending succession of alternating trends by air–sea interaction in the equatorial belt" as the cause of the Southern Oscillation, but he was uncertain about the mechanisms that cause a turnabout from its warm to its cold phase.

No element of Bjerknes' synthesis was entirely new but he organized the intertwined elements of the Southern Oscillation, El Niño, and large-scale air–sea interactions into a new conceptual framework supported by plausible dynamic and thermodynamic reasoning. He allowed that some of his arguments were tenuous, but research since the early 1970s has given them a firm basis and has explored new aspects not considered by Bjerknes. Of particular importance is the recent development of models that simulate the interactions between the ocean and atmosphere. Studies with these models reveal that the interactions can support modes of oscillation that are features strictly of the coupled ocean–atmosphere system. Many of these modes are interannual fluctuations between warm El Niño and cold La Niña conditions and hence correspond to possible Southern Oscillations. In simple models that capture one of these modes, the simulated Southern Oscillation is periodic, and hence is perfectly predictable, provided there is no "weather," in other words no "random," high-frequency atmospheric disturbances unrelated to sea surface temperature changes. The introduction of weather into such a model can disrupt the regular oscillation, thus causing a realistically irregular, interannual Southern Oscillation.

If this were all there was to the story, then it would end on a rather discouraging note because the unpredictability of weather beyond a few days would imply that phenomena such as El Niño and La Niña are unpredictable. There is more to the story, however. The strength of the coupling between the ocean and atmosphere, and hence the amplitude of the interannual ocean–atmosphere mode, varies with time and this causes the effect of weather on the mode to vary with time. The extent to which a regular mode can be disrupted depends on the phase of the seasonal cycle and, for a given month, can be different in different years. Models that succeed in predicting El Niño succeed in identifying those occasions when interactions between the ocean and atmosphere are so strong that random atmospheric disturbances are unlikely to affect an ocean–atmosphere mode. At the moment the models are sufficiently accurate to anticipate when El Niño will evolve, but they are too crude to predict how it will evolve or what amplitude it will attain. More sophisticated models that are now being developed will be available shortly for detailed predictions.

The fascinating story of how independent lines of oceanographic and meteorological research elegantly converged to reveal interactions between the ocean and atmosphere has not ended yet. Many aspects of the Southern Oscillation are poorly understood. Further studies are needed to determine whether it does indeed correspond to a periodic ocean–atmosphere mode which, in the presence of weather, becomes irregular. Theory indicates that such modes are impossible in ocean basins as small as the Atlantic. A phenomenon very similar to El Niño nonetheless occurs in that ocean. Much remains to be explained. To promote studies of the Southern Oscillation and, more generally, to further investigations of the seasonal and interannual variability of the oceans and atmosphere, a ten-year, international research program called TOGA[10] was launched in 1985. One of its goals is the development of coupled ocean–atmosphere models for the routine prediction of Southern Oscillation conditions. It is hoped that success in understanding, simulating, and predicting this phenomenon will lead to similar success with other aspects of global climate variability.

Notes

1. These are remarks from Señor Federico Alfonso Pezet's address to the Sixth International Geographical Congress in Lima, Peru in 1895.

2. The ocean off the western coast of South America is one of the most productive regions of the world ocean. The prevailing equatorward winds drive the surface waters offshore and cause the upwelling of deep, cold water that is rich in inorganic plant nutrients such as nitrate, phosphate, and silicate. This continuous injection of nutrients into the surface layers, where optimal light conditions prevail, sustains the high rate of primary production. Abundant phytoplankton are eaten by herbivores and, through grazing and predation, the organic matter passes up the food chain.

During El Niño there is decreased primary production, which disrupts the food chain and contributes to the reproductive failure of some species. The reason, surprisingly, is not cessation of upwelling. The problem is that the upwelled water comes from relatively shallow depths. Normally the cold, nutrient-rich water is close to the surface, but during El Niño the depth of the surface layer of warm, nutrient-poor water increases so that the deeper, richer water is inaccessible. Gardiner-Garden (1987) notes that the depth from which the upwelled water comes depends on the offshore shear of the wind. During El Niño the winds close to the coast remain strong but offshore the winds relax (Enfield, 1981a). Such a shear in the wind induces shallow upwelling. If the wind had no shear, then the depth of the warm surface layer would not be a factor; upwelling would always bring deep water to the surface. [Barber and Chavez (1986) provide further references on the biology of the coastal zone.]

3. In March 1925, 394.4 mm of rain (over 200 mm in one 3-day period) fell on Trujillo (8°S, 70°W), which had received only 17.9 mm in the preceding five years. The heavy rains are generally restricted to the northern part of Peru. It is very unusual for Lima (12°S, 78°W) to receive heavy rainfall even during a major El Niño episode, but it can experience flooding because the rains can cause the level of the Rimak River to rise.

4. See Note 1.

5. These quotations are from letters written by Mr. S. M. Scott of Florence, Italy, in April 1925 and by Mr. H. Twiddle in 1922, in which they recall conditions they witnessed in Peru in 1891. The letters are quoted by Murphy (1926) in his account of the 1925 *año de abundancia.*

6. Walker introduced the name Southern Oscillation to avoid confusion with the North Atlantic Oscillation (which is a tendency for pressure to be low near Iceland in winter when it is high near the Azores and over southwestern Europe) and the North Pacific Oscillation (which is the tendency for pressure variations over Hawaii to be opposed to those over Alaska and Alberta).

7. Several factors contribute to the decimation of the bird and marine life along the coast of South America during El Niño. One is reduced primary production, which affects the entire food chain (Note 2). As El Niño evolves, nutrient-rich, cool water occupies a smaller and smaller volume, usually close to the coast. This causes some species to redistribute themselves spatially. For example, Peruvian anchovy (*Engraulis ringens*), which seek out and remain in relatively cool water of 16°C to 18°C, a temperature at which phytoplankton are abundant, become more concentrated in small pockets of cold water. Natural predators, especially humans, can then reduce their numbers significantly. Some will survive a mild El Niño, but during a severe El Niño, such as that of 1982–1983, the cool pockets disappear altogether and huge numbers of anchovies perish. The anchovy catch off Peru in 1983 was less than 1% of the catch a decade earlier.

Not all species are adversely affected by the change in environment during El Niño. Along the coast of Peru both shrimp and the scallop *Argopecten purpuratos* rapidly increased in abundance during El Niño of 1982–1983. The scallops were local and probably increased because of excellent growing conditions; the shrimps were probably carried southwards by the currents [Barber and Chavez (1986)].

Ecological damage during El Niño is not confined to the coast of South America. In 1982–1983 relatively warm, nutrient-poor water appeared off the coast of California and contributed to a reproductive failure of the northern anchovy (*Engraulis mordax*) and other species (Fiedler, 1984). In the central equatorial Pacific much of the usual marine life disappeared and the bird populations of several islands were decimated. Not only the food of the birds disappeared but nests in sand bars were flooded by the heavy rains (Schreiber and Schreiber, 1984). Farther west, a severe drought in Indonesia contributed to a fire that ravaged vast tropical forests in 1982. Caviedes (1984) describes in detail the ecological and economic damage wrought in South America by El Niño of 1982–1983.

8. The *New York Times* of 2 August 1983 and the *Los Angeles Times* of 17 August 1983 include detailed estimates of the worldwide economic impact of El Niño of 1982–1983. The country that suffers the most from El Niño is probably Peru. Before the onset of El Niño of 1972 it supplied 38% of the world's fish meal, which is used in animal feed; its catch then shrank from 10.3 to 1.8 million metric tons in just two years. Without fish to eat, many seabirds died, which was a disaster to the guano industry. In the United States the price of soybeans, used as a fishmeal substitute, more than tripled in 1972. The soaring cost of feed in turn contributed to an increase in the retail price of chicken. Another example of how El Niño

affects the world economy is the considerable increase in the price of coconuts in 1983 because of the drought in the Philippines. This increased the cost of making soaps and detergents.

9. The unfortunate term *Anti-El Niño* is sometimes used for La Niña. In Spanish, *el niño* and *la niña* mean "the boy" and "the girl," respectively. The term *El Niño* refers to the Child Jesus, as is evident from the quotation at the beginning of this introduction. The term *Anti-El Niño* that is sometimes used for La Niña is clearly an unfortunate if not apocalyptic choice. The term *El Viejo* (the old man) (Butler, 1988), instead of La Niña, will appeal to those who believe in reincarnation.

10. TOGA is an acronym for tropical oceans and global atmosphere. The acronym ENSO is often used for El Niño and the Southern Oscillation.

Chapter 1 The Southern Oscillation: Variability of the Tropical Atmosphere

1.1 Introduction

Towards the end of the nineteenth century Hildebrandsson (1897) noticed that the atmospheric pressure fluctuations at Sydney, Australia are out of phase with those at Buenos Aires, Argentina. A few years later Lockyer and Lockyer (1902a), father and son, confirmed this and estimated the period of the oscillation to be approximately 3.8 years. Their analyses of additional data from 95 stations around the world revealed that the oscillation was almost global in extent (Lockyer and Lockyer, 1902b, 1904). The map of the pressure fluctuations that appears in their paper of 1904 is, in its gross aspects, very similar to that in Fig. 1.1, which shows that the oscillation has two centers of action, one over the western tropical Pacific and eastern Indian Ocean and the other over the southeastern tropical Pacific. Fluctuations at these two centers, which are thousands of kilometers apart, are remarkably coherent and are out of phase. This is evident in Fig. 1.2, which also shows that the interannual fluctuations are very irregular in time. Sir Gilbert Walker named these fluctuations the Southern Oscillation. In collaboration with Bliss and others he established that the Southern Oscillation involves far more than a seesaw in the surface pressure difference across the Pacific Ocean. It is associated with major changes in the rainfall patterns and wind fields of the tropical Indian and Pacific Oceans and is correlated with meteorological fluctuations in other parts of the globe (Walker, 1923, 1924, 1928; Walker and Bliss, 1930, 1932, 1937). The important relation between the Southern Oscillation and sea surface temperature variations in the tropical Pacific was not discovered until the 1960s

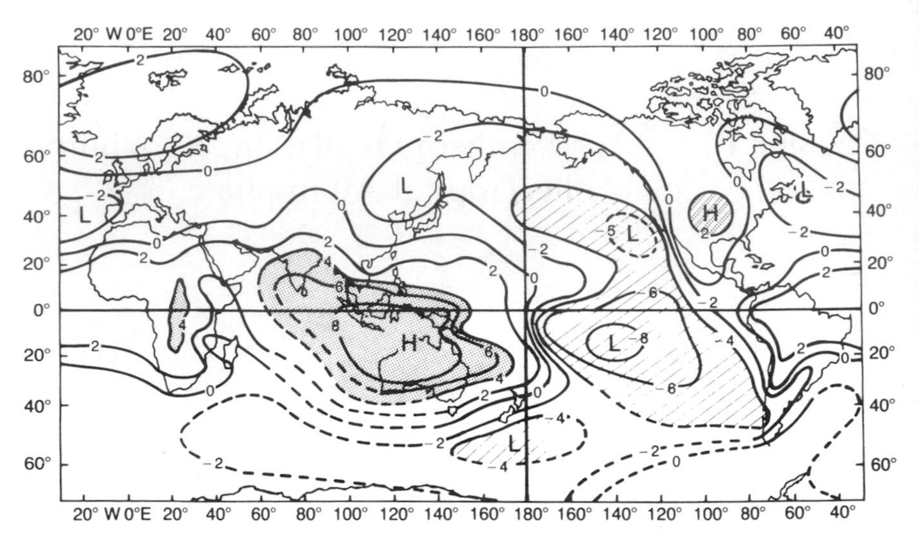

Figure 1.1. Correlations (×10) of annual mean sea level pressure with the pressure at Darwin. Correlations exceed 0.4 in the shaded regions and are less than −0.4 in the regions with dashed lines. [From Trenberth and Shea (1987).]

in studies by Ichiye and Petersen (1963), Berlage (1966), and Doberitz (1968). The correlations between the various parameters establish that high surface pressure over the western and low surface pressure over the southeastern tropical Pacific coincide with heavy rainfall, unusually warm surface waters, and relaxed trade winds in the central and eastern tropical Pacific. This phase of the Southern Oscillation is known as El Niño. Although some

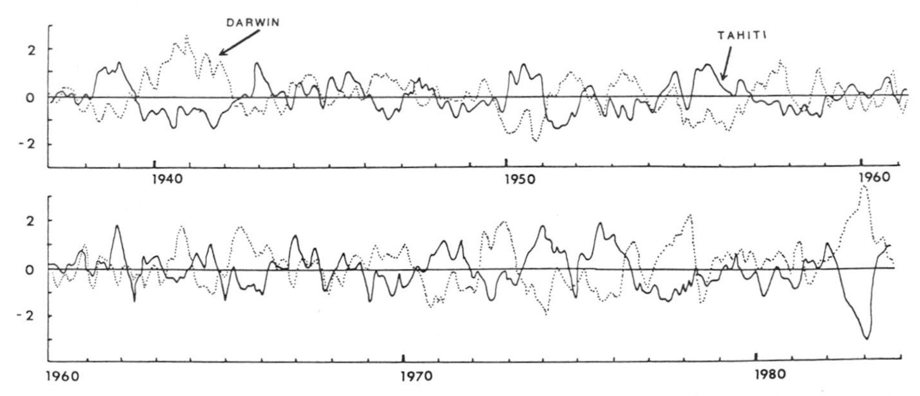

Figure 1.2. Sea level pressure fluctuations between 1937 and 1983 at Tahiti (solid line) and Darwin (dotted line) in units of standard deviations for the respective stations. (Data provided by the Climate Analysis Center, NOAA, Washington, D.C.)

descriptions give the impression that El Niño is a temporary departure from some "normal" condition of the tropical Pacific, this is inaccurate. Normal conditions can be defined statistically, but it is clear from Fig. 1.2 that the Pacific is usually not in a "normal" state. It is either in one phase of the Southern Oscillation, known as El Niño, or in the complementary phase for which the term *La Niña* is apposite. During La Niña, surface pressure is high over the eastern but low over the western tropical Pacific, while trades are intense and the sea surface temperatures and rainfall are low in the central and eastern tropical Pacific.

The terms *El Niño* and *La Niña* cover a wide range of conditions. For example, it is evident in Fig. 1.2 that the amplitudes of different El Niño episodes vary enormously. This prompted Quinn *et al.* (1978) to introduce four El Niño categories—strong, moderate, weak, and very weak[1]—but there are still considerable differences within each category. The strong El Niño of 1972 was more intense than that of 1976 but it was of shorter duration; it lasted for approximately 18 months whereas warm conditions lingered for several years after 1976. In 1972 and 1976 El Niño evolved in a similar manner, which is described in Section 1.4, but the exceptionally intense El Niño of 1982–1983 evolved very differently, as did the one of 1941. The term *El Niño* clearly covers diverse phenomena, as do *La Niña* and the *Southern Oscillation*.

There are relatively brief periods when none of these terms adequately describes conditions in the tropical Pacific. For example, the pressure fluctuations at Darwin and Tahiti are sometimes uncorrelated. It is possible for an increase in surface pressure at Darwin and a simultaneous decrease at Tahiti not to coincide with the appearance of unusually warm surface waters off Peru and with heavy rainfall in the central equatorial Pacific. It is pointless to debate whether or not El Niño occurred in such years. This problem is a consequence of the imperfect correlations between various parameters in the tropical Pacific. Table 1.1 shows that the values of the correlation coefficients, though high, are not equal to one. It follows that a definition of the Southern Oscillation in terms of the pressure difference between Darwin and Tahiti, for example, will differ from definitions in terms of sea surface temperature[2] or rainfall. Any one definition is of course unambiguous, but a multitude of definitions serves no purpose. It is more practical to avoid strict definitions and to accept that the terms *Southern Oscillation*, *El Niño*, and *La Niña* are general and qualitative. They are useful in the same way that the term *winter* is useful even though each winter is distinct. The features that are common to different El Niño episodes need to be identified—this is done in Section 1.4—because they provide a focus from which to explain the phenomenon. There are, however, many purposes for which this focus is too diffuse. Thus the term

Table 1.1

Matrix of Contemporaneous Correlations between Various Parameters in the Tropical Pacific[a]

	SST index	SLP index[b]	200-mbar index[c]	PNA index[d]	Tarawa rainfall	Canton rainfall	Christmas rainfall
SST index							
SLP index	−83						
200-mbar index	80	−68					
PNA index	46	−31	57				
Tarawa rainfall	78	−71	63	32			
Canton rainfall	82 (23)	−65 (23)	61 (23)	40 (23)	60 (23)		
Christmas rainfall	64 (24)	−49 (24)	57 (24)	40 (24)	55 (24)	82 (19)	
Fanning rainfall	79 (23)	−60 (23)	69 (23)	38 (23)	72 (23)	85 (18)	84 (23)

Source: Horel and Wallace (1981).

[a] Correlation coefficients are ×100. Correlations are between time series of the sea surface temperatures (SST) index for the tropical Pacific Ocean, the sea level pressure (SLP) index, the 200-mbar index, the Pacific North American (PNA) index, and rainfall indexes on various islands. The time series are for the winter seasons of at least 28 years (1951 to 1978) unless it is indicated, in parentheses, to be shorter.

[b] The SLP index measures the Tahiti–Darwin normalized sea level pressure.

[c] The 200-mbar index measures the height of the 200-mbar surface in the tropics and is proportional to the averaged temperature of the tropical troposphere.

[d] The PNA index is drawn from the teleconnection pattern in Fig. 1.26, which is based on the 700-mbar height at the four points indicated.

El Niño is no substitute for a detailed description of how different parameters varied during a certain period. A prediction that El Niño will occur is of limited practical value unless it specifies how surface pressure, rainfall, sea surface temperature, and other parameters will vary over a given period.

The Southern Oscillation at first appears complex because the large number of correlations between various parameters in different parts of the globe presents a bewildering picture. A physically plausible framework for a discussion of the correlations emerged from the seminal papers by Bjerknes (1966, 1969). The principal result is that large-scale atmospheric motion in the tropics, on time scales of weeks and longer, corresponds to direct thermal circulations. In such circulations, moisture-laden air converges onto the warmest regions of the earth's surface where the air rises and condenses, causing those regions to have widespread cloudiness and heavy precipitation. Elsewhere subsidence of dry air from the upper troposphere forms a lid on the planetary boundary layer and prevents small cumulus clouds from growing to a size that can produce substantial rainfall. The monsoons that bring heavy rainfall to the Indian subcontinent during the summer when that region is warmer than the surrounding ocean are an example of a direct thermally driven circulation. Other examples include the meridional Hadley Circulation, in which air rises near the equator and sinks in higher latitudes, and the zonal Walker Circulation, in which air rises over the warm western tropical Pacific and sinks over the cold eastern tropical Pacific. The Southern Oscillation is a perturbation to these direct thermal circulations and is associated with fluctuations in the intensity and the positions of the regions of rising, moist air. The factors that influence the interannual movements of the convective zones—variations in sea surface temperature patterns and variations in the heating of the continents—also influence the seasonal movements of the convective zones. The Southern Oscillation and the seasonal cycle therefore have much in common.

1.2 The Seasonal Cycle

Satellite photographs such as that in Fig. 1.3 clearly show that certain regions in the tropics are bright and highly reflective. These cloudy regions, where moist air rises and condenses, are convective zones and occur principally over equatorial Africa, Central and South America, and the "Maritime Continent" of the western Pacific and southeastern Asia. Important extensions over the oceans include the east–west Intertropical Convergence Zone (ITCZ) north of the equator and the South Pacific Convergence Zone, which slopes southeastward from the western equatorial Pacific and which has a weak counterpart in the Atlantic Ocean. Measurements of the outgoing long-wave radiation emitted by the tops of deep convective clouds, which are at much lower temperatures than areas of clear skies, quantify the convective activity[3] in different regions as shown in Fig. 1.4.

The seasonal movements of the regions of intense convection over the Indian Ocean sector and western Pacific are along a line that runs in a southeastward direction from northern India to the southwestern tropical Pacific. During the summer of the Northern Hemisphere the South Pacific Convergence Zone is poorly developed but convection is intense over northern India (Fig. 1.4). At that time the ITCZ is strong all across the Pacific. During the transition to the northern winter, the region of most intense convection moves southeastward so that by January the South Pacific Convergence Zone is well developed while the ITCZ over the western Pacific has become weak. From April onwards the intense convection moves back northwestward, the ITCZ strengthens over the western Pacific, and the South Pacific Convergence Zone weakens as it moves westward and equatorward (Trenberth, 1976; Meehl, 1987). Figure 1.5 depicts the seasonal rainfall and sea level pressure changes associated with the movements of the convergence zones. The stations in this figure are arranged to follow the convective maximum southeastward from India into the Pacific. Stations northeast of the South Pacific Convergence Zone are farthest to the right. Rainfall and sea level pressure are highly correlated and their maxima occur during the local summer.

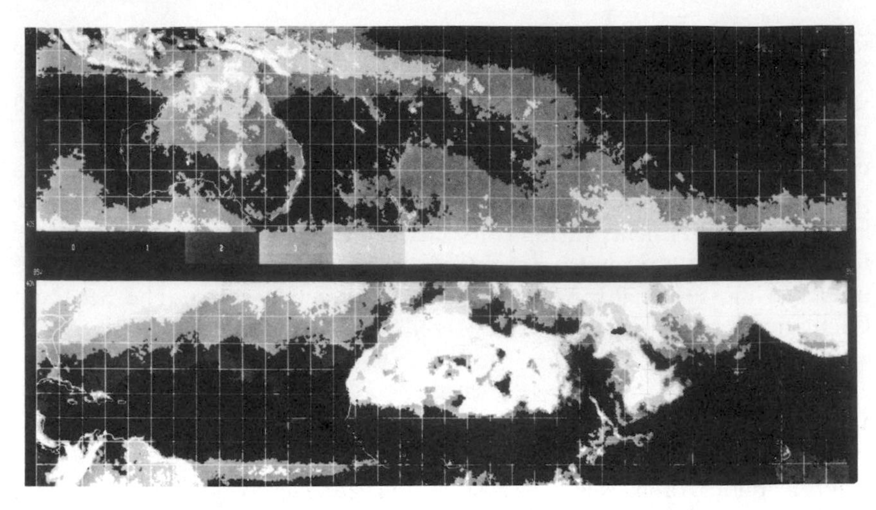

Figure 1.3. The geographical distribution of reflectivity for February on the basis of satellite measurements during the years 1967 to 1970. The Intertropical Convergence Zone (ITCZ) is the zonal band of cloudiness just north of the equator in the Pacific and Atlantic Oceans. The South Pacific Convergence Zone (SPCZ) stretches southeastward across the southwestern tropical Pacific. Most bright areas are characterized by persistent cloudiness and heavy precipitation, but there are two exceptions: the surface of the earth is highly reflective in deserts such as the Sahara, which therefore appear bright, and regions of low sea surface temperature (off southwestern Africa, for example) often have low, nonprecipitating cloud decks. The Mercator grid lines are spaced at intervals of 5° of latitude and longitude; the tick marks along the side indicate the equator. [From U.S. Air Force and U.S. Department of Commerce (1971). "Global Atlas of Relative Cloud Cover, 1967–1970." Washington, D.C.]

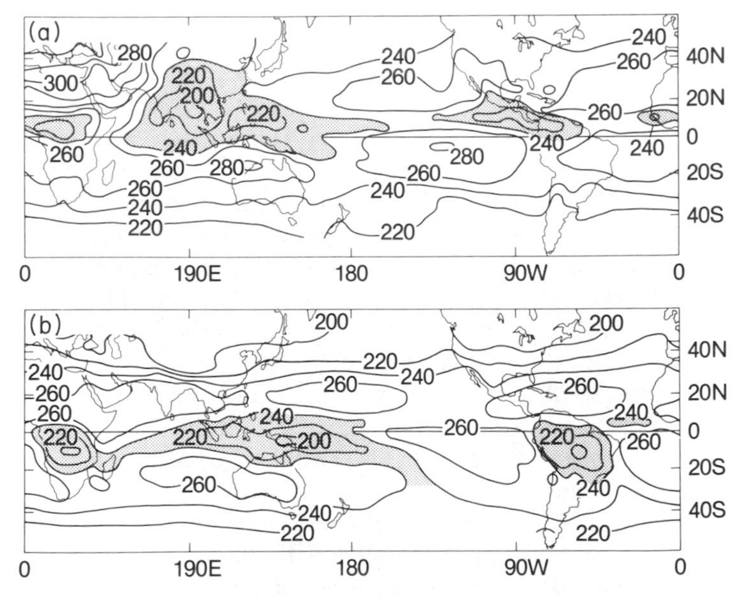

Figure 1.4. Mean outgoing long-wave radiation (watts per square meter) for (a) June to August and (b) December to February as measured by polar orbiting satellites. Tropical regions with values less than 240 W/m^2 have strong convection and heavy rainfall.

The eastern tropical Pacific and Atlantic Oceans are arid except for the ITCZ whose seasonal movements dictate seasonal rainfall variations. The east–west zone of maximum annual mean rainfall in Fig. 1.6 coincides with the mean position of the ITCZ. Whereas the convective zones over the continents, including the Maritime Continent, are always in the summer hemisphere and therefore cross the equator twice a year, the ITCZ is predominantly in the Northern Hemisphere: it migrates seasonally between the neighborhood of the equator in March and April and the latitude 12°N approximately in August and September (Horel, 1982). Equatorial regions whose rainfall depends on the presence of the ITCZ—the equatorial Atlantic, the central and eastern Pacific, and the adjacent coastal regions—have a single pronounced rainfall maximum in March and April when the ITCZ is in its southernmost position. Between 10°N and 15°N the maximum is in August and September when the ITCZ is farthest north. The ITCZ has a very small latitudinal extent so that the north–south gradients of the associated rainfall are sharp. For example, annual mean rainfall along the West African coast is 1939 mm at Bissau (12°N), 542 mm at Dakar (15°N), and 123 mm at Nouakchott (18°N). These statistics indicate that even small variations in the position of the ITCZ can have a major effect on rainfall in certain regions. Another example is the coast of Peru, where the increase in rainfall during El Niño is associated with a southward displacement of the ITCZ. Trujillo at 8°S usually has torrential rains during El Niño, but it is unusual for Lima at 12°S to receive heavy rainfall even during a major El Niño.

Moist air rises in the convective zones where surface pressure is low and rainfall is high. Dry air subsides over the arid high-pressure zones of the

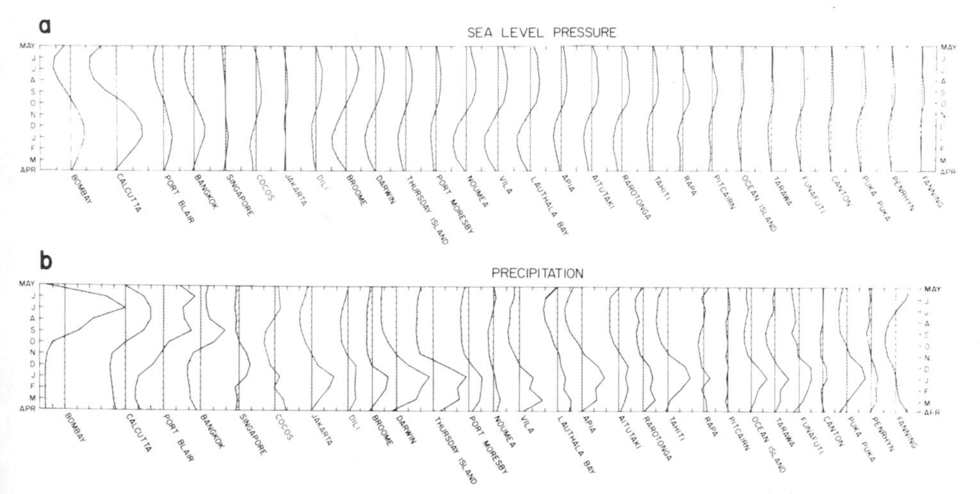

Figure 1.5. The annual cycle of (a) sea level pressure and (b) precipitation plotted as deviations of the three-month running means from the annual mean. Each tick mark on the horizontal axes indicates (a) 5.0 mbar and (b) 100 mm/month. [From Meehl (1987).]

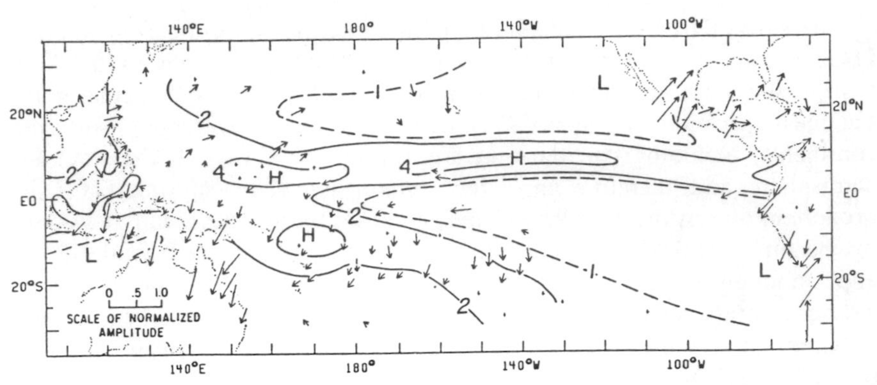

Figure 1.6. The annual mean rainfall of Taylor (1973) and the first harmonic of the annual cycle in rainfall. The contour interval for the annual mean is 2 m. The amplitudes of the first harmonic have been divided by the annual mean rainfall. The normalized amplitudes are denoted by the lengths of the arrows according to the scale in the figure. A plus sign is plotted at stations where the amplitudes are less than one-tenth of the annual mean. The phase is indicated by the direction of the arrow. An arrow that points downwards means maximum rainfall on 1 January; a horizontal arrow pointing left means maximum rainfall on 1 April. [From Horel (1982).]

tropics and subtropics. Figure 1.7 shows that the major high pressure zones are over the northeastern and southeastern tropical Atlantic and Pacific Oceans and over the southern Indian Ocean. The trade winds and monsoons generally blow from the high-pressure zones towards the convective regions where the surface pressure is low. (Allowance must be made for

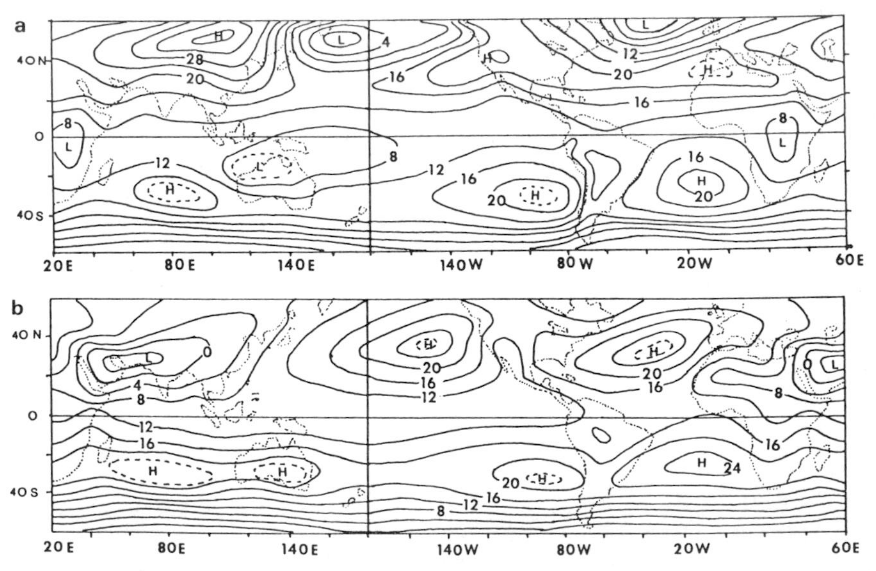

Figure 1.7. The mean sea level pressure (−1000 mbar) for January (a) and July (b) based on data for the period 1961 to 1976. [From Godbole and Shukla (1981).]

effects associated with the rotation of the earth.) The seasonal movements of the convective zones therefore imply seasonal changes in the surface winds. As shown in maps of these winds (Fig. 1.8), the most dramatic changes are over the Indian Ocean, where the winds reverse direction seasonally: the southwest monsoons prevail from May until October and the northeast monsoons prevail from December until March. April and November are months of transition. The southwest monsoons are most intense in a low-level jet along the East African coast.[4] Seasonal changes in the trade winds over the Atlantic and Pacific are associated with the movements of the ITCZ onto which the trades converge. On average, there is a northward surface flow across the equator—the southeast trades penetrate into the Northern Hemisphere—because the ITCZ is predominantly in the Northern Hemisphere. The trade winds attain their maximum speeds around 15°N and S. Equatorward of these latitudes the fluctuations in the intensity of the northeast and southeast trades are out of phase: the southeast trades are intense and the northeast trades are relaxed in September when the ITCZ is farthest north. The situation is reversed in March and April when the ITCZ is close to the equator.

The meridional component of the circulations in the tropics, known as the Hadley Circulation, can be isolated by taking zonal averages to filter out longitudinal variations. The Hadley Circulation has two meridional cells with moist air rising in the ITCZ, then diverging poleward in the upper troposphere, and descending over regions to the north and south of the ITCZ. The more intense cell has subsiding motion in the winter hemisphere where latitudinal thermal gradients are larger. In other words, seasonal variations involve changes in both the location and intensity of each of the meridional cells. Superimposed on this meridional circulation are zonal circulations. For example, there is ascending motion over the western tropical Pacific but descending motion over much of the eastern side of this ocean basin. The trade winds at the surface link these two regions. Bjerknes (1966) introduced the term Walker Circulation for this zonal cell, which is weak in March and April when sea level pressure gradients are small. The Hadley and Walker Circulations refer to the meridional and zonal components of the motion and are abstractions that are shown schematically in Fig. 1.9a. Figure 1.9b shows a more realistic state of affairs. A parcel of air that moves westward in the surface layers over the equatorial Pacific is unlikely to remain in the equatorial plane and complete a Walker Circulation because the parcel also has a meridional velocity component and will make latitudinal excursions as part of the Hadley Circulation.

A comparison of the sea surface temperature patterns in Fig. 1.10 with the convective patterns in Fig. 1.4 reveals that, over the oceans, the moist air rises where the sea surface temperatures are highest while dry air subsides where surface waters are cold. The warmest waters are in the western tropical Pacific and in a band of latitudes just north of the equator

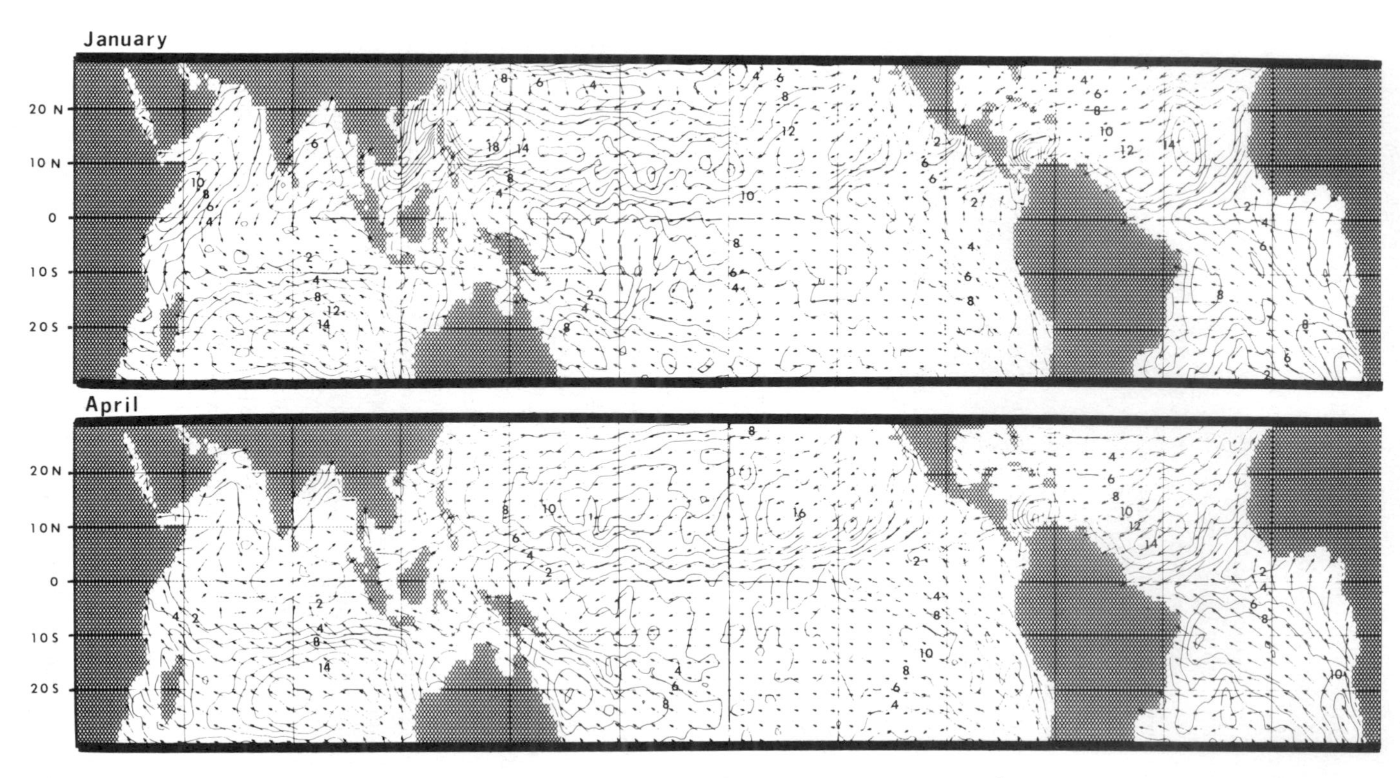

Figure 1.8. The surface windstress in January, April, July, and October in units of 0.1 dyne/cm^2. The contours are lines of constant amplitude. The arrows indicate direction and are drawn as if scales in the zonal and meridional directions are equal so that an arrow at 45° to the vertical indicates a wind from the southwest. [From Hellerman and Rosenstein (1983).]

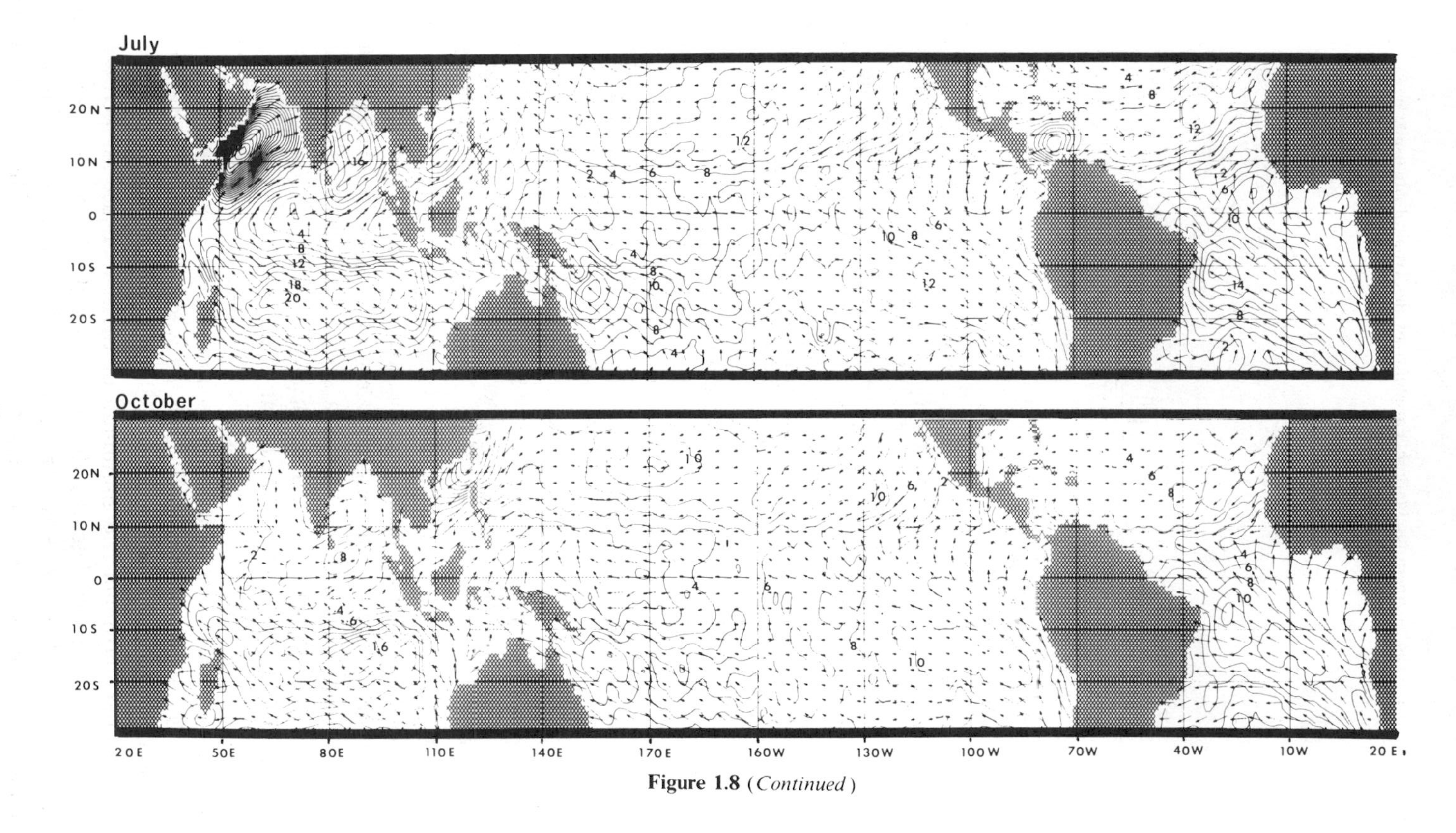

Figure 1.8 (*Continued*)

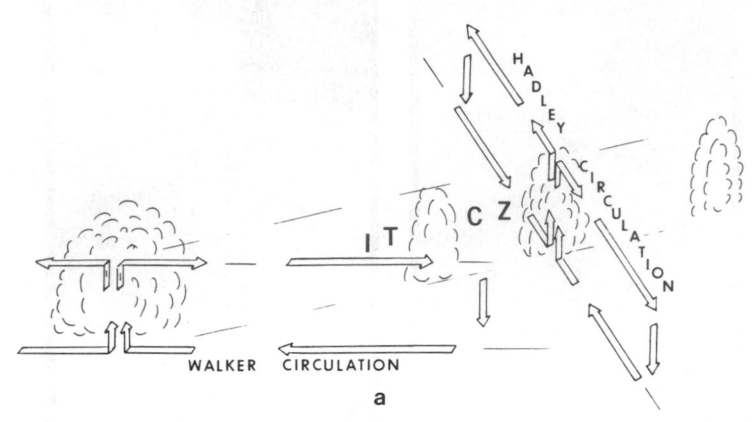

Figure 1.9. (a) A schematic diagram of the zonal Walker Circulation and meridional Hadley Circulation. (b) The horizontal portion of the diagram illustrates the January mean 1000-mbar streamline pattern for the years 1979 to 1981. The superimposed dashed line is the approximate position of the South Pacific Convergence Zone. In the vertical portion of the diagram, which is a section along 5°S, solid lines indicate the vertical motion w (in units of 10^{-4} mbar/sec with regions of upward motion shaded) and dashed lines indicate the zonal velocity. [From Kousky *et al.*, (1984).] (*Figure continues.*)

where the ITCZ is located. A sea surface temperature of 27.5°C seems to be a threshold for organized atmospheric convection[5], but it is not a sufficient condition because convection is absent from many areas where sea surface temperatures exceed 27.5°C (Gadgil *et al.*, 1984; Graham and Barnett, 1987). The large-scale convection over the warmest surface waters is associated not with increased local evaporation from the ocean but with the convergence of moist air onto those regions (Cornejo-Garrido and Stone, 1977; Ramage and Hori, 1981). In other words, the air that converges onto the convective zones carries the fuel that stokes the atmospheric engine, thus ensuring continued convergence of moist air onto the convective zone.

The movements of the atmospheric convergence zones are such that they always remain over the warmest surface waters. Thus the ITCZ is farthest south, near the equator, when the sea surface temperatures there are at a maximum. It moves northward towards warmer water when sea surface temperatures at and south of the equator start to fall, usually in May. The convective zone over the western tropical Pacific similarly remains over the warmest surface waters as it moves seasonally from hemisphere to hemisphere. In the eastern tropical Pacific the westward phase propagation of the annual harmonic of the sea surface temperature—Fig. 1.11 shows that the seasonal maximum and minimum of the sea surface temperature are attained progressively later as distance from the South American coast increases—is echoed by westward phase propagation of the annual harmonic of the surface wind convergence (Horel, 1982). These observations

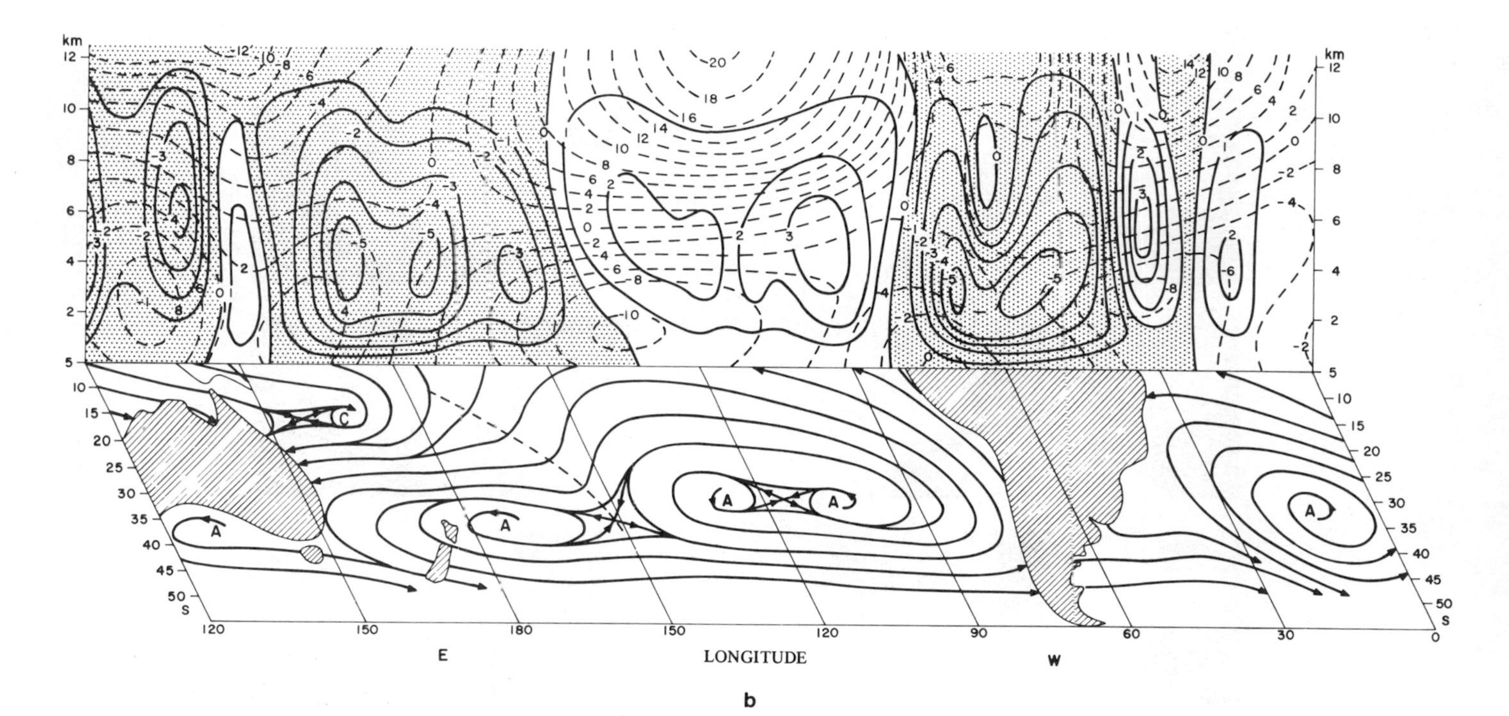

Figure 1.9 (*Continued*)

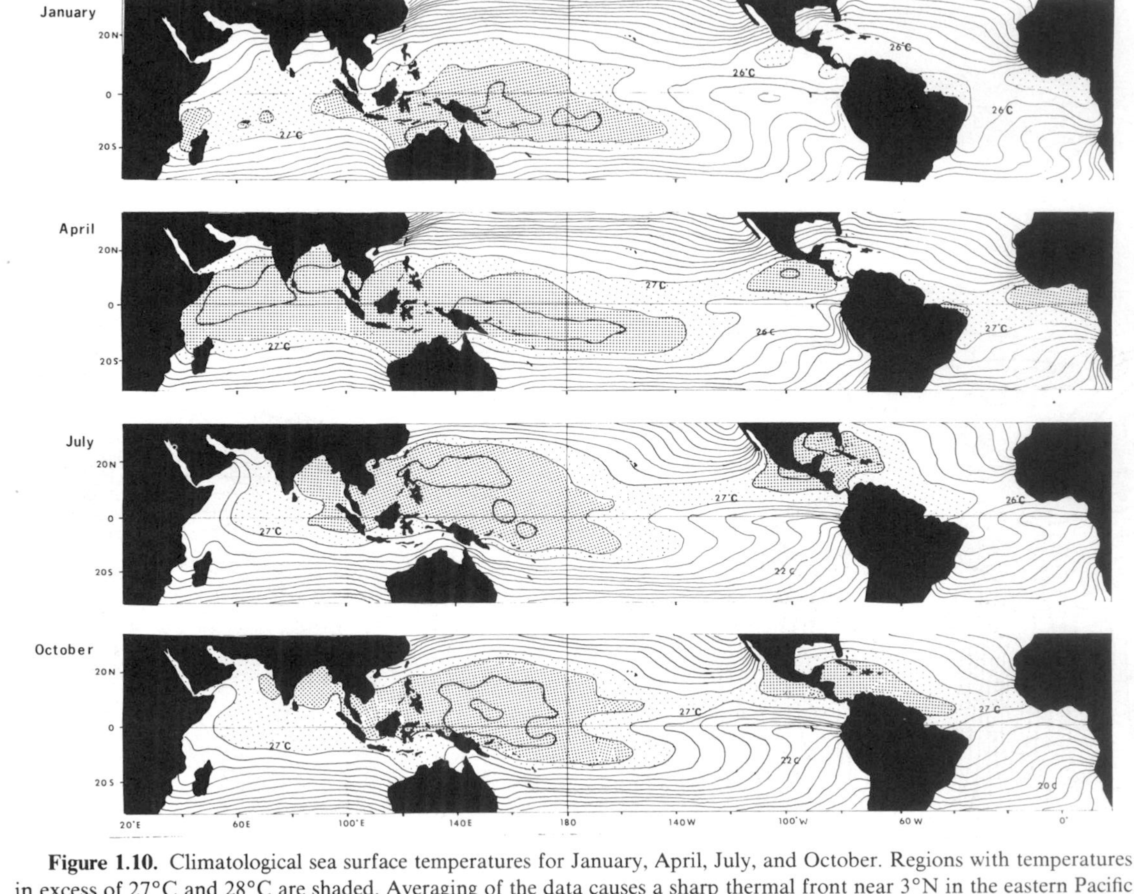

Figure 1.10. Climatological sea surface temperatures for January, April, July, and October. Regions with temperatures in excess of 27°C and 28°C are shaded. Averaging of the data causes a sharp thermal front near 3°N in the eastern Pacific and Atlantic in July and October to be diffuse. A wedge of cold surface waters off northeastern Africa in July is absent for similar reasons. (The data were provided by R. Reynolds of the Climate Analysis Center, NOAA, Washington, D.C.)

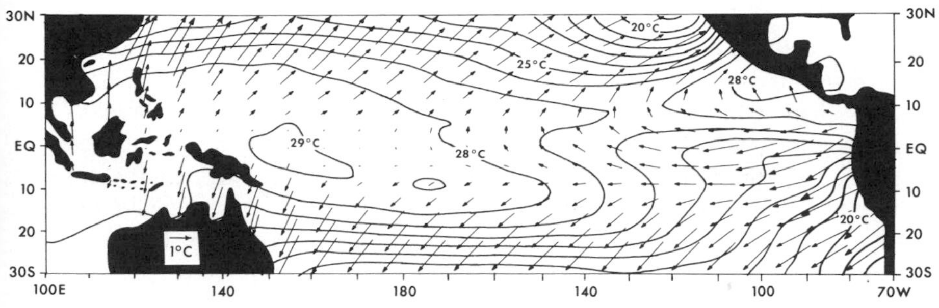

Figure 1.11. Mean sea surface temperatures (solid lines) and the amplitude (length of arrow) and phase (direction of arrow) of the annual cycle of sea surface temperature in the tropical Pacific Ocean. Maximum temperature is in January for an arrow that points downwards and in April for an arrow that points to the left.

give the impression, which is confirmed by theoretical studies, that sea surface temperatures influence the movements of the convergence zones and hence affect the surface windstress patterns. The surface winds in turn strongly influence the sea surface temperatures. (This matter is discussed in Chapters 2 and 4.) It follows that interactions between the ocean and atmosphere determine the low-frequency variability in the two media.

1.3 Interannual Variability

The dominant mode of interannual variability in the tropics is the Southern Oscillation, which has its largest amplitude over the Indian and Pacific Oceans. Its spatial structure depends on the parameter under consideration; each parameter has a core region where the Southern Oscillation accounts for the major part of its variance. Surface pressure has two core regions as was shown in Fig. 1.1. One is over the southeastern tropical Pacific and the other is over the Maritime Continent. Rainfall has a single core region that coincides with neither of the pressure core regions. It is close to the equator and extends approximately from 160°E to 150°W (Wright, 1984b; Reiter, 1978). For sea surface temperature the core region is also close to the equator but it is confined to the central and eastern Pacific (Weare *et al.*, 1976; Wright, 1984a). An index that measures the fluctuations of a variable in its core region effectively monitors the Southern Oscillation. Some parameters, surface pressure, for example, have considerable variability on time scales of the order of a month that is unrelated to the Southern Oscillation. An effective index then requires averages over a season or longer (Trenberth, 1984). Table 1.1 describes the remarkably high correlations between different Southern Oscillation Indices.

The time scale of the Southern Oscillation is of the order of three years but the oscillation is so irregular that spectra of meteorological variables in the tropics have a broad peak in the range from two to ten years (Berlage, 1957; Trenberth, 1976; Julian and Chervin, 1978). Furthermore, there is a tendency for peaks to shift within this range depending on the period being analyzed (Barnett, 1985a, b). In other words, El Niño episodes are more frequent during some decades than others. This indicates considerable variability at frequencies much lower than that of the Southern Oscillation but very little is known about fluctuations at those long periods.

Figure 1.1 gives the impression that the Southern Oscillation is a standing mode, but this is not the case. Spectral analysis and cross-correlations reveal that interannual sea level pressure variations over the South Pacific Ocean tend to precede, by several months, variations of the opposite sign in the Australian–Indonesian Low-Pressure Zone.[6] This has prompted suggestions that El Niño episodes are initiated by developments in the South Pacific (van Loon and Shea, 1985; Trenberth and Shea, 1987).

The interannual movements of the atmospheric convective zones in the tropics are influenced by sea surface temperature variations. The convective zones over the oceans, as noted in Section 1.2, occur over surface waters warmer than 27.5°C. In Fig. 1.12, which depicts the interannual movements of isotherms along the equator, warm surface waters are seen to expand eastward during El Niño and to contract back westward during La Niña. Figure 1.13 shows how much the area and location of the pool of warm surface waters can change interannually. Analysis using empirical orthogonal functions indicates that the amplitude of the interannual sea surface temperature variations has a maximum in the central and eastern equatorial Pacific, between approximately 10°N and S (Newell and Weare, 1976). Sea surface temperature variations in this region are far larger than those west of the date line. Furthermore, changes to the east and west of the date line have opposite signs: sea surface temperatures tend to decrease slightly in the western equatorial Pacific during El Niño when they increase significantly in the east. In the west, where the local flux of heat across the ocean surface has a strong influence on the sea surface temperature, more intense local winds and hence enhanced evaporation reduce this flux so that the sea surface temperature decreases (Meyers *et al.*, 1986). The increase in sea surface temperatures in the eastern tropical Pacific during the early stages of El Niño (usually during the early calendar months of the year) is caused by a local weakening of the winds and hence reduced evaporation. As El Niño evolves, eastward advection of warm surface waters maintains the high sea surface temperatures in the east in the face of an increase in evaporation caused by the seasonal intensification of the winds. This latent heat loss during the later stages of El Niño exceeds the gain during the early

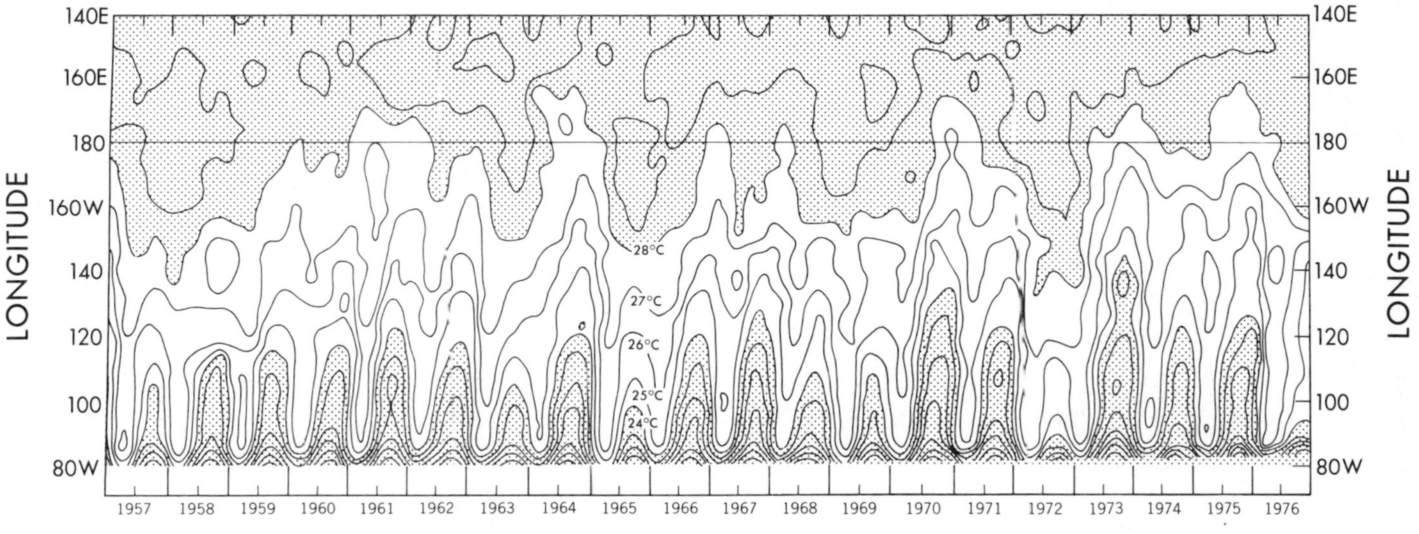

Figure 1.12. Sea surface temperatures along the equator between 140°E and 100°W and along a line from 0°N, 100°W to the South American coast at 10°S. Regions with temperatures less than 24°C and greater than 28°C are shaded.

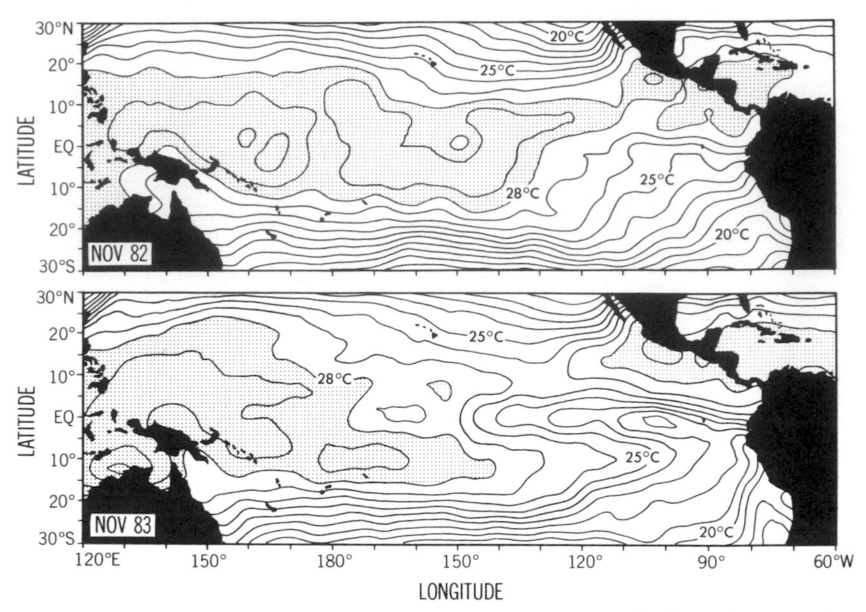

Figure 1.13. Sea surface temperatures in November 1982 during El Niño and one year later during La Niña.

stages. The net result is that the tropical Pacific loses more latent heat during El Niño than it does during La Niña (Weare, 1983; Ramage and Hori, 1981). These results suggest that there are interannual variations not only in the location but also in the intensity of the atmospheric heating in the tropics.

During La Niña, water warmer than 27.5°C covers relatively small areas in the western tropical Pacific and north of the equator in the east. The area of warm surface waters expands into the central and eastern tropical Pacific during El Niño when the tongue of cold surface waters along the equator contracts eastward. Figure 1.13 illustrates how much sea surface temperature patterns can change between El Niño and La Niña. The relatively uniform sea surface temperatures during El Niño are associated with a merger of the atmospheric convergence zones: the ITCZ moves equatorward, the South Pacific Convergence Zone moves northward, and the convergence zone over the western Pacific moves eastward (Pazan and Meyers, 1982). The zonal Walker Circulation is weak during these periods but the meridional Hadley Circulation intensifies. The opposite happens during La Niña when the cold surface waters of the southeastern tropical Pacific stretch far westward in a tongue along the equator: the convergence zones are separated from each other as the ITCZ moves northward and the convergence zone over the western Pacific moves westward. Figure 1.14 depicts zonal movements of the convergence zones in the equatorial Indian and Pacific Oceans. In 1975 when intense La Niña conditions prevailed, the equatorial Pacific was essentially cloud free east of the date line. This is in sharp contrast to what happened in 1982–1983 when the convective zone moved across the full width of the Pacific until it reached the Americas.

Rainfall variations during El Niño reflect the merger of the convective zones as shown schematically in Fig. 1.15. The season during which there is usually precipitation, although it can be shortened or lengthened, does not change in any of the affected regions. In other words, the phase of the annual precipitation cycle stays essentially the same but the amplitude changes. The months indicated in Fig. 1.15 are the months during which rainfall is influenced during the dominant El Niño year (0) and during the subsequent year (+). In the Pacific there is heavy rainfall in the central and eastern equatorial regions but droughts in the surrounding anvil that covers Indonesia, the Philippines, Hawaii, eastern Australia, and the area of Fiji and New Caledonia. In the western part of South America, the equatorward displacement of the ITCZ brings heavy rainfall to Ecuador and northern Peru but droughts to parts of Bolivia and Central America. In the Indian sector there is a strong tendency for the monsoons to fail and for southeastern Africa to suffer droughts, while Sri Lanka and equatorial East Africa often have increased rainfall during El Niño (Ropelewski and Halpert, 1987; van Loon and Shea, 1985; Shukla and Paolino, 1983; McBride and Nicholls, 1983). The sign of these changes is generally reversed during La Niña.

From Fig. 1.16 it is evident that factors other than those associated with the Southern Oscillation also affect rainfall. For example, India can have droughts in the absence of El Niño and can have wet seasons even when El Niño occurs. Sea surface temperature variations in the Indian Ocean are poorly correlated with rainfall over India and do not appear to have a strong influence on the southwest monsoons. During exceptionally wet monsoons, surface waters are initially slightly warmer than normal in the Arabian Sea (Shukla and Misra, 1977) but the waters then cool so that temperatures are lower than normal during and after the wet season (Weare, 1979; Shukla, 1986), possibly because of unusual evaporative cooling associated with the intense winds. These results suggest that, during the northern summer, sea surface temperature variations in the Indian Ocean may be a consequence rather than a cause of variations in the intensity of the monsoons.

Matters may be different during the southern summer. El Niño brings enhanced rainfall to equatorial eastern Africa but diminished rainfall to southeastern Africa (Fig. 1.15). These tendencies, during the southern

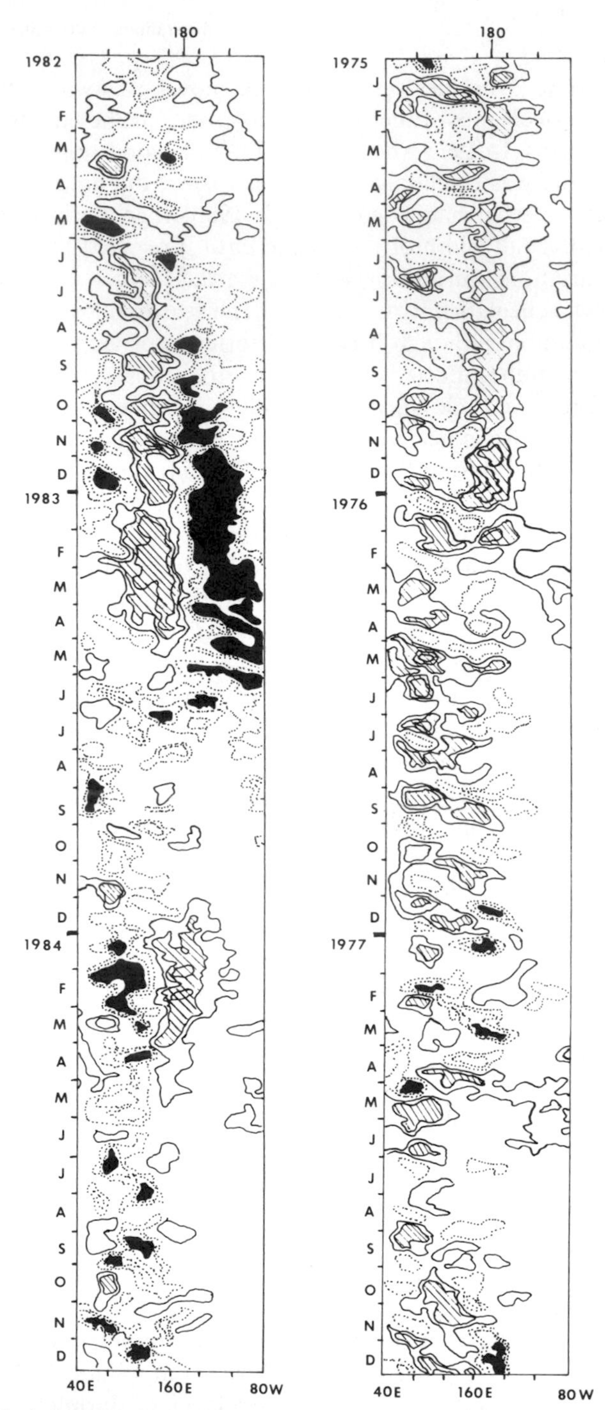

Figure 1.14. A time–longitude plot of 5-day mean outgoing long-wave radiation anomalies (OLR) averaged between 5°S and 5°N for the periods 1975 to 1977 and 1982 to 1984. Convection is enhanced in regions enclosed by dotted lines and is at a maximum (OLR < -30 W/m^2) in solid areas. Convection is suppressed in regions enclosed by solid lines and is at a minimum (OLR > 10 W/m^2) in hatched areas. [From Lau and Chan (1986).]

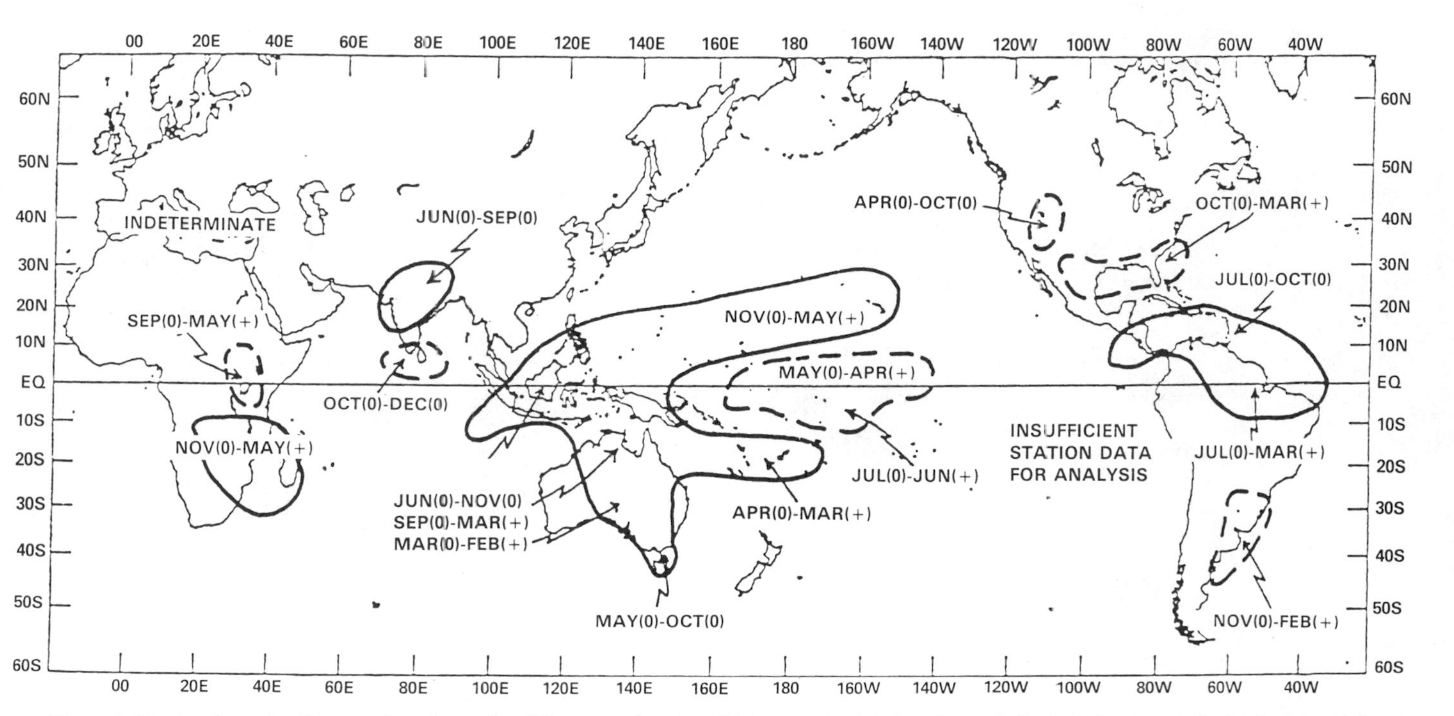

Figure 1.15. A schematic diagram that shows the different regions in which precipitation is enhanced (dashed lines) and diminished (solid lines) during El Niño episodes. The months, which indicate when the regions are affected, generally coincide with the local rainy season. The year in which anomalously high sea surface temperatures first appear and then amplify in the tropical Pacific is denoted by (0); (+) refers to the subsequent year. [From Ropelewski and Halpert (1987).]

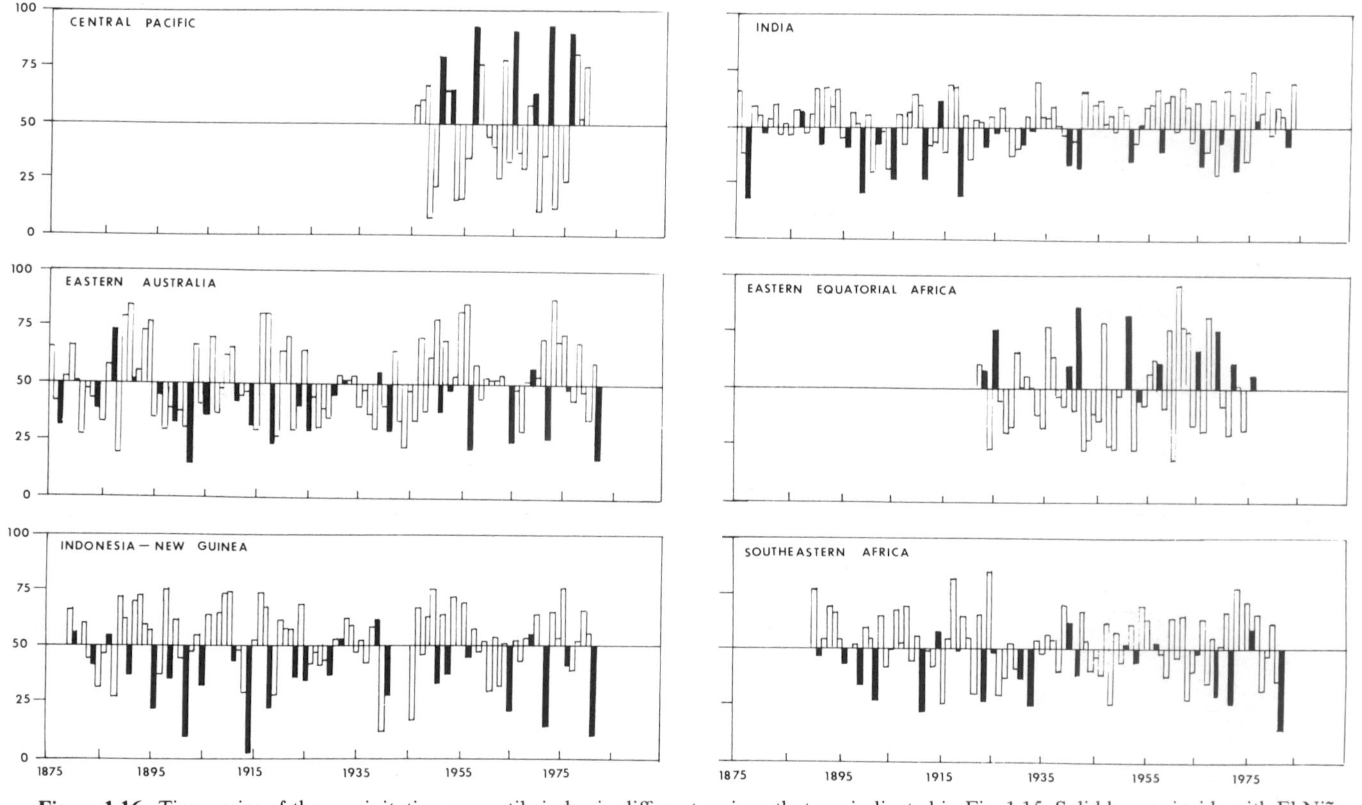

Figure 1.16. Time series of the precipitation percentile index in different regions that are indicated in Fig. 1.15. Solid bars coincide with El Niño years. For each station the data in each of n years are ranked from 1 (the year with lowest precipitation) to n (the year of maximum precipitation). The percentile index for rainfall in the year with ranks m is $100m/n$. The value for a region is the average of the percentiles for each station in the region. This procedure copes with inhomogeneous records and unrepresentative individual stations. [From Ropelewski and Halpert (1987).]

summer, suggest that the convective zone over the southwestern Indian Ocean and adjacent land is displaced equatorward during El Niño. To determine whether sea surface temperature variations in the Indian Ocean induce this shift, further studies of the northeast monsoon season are needed. Such studies may also shed light on the factors that influence the termination of El Niño. In early 1973, for example, easterly surface winds replaced westerly winds over the western tropical Pacific and signaled the end of El Niño, even though surface waters in the central and eastern tropical Pacific were unusually warm at the time. Could the change in the winds have been in response to conditions over the Indian Ocean?

Movements of the atmospheric convective zones affect the sea level pressure as shown in Figs. 1.1 and 1.2. Variations in the pressure gradients in turn are correlated with variations in the trade winds over the Pacific Ocean and the monsoons over the Indian Ocean. The areal extents of these wind systems expand and contract out of phase. During the warm (El Niño) phase of the Southern Oscillation the monsoons strengthen and expand into the Pacific Ocean almost as far as the date line, while the trade winds weaken and retreat eastward. During the cold La Niña phase the reverse happens. The variability of the zonal component of the surface wind has its maximum amplitude west of the date line but the meridional component has its further east (Barnett, 1983). Exceptions to these large-scale coherent fluctuations occur over the Arabian Sea, where the southwest monsoons can be weak during El Niño, and along the coast of Peru, where the intensity of the southeasterly winds can be unabated during El Niño. The latter case is associated with a coastal atmospheric jet, partially in response to the temperature difference between the ocean and the adjacent desert. During El Niño the large-scale trade winds weaken but within tens of kilometers of the coast the winds remain intense (Enfield, 1981a and b).

In the lower atmosphere the winds converge onto the convective zones; in the upper troposphere the winds diverge from these zones. The velocity potential at 200 mbar is largely the mirror image of the potential near the surface. Matters are different for the rotational part of the flow. The dipole that characterizes the geopotential heights on a 1000-mbar map—high values over the southeastern tropical Pacific and low values over the eastern Indian Ocean and western tropical Pacific—attenuates with increasing altitude and at 200 mbar streamlines have considerable east–west uniformity, as shown in Fig. 1.17. The Southern Oscillation is associated with interannual fluctuations in the height of the 200-mbar surface. These fluctuations are of the same sign throughout the tropics, are correlated with variations in the thickness of the layer between 1000 and 200 mbar, and are also correlated with the average temperature of the tropical troposphere. This temperature can be 1 K higher during El Niño than during La Niña

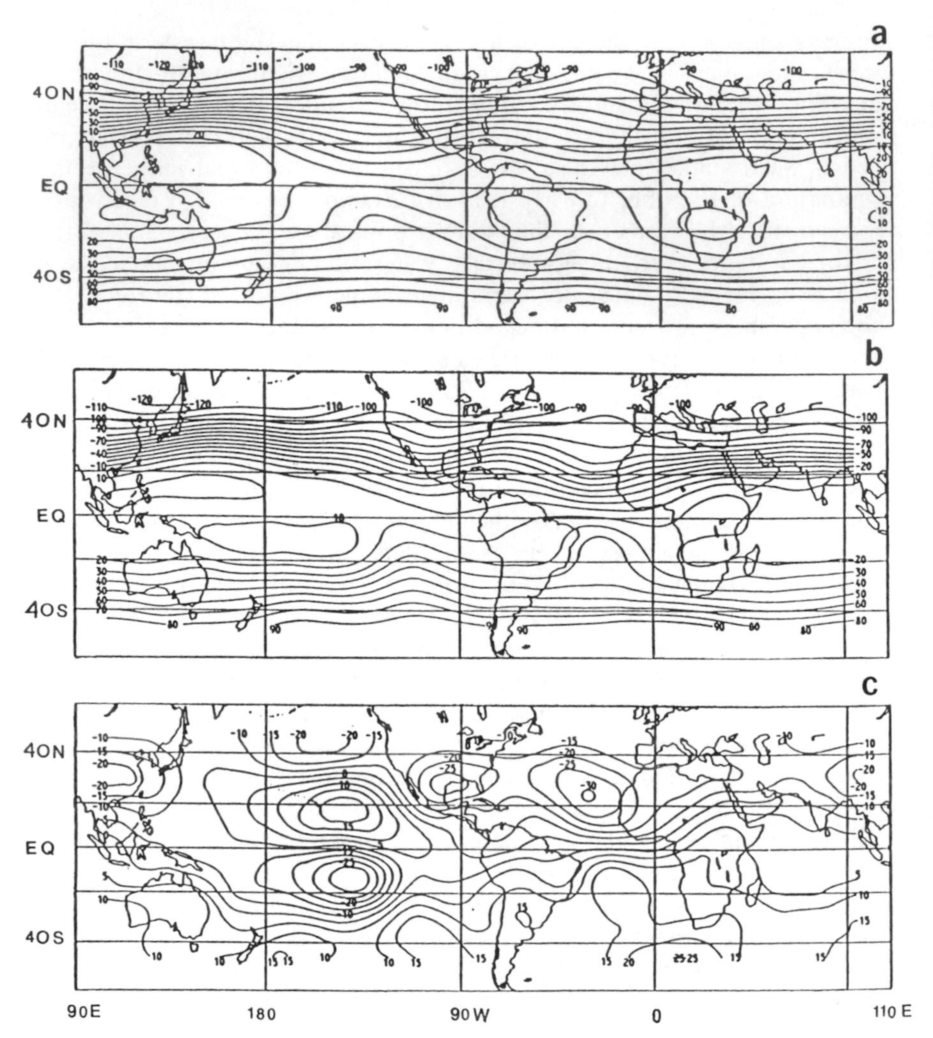

Figure 1.17. The mean stream function for the December, January, and February quarter computed from the 200-mbar winds. The contour interval is 10^7 m^2/sec. (a) Climatological conditions, (b) conditions in 1982–1983, and (c) the anomaly pattern for 1982–1983. [From Arkin (1983).]

(Angell and Koshover, 1978; Pan and Oort, 1983; Horel and Wallace, 1981; Barnett, 1985c). The probable source of heat for this warming is the increase in the flux of latent heat from the ocean to the atmosphere during El Niño.

Striking features of the anomalous conditions in the upper troposphere during El Niño are two anticyclones, shown in Fig. 1.17, that straddle the equator near the date line, where there is upper-level divergence from the convective zone. On their equatorward side the anticyclones have anomalous easterly winds that are opposite in direction to the anomalous equatorial westerly winds near the surface. On their poleward flanks the anticyclones have westerly wind anomalies and thus intensify the subtropical jets in upper levels especially in the winter hemisphere. Bjerknes (1966) argued that this happens because the intensified Hadley Circulation increases the poleward transport of angular momentum in the upper troposphere during El Niño.

The phase of the Southern Oscillation is closely related to that of the seasonal cycle not only in the case of rainfall, as mentioned earlier, but for other parameters too. On many occasions, some of which are depicted in Fig. 1.18, the onset of the warm El Niño phase of the Southern Oscillation in the eastern tropical Pacific coincides with the warm phase of the seasonal cycle so that El Niño initially amplifies this phase of the seasonal cycle. This means that exceptionally high sea surface temperatures and rainfall often appear at a time when these variables have their seasonal maxima. Such conditions are associated with an unusual southward displacement of the ITCZ during the early calendar months of the year when the seasonal migrations of the ITCZ take it to its southernmost position. Subsequently the seasonal cycle continues to modulate the development of El Niño. For example, the ITCZ next moves northward, as it does each year, but this movement is inhibited during El Niño (Ramage and Hori, 1981). In the western Pacific and Indian Ocean sector there is a similar close relation between the seasonal cycle and the Southern Oscillation. There the principal movement of the region of intense convection is from the northwest (over India) to the southeast (over the southwestern Pacific) and back again. There is evidence for a biennial variation in the amplitude of this oscillation so that high rainfall and low sea level pressure precede the opposite tendencies in the following year. Extreme amplitudes are attained during El Niño and La Niña (Meehl, 1987). Sea surface temperature fluctuations in the eastern Indian and western equatorial Pacific play a role in the low-frequency variability of that region (Nicholls, 1983, 1984a and b).

The close relation between the seasonal cycle and the Southern Oscillation is evident in Fig. 1.18, which shows how, in the eastern Pacific, different El Niño episodes all start at the same time of the year, and in Fig.

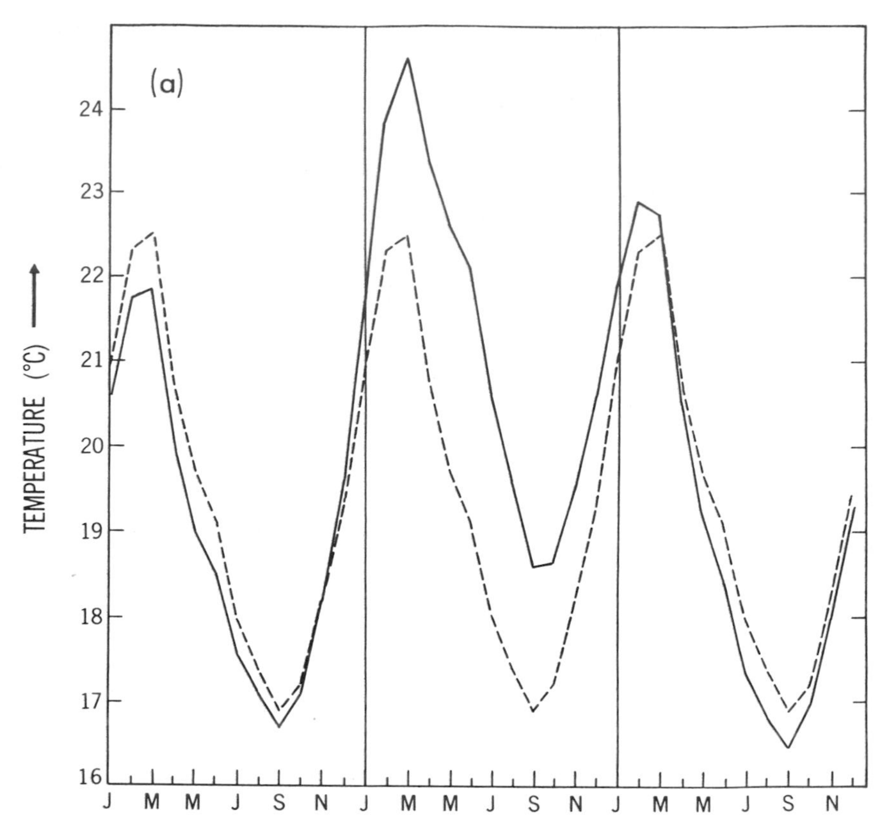

Figure 1.18. (a) The seasonal sea surface temperature changes (dashed line) and the changes during a typical El Niño episode as measured along a ship track 100 km off the coast of South America between 3°S and 12°S. (b) Sea surface temperature departures from the seasonal cycle, along the track mentioned in (a), during El Niño episodes in 1951, 1953, 1957, 1963, 1965, 1969, and 1972. The figure also shows variations during the preceding and following years. The solid line in (a) is obtained by averaging the curves in (b). [From Rasmusson and Carpenter (1982a).] (*Figure continues.*)

2.12, which shows how, in the western Pacific, different El Niño episodes all terminate at the same time of the year. Reasons for this close relation are discussed in Chapter 6. Here we exploit one of the fortunate consequences of this relation. Because many different El Niño episodes evolve in a similar manner, the paucity of the available data can be mitigated by analyzing a composite episode. In general, the similarities between different El Niño episodes are greater than the similarities between different La Niña episodes, to such an extent that compositing has been attempted only for El Niño

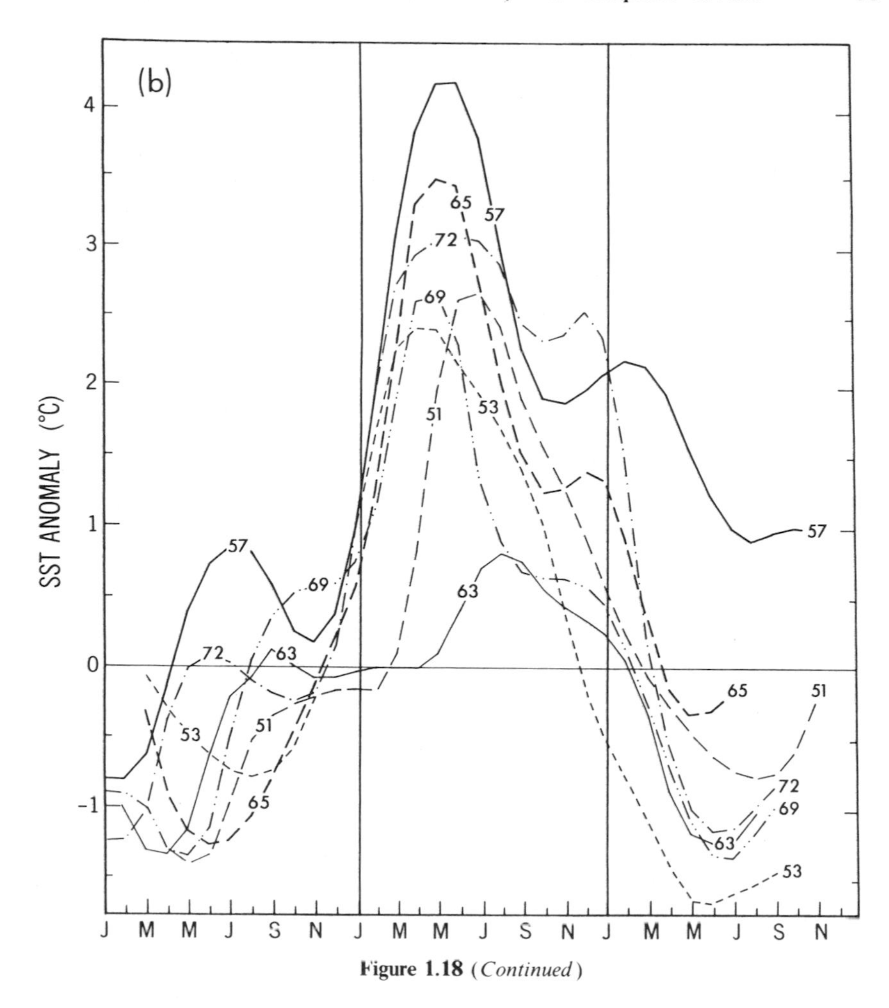

Figure 1.18 (*Continued*)

and not for La Niña. This difference is probably related to the tendency towards spatially homogeneous conditions during El Niño, when convergence zones merge, and the opposite tendency during La Niña. (There is only one spatially homogeneous state and many inhomogeneous states.)

1.4 A "Composite" El Niño

The amplitudes of different El Niño episodes can vary significantly but, as shown in Fig. 1.18, the phases of different episodes can be remarkably similar. Because of this, composite analyses are possible. The following is a description of such an analysis by Rasmusson and Carpenter (1982a) for the six warm episodes between 1950 and 1976.

Towards the end of a year preceding a warm event, the end of year (−1), sea level pressure in the central and southeastern tropical Pacific starts to decrease, and the easterly trade winds weaken to the west of the date line, where sea surface temperatures become higher than normal. (None of these changes is a reliable precursor of El Niño; similar changes can occur when no El Niño is imminent.) Next there is an amplification of the warm phase of the seasonal cycle in the eastern tropical Pacific during the early calendar months of year (0) so that sea surface temperatures and rainfall in the east attain exceptionally high values at the time of their seasonal maxima. Sea surface temperatures were exceptionally high at the start of 1975 and 1986 but those conditions subsequently attenuated and did not develop into El Niño. In the case of a composite El Niño, however, these conditions persist in the east for several months while the seasonal northward migration of the ITCZ is inhibited. Not only is the northward movement of the ITCZ absent but so is the usual westward propagation of the cooling phase of the seasonal cycle (Fig. 1.11). The persistence of high sea surface temperatures therefore implies that sea surface temperature anomalies, relative to the climatology, propagate westward. The speed of propagation is between 50 and 100 cm/sec. On the basis of data along the coast of Peru, it is therefore possible to predict when, in the central equatorial Pacific, the cold season will fail to develop and the warm surface waters will persist for several more months (Barnett, 1981a).

While warm, wet conditions prevail in the eastern equatorial Pacific, the convective zone over the western Pacific moves eastward starting in April of year (0) so that rainfall decreases and sea level pressure increases over Indonesia. Farther east, over a region that extends from Nauru and Ocean Islands (near 165°E) to the South American coast, there is now heavy rainfall and northerly wind anomalies indicative of a southward displacement of the ITCZ. By July of year (0), anomalous conditions in the eastern equatorial Pacific have peaked, although a secondary maximum can appear late in year (0) and early in year (+1). These developments in the east are different from those in the western and central Pacific, where anomalous conditions continue to amplify until the end of year (0). Rainfall at the Line Islands near 157°W, for example, has its strongest enhancement towards the end of year (0). By contrast, anomalous rainfall at the Galapagos Islands near 90°W has its maximum in March, April, and May of year (0). Fig. 1.19 indicates that not only rainfall but also sea surface temperature anomalies attain their maximum amplitude along the coast of Peru long before they do so in the central Pacific.

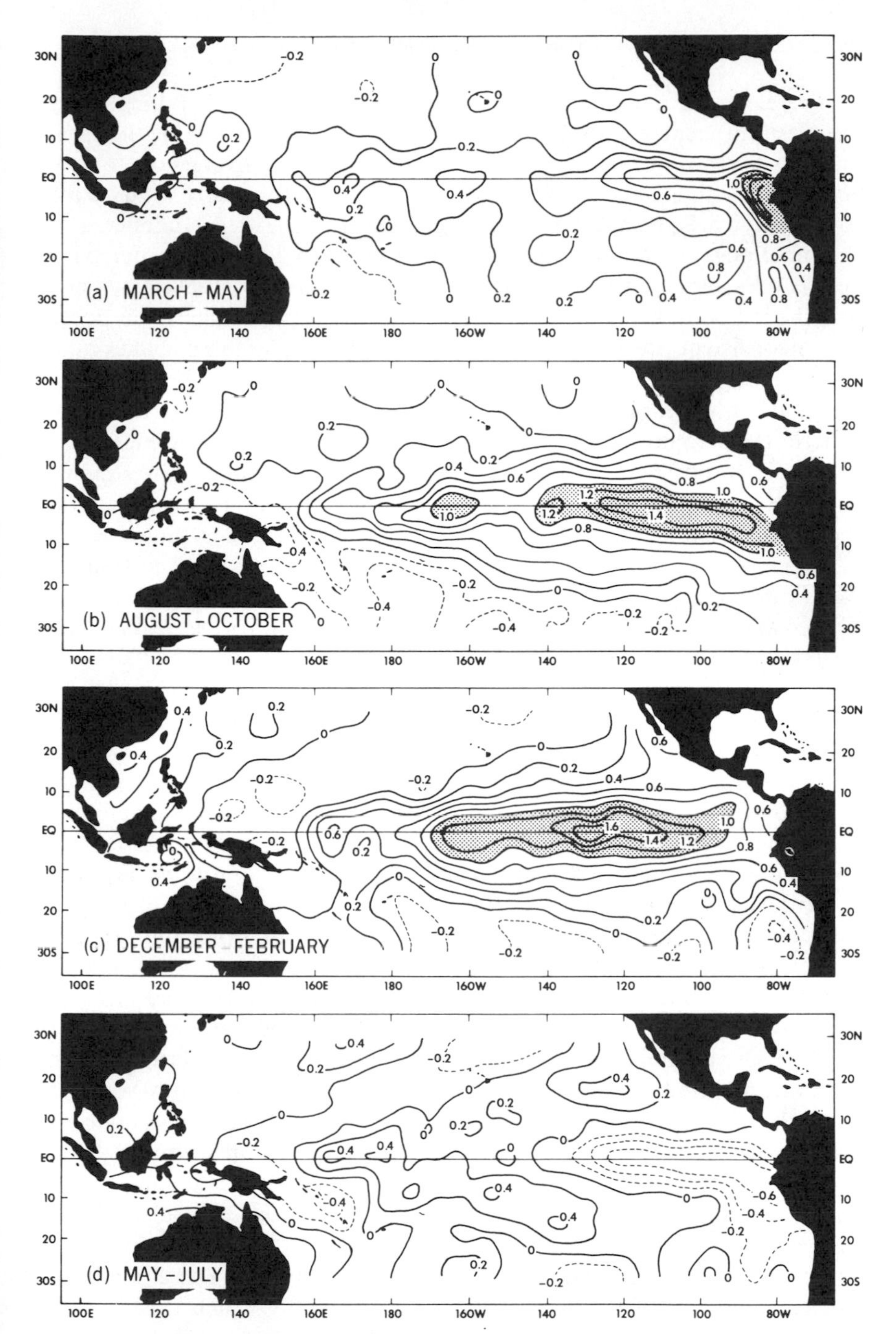

Figure 1.19. Sea surface temperature anomalies (in °C) during a typical El Niño obtained by averaging data for the episodes between 1950 and 1973. (a) March to May after the onset, (b) the following August to October, (c) the following December to February, and (d) May to July more than a year after the onset. [From Rasmusson and Carpenter (1982a).]

The eastward movement of the convergence zone that is normally over the western equatorial Pacific, the equatorward displacement of the ITCZ, and the northward movement of the South Pacific Convergence Zone are associated with a collapse of the trade winds so that westerly winds extend to 160°E by September of year (0). Farther west, over Indonesia, for example, negative rainfall anomalies peak between July and September of year (0). Eastern New Guinea, which is between drought-stricken Indonesia and the region of heavy rainfall in the central equatorial Pacific, has a weak and inconsistent precipitation anomaly pattern. The large anomalies attain their maximum areal extent towards the end of year (0) and early in year (+1), the mature phase of El Niño. This is shown in Fig. 1.19. Tropospheric temperatures are now above normal throughout the tropics and in the upper troposphere the anticyclonic couplet that straddles the region of enhanced precipitation in the central equatorial Pacific is associated with the pronounced teleconnection pattern to higher latitudes shown in Fig. 1.26. A precursor of the end of El Niño is the appearance of cold surface waters in the eastern equatorial Pacific towards the middle of year (+1). These low sea surface temperatures spread westward and inaugurate La Niña. The duration of El Niño is of the order of 18 months.

1.5 El Niño of 1982–1983

El Niño of 1982–1983 was exceptional because of the very large amplitude it attained and because of the unusual way in which it evolved. The anomalous movements of the convergence zones occur in the following order during a composite El Niño: first the ITCZ is displaced equatorward early in the calendar year and a few months later, in April and May, the convergence zone over the western Pacific starts to move eastward. El Niño of 1982–1983, and probably that of 1940–1941, was exceptional in having this order reversed. In the western Pacific the evolution of El Niño of 1982–1983 was similar to that of a composite El Niño and started with an eastward movement of the convergence zone in that region from May 1982 onwards. These developments, however, were not preceded by anomalous warm, wet conditions in the eastern equatorial Pacific. Instead the western convergence zone continued its gradual eastward migration across the Pacific during most of 1982. Figure 1.20 shows that this coincided with an eastward expansion of warm surface waters and with a collapse of the trade winds as westerly wind anomalies propagated eastward. These conditions reached the coast of South America in late 1982, by which time westerly winds prevailed over much of the western and central equatorial Pacific. Anomalies attained their peak amplitude in February 1983, by which time

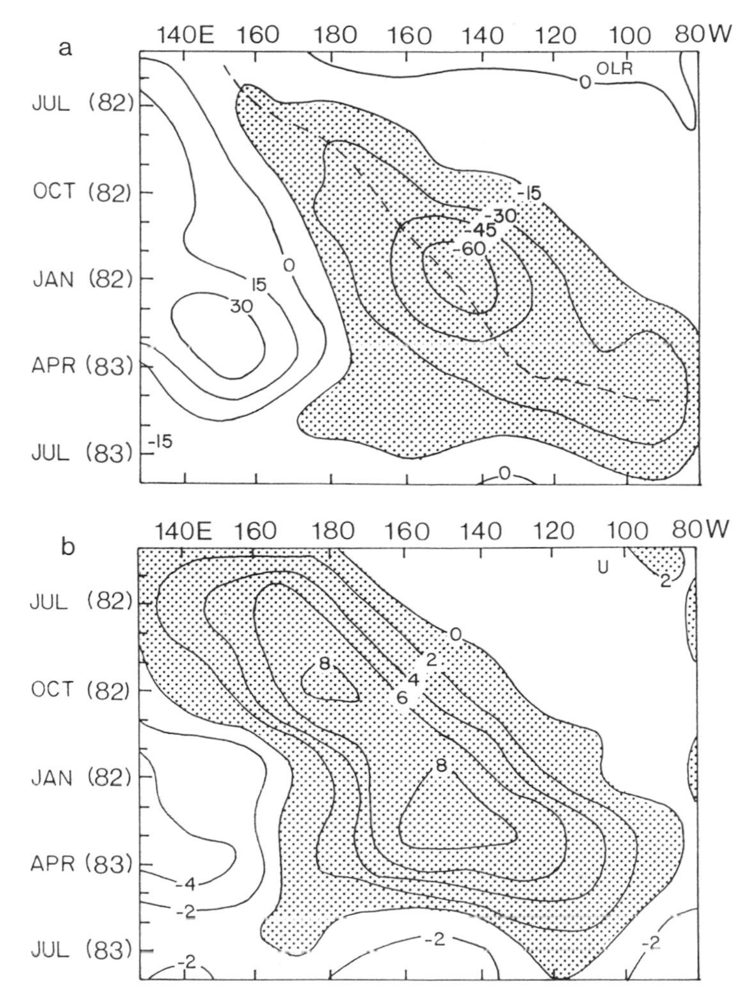

Figure 1.20. A time–longitude plot along the equator of monthly mean values, for the period 1982–1983, of (a) outgoing long-wave radiation anomaly (OLR) in watts per square meter, (b) westerly wind anomaly (5°S to 5°N) in meters per second, and (c) sea surface temperature (4°S to 4°N) in degrees Celsius. The dashed lines in (a) and (c) follow the peak OLR anomaly. [Reprinted by permission from Gill and Rasmusson (1983). *Nature* (*London*) **305**, 229–234. Copyright ©1983 Macmillan Journals Limited.] (*Figure continues.*)

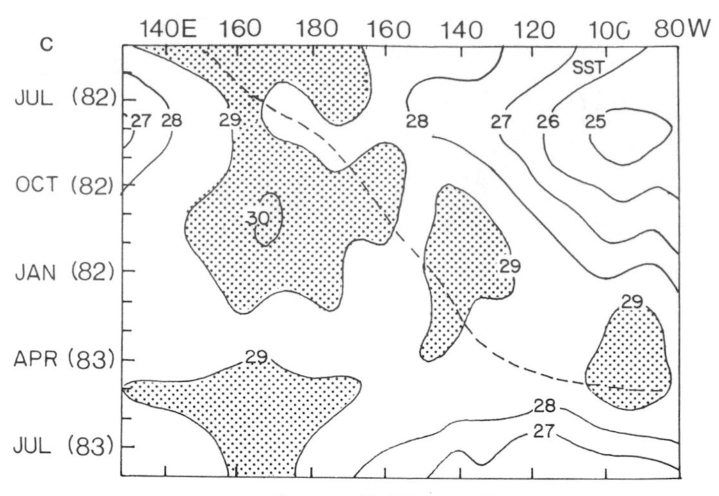

Figure 1.20 (*Continued*)

they had the structure shown in Fig. 1.21. By then the eastern tropical Pacific had become so exceptionally warm that the region of tropical storm genesis was moved far eastward. French Polynesia, noted for its benign climate, was battered by five full-blown hurricanes during El Niño of 1982–1983 (De Angelis, 1983; Rasmusson and Wallace, 1983). The region of drought, which covered parts of the Indian Ocean and the western tropical Pacific as in previous El Niño episodes, now extended over the much larger area shown in Fig. 1.21 and affected the Philippines and Hawaii.

Anomalous conditions in the western Pacific started to attenuate in early 1983 as sea level pressure at Darwin began to decline, the drought over Australia and Indonesia gradually eased, and the westerly surface winds west of the date line disappeared. In the central and eastern tropical Pacific, El Niño conditions persisted for several more months. The equatorward displacement of the ITCZ from its normal position was premature and occurred in late 1982 when unusually warm surface waters arrived in the east. Heavy rainfall in Ecuador and northwestern Peru therefore started several months in advance of the normal wet season and continued far longer than normal. Rainfall records were shattered month after month from November 1982 until June 1983. For example, Guayaquil, Ecuador, recorded a total of 2636 mm between November 1982 and April 1983, compared with a previous record of 1670 mm during El Niño of 1972. (The annual mean for the decade 1951 to 1960 was 1022 mm, of which 942 mm fell between November and April.) Ironically, southern Peru and Bolivia were experiencing serious droughts at this time. Only in June 1983 did sea

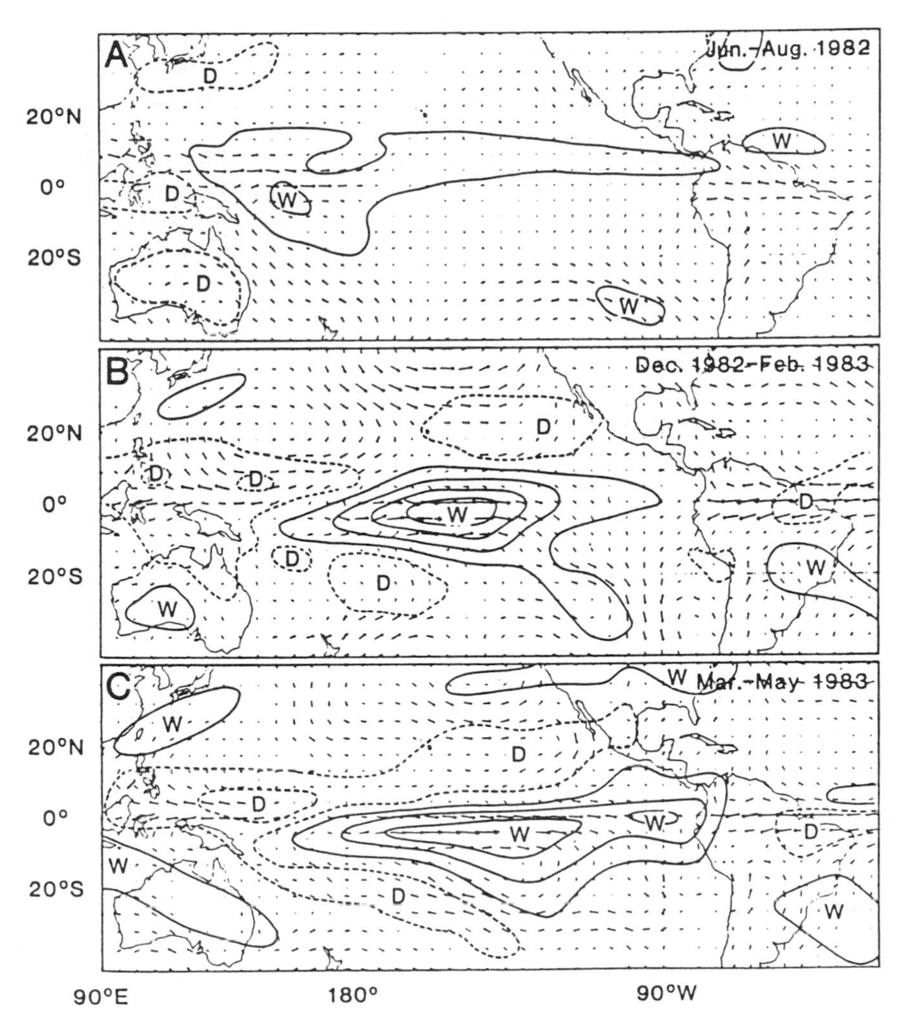

Figure 1.21. Anomalies in satellite-sensed outgoing long-wave radiation (OLR) (contours at intervals of 20 W/m^2) and wind at 850 mbar (arrows) for three seasons during El Niño of 1982–1983. Negative anomalies in OLR, indicated by the solid contours and labeled W for "wet," correspond to regions of enhanced precipitation and vice versa (D for "dry"). The longest arrows correspond to wind anomalies on the order of 10 m/sec. [From Rasmusson and Wallace (1983).]

surface temperatures start to fall, first in the central equatorial Pacific and along the coast of Peru. The southeast trades intensified rapidly, the ITCZ moved northward, and by September an intense La Niña episode was under way.

In the western tropical Pacific the evolution of El Niño of 1982–1983 was very similar to that of the composite El Niño described in Section 1.3, but its amplitude was exceptionally large. In the eastern tropical Pacific, developments were also qualitatively similar to those of a composite El Niño except that the episode started several months earlier than usual and continued longer than usual. During a composite El Niño, developments in the east come first and are followed by those in the west. This order was reversed in 1982–1983. A further difference concerns the evolution of sea surface temperature patterns. During a composite El Niño the high sea surface temperatures of the warm phase of the seasonal cycle persist when the cold phase in the east fails to develop fully. Isotherms do not propagate eastward along the equator as the event evolves. During 1982, however, isotherms did propagate eastward as shown in Fig. 1.20.

1.6 Interannual Variability in the Atlantic Sector

In June 1983, towards the end of the exceptionally intense El Niño, sea surface temperatures were high in the eastern tropical Pacific but were unusually low in the tropical Atlantic Ocean. Fig. 1.22 shows that one year later, in June 1984, the situation was reversed. The Pacific experienced an intense La Niña with low sea surface temperatures and intense trades while conditions in the tropical Atlantic resembled El Niño. Is there in general a negative correlation between variations in the tropical Atlantic and Pacific?

In 1982 and 1983 the convective zone that is normally over the western tropical Pacific moved far eastward. Winds converged onto this convective zone from both the east and west so that westerly winds prevailed over much of the Pacific while the easterly winds were intensified to the east of the convective zone. Hence the strong trade winds and associated low sea surface temperatures in the tropical Atlantic during 1983 can in part be attributed to the unusual presence of a convective zone over the eastern tropical Pacific. An associated change is increased atmospheric subsidence and reduced rainfall over a wide region centered on the mouth of the Amazon and shown in Fig. 1.15. In general, interannual rainfall variations in this region are strongly influenced by El Niño episodes in the Pacific Ocean. Out of 17 such episodes between 1906 and 1985, 16 coincided with dry years near the mouth of the Amazon (Ropelewski and Halpert, 1987). However, dry years are possible in that region when there is no El Niño; factors other than the Southern Oscillation also influence rainfall variations

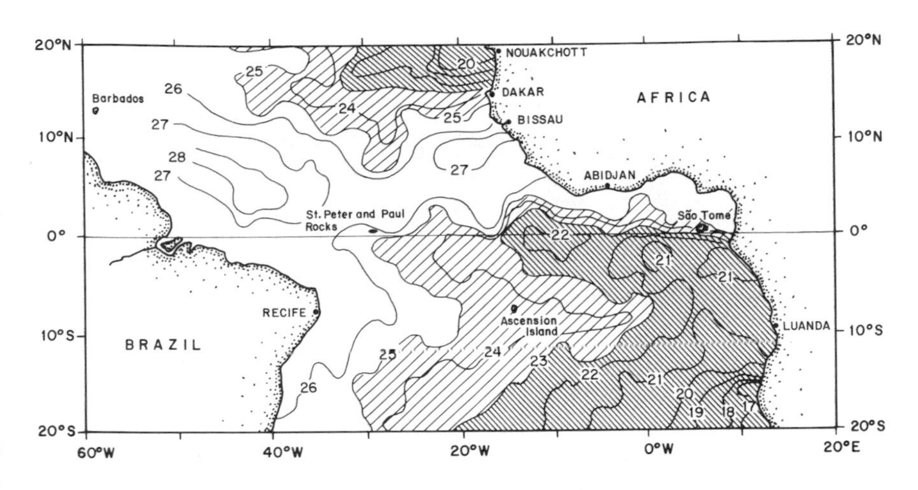

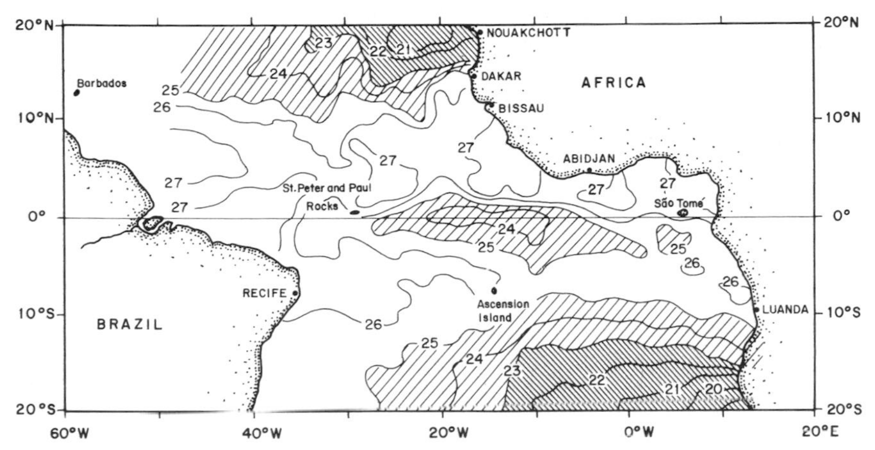

Figure 1.22. Sea surface temperatures (°C) in the tropical Atlantic in June 1983 (top) and June 1984 (bottom).

in the Amazon basin. Further evidence is the relatively low correlations, −0.32 at most, between a Southern Oscillation Index in the Pacific and rainfall in northeastern Brazil, the region centered on Recife (Kousky *et al.*, 1984; Hastenrath and Heller, 1977; Caviedes, 1973; Walker, 1928). This result is statistically significant at the 95% level but it indicates that El Niño episodes in the Pacific account for only 10% of the rainfall variance over northeastern Brazil.

Much of the interannual variability over the tropical Atlantic Ocean is independent of the Southern Oscillation. The Atlantic variability can nonetheless be expected to resemble the Southern Oscillation in many respects because time-averaged conditions and the annual cycle in the tropical Atlantic and Pacific Oceans are remarkably similar. Both regions have convective zones over the continents that bound them in the east and west and both have an ITCZ. The seasonal movements of these convergence zones are similar in the two regions so that rainfall patterns are similar. The arid coastal zone of southwestern Africa corresponds to that of Peru and Chile; the rain forests of the Amazon correspond to those of the western tropical Pacific. The zonal extent of the Atlantic is far smaller than that of the Pacific so that arid northeastern Brazil is the counterpart of the central equatorial Pacific south of the ITCZ. The low-pressure, convective zone over the Amazon basin can be viewed as the rising branch of an Atlantic-Walker Cell (Kidson, 1975). Subsidence occurs to the east, over northeastern Brazil and the South Atlantic high-pressure zone. The seasonal variations in the trade winds, which link the regions of ascending and descending motion, are seen to be very similar in the Atlantic and Pacific in Fig. 1.8. These winds strongly influence the sea surface temperature patterns, which are essentially the same for the two oceans (Fig. 1.10).

Given that conditions in the tropical Atlantic resemble those in the Pacific in many respects, it is not surprising that a phenomenon similar to El Niño occurs in the Atlantic. The best documented episode in the Atlantic, one that had an unusually large amplitude, happened in 1984. Sea surface temperatures in the southeastern tropical Atlantic were anomalously high, the trade winds were weak, the ITCZ was displaced southward from its usual position, rainfall was heavy over the normally arid coastal zone of southwestern Africa and over northeastern Brazil, and oceanic upwelling along the coast of southwestern Africa was inhibited[7] (Katz *et al.*, 1986; Weisberg and Colin, 1986; Hisard *et al.*, 1986; Horel *et al.*, 1986; Shannon *et al.*, 1986). Conditions in the tropical Atlantic in 1984 clearly resembled El Niño in the Pacific, however, there were also important differences between these episodes. The main difference concerns zonal gradients. In the tropical Pacific, interannual variations in the east and west are negatively correlated. For example, rainfall increases and surface pressure decreases in the eastern tropical Pacific during El Niño while the opposite happens in the west. Variations in the tropical Atlantic, on the other hand, have the same sign on the two sides of that ocean basin and are relatively uniform in the east–west direction. In 1984, surface pressure was low across the entire tropical Atlantic and rainfall increased in both southwestern Africa and northeastern Brazil. Interannual variability in both the tropical Atlantic and Pacific involves unusual meridional displacements of the ITCZ. The Southern Oscillation, however, has an additional feature with no counterpart in the Atlantic, namely, an east–west movement of the convergence zone over the western tropical Pacific.

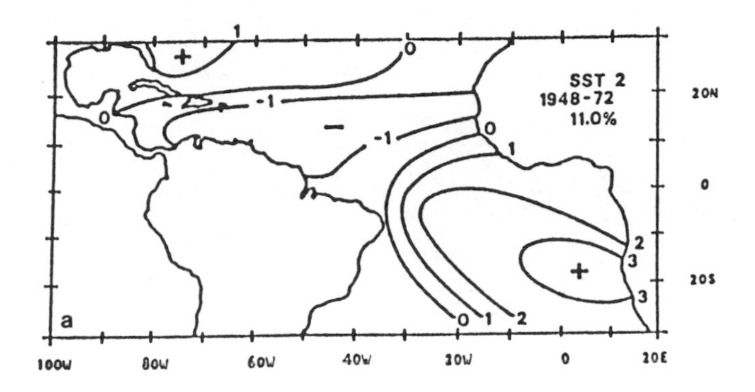

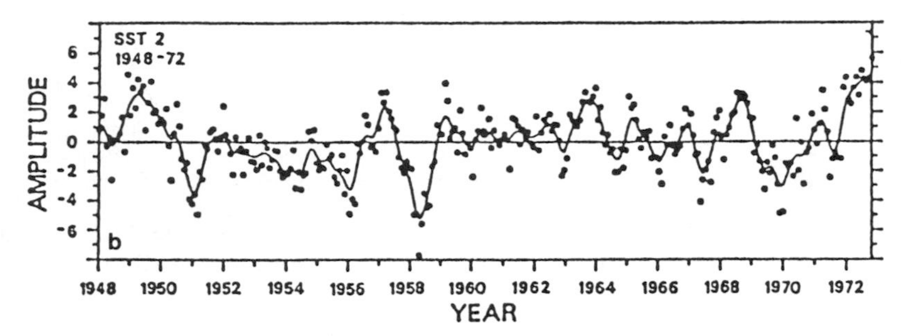

Figure 1.23. The structure of the second empirical orthogonal function for sea surface temperature variations in the Atlantic. (a) The loading (multiplied by a factor of 10) and (b) the amplitude. The curve in (b) was obtained with a 13-point low-pass filter. [From Lough (1986).]

Analyses of long time series confirm that an important mode of Atlantic interannual variability involves meridional displacements of the ITCZ and is antisymmetrical about the equator. This is evident in Fig. 1.23, which shows the structure of the second empirical orthogonal function for sea surface temperature variations in the tropical Atlantic. (The first function describes a warming and cooling that is symmetrical about the equator.) In Fig. 1.23, fluctuations south of approximately 3°N are out of phase with those farther north. The amplitude of these fluctuations is generally modest in comparison with that of the seasonal cycle, but the fluctuations nonetheless influence the position of the ITCZ and affect rainfall variations. Northeastern Brazil, for example, has unusually heavy rainfall when sea surface temperatures are high south of the equator but are cold farther north. Rainfall over southwestern Africa is correlated with that over northeastern Brazil (Berlage, 1966); the correlations between rainfall in the latter

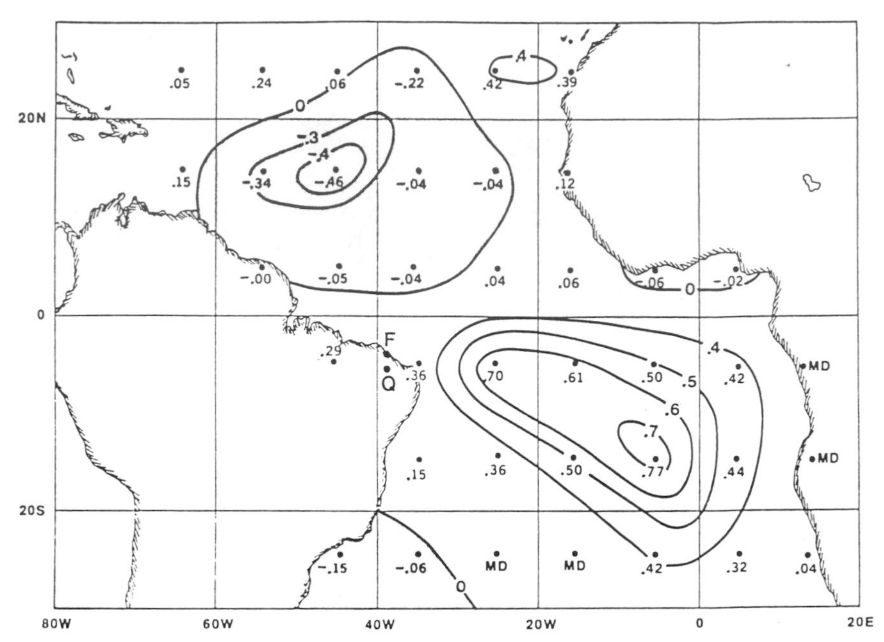

Figure 1.24. The correlation between March sea surface temperature anomalies for the 10° squares that are indicated and the March to May rainfall at the points F and Q. [From Moura and Shukla (1981).]

region and sea surface temperatures over the tropical Atlantic are shown in Fig. 1.24. Fluctuations in the surface winds are also correlated with sea surface temperatures: warm surface waters south of the equator generally coincide with relaxed trade winds (Servain and Legler, 1986).

The interannual fluctuations of the Atlantic, like the Southern Oscillation, are closely related to the seasonal cycle. Precipitation over northeastern Brazil and southwestern Africa, for example, is enhanced during the normal rainy season of those regions, the early months of the calendar year. In a regular seasonal cycle the ITCZ starts to move northward in April and May when sea surface temperatures start to fall and the trade winds intensify. In some cold, dry years this northward movement can begin as early as February and in some warm, wet years it can begin as late as June. In 1984, the ITCZ did not start to move northward until July.

During El Niño in the Pacific Ocean the equatorward displacement of the ITCZ early in the year is usually followed, a few months later, by an eastward movement of the western Pacific convergence zone. The absence of such an eastward displacement from the Atlantic—apparently the convergence zone over the Amazon basin and Gulf of Mexico cannot be easily

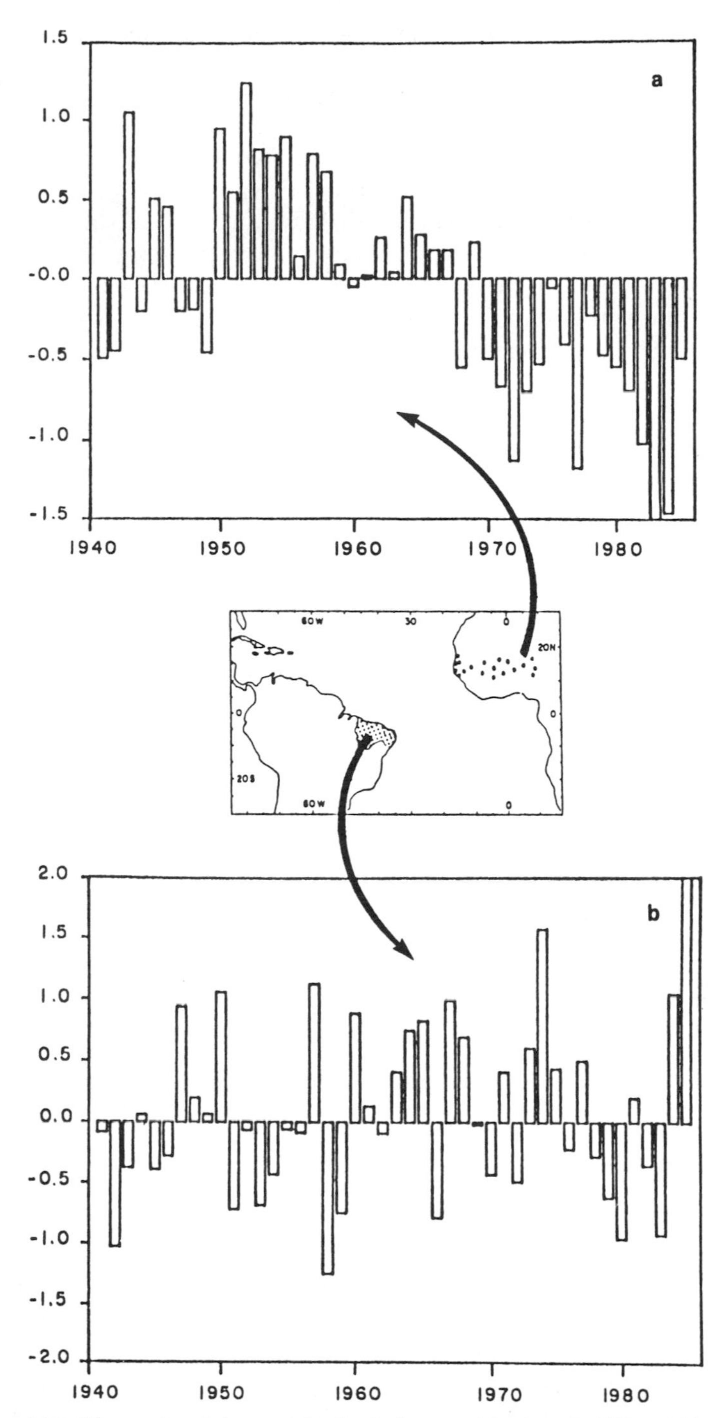

Figure 1.25. Time series of the precipitation index over (a) subtropical West Africa and (b) northeastern Brazil according to Lamb *et al.* (1986). The index for year j is $X_j = N_j^{-1}\Sigma_i(r_{ij} - \bar{r}_i)/\sigma_i$, where r_{ij} is that year's rainfall total at station i, $\bar{r}_i$ and σ_i are respectively the mean and standard deviation of station i's yearly totals, and N_j is the number of stations with complete records in year j. [From Servain and Seva (1987).]

dislodged—is part of the reason why warm episodes in the Atlantic tend to terminate early in the second half of the year. This is also why rainfall variations south of the equator are significantly different from those in the Sahel, a region south of the Sahara Desert (Fig. 1.25). The Sahel gets rain during the summer of the Northern Hemisphere when the ITCZ is a farthest north. That is almost six months after the rainy season of the coastal zones south of the equator, and is at a time when anomalous sea surface temperatures south of the equator have started to disappear. Apparently sea surface temperature variations in the tropical Atlantic have a much stronger influence on the convective zones, especially the ITCZ, during the first half of the year than later in the year. Rainfall variations in the Sahel are influenced not only by changes in the position of the ITCZ but also by changes in the moisture flux convergence into the West African sector of the Sahel (Newell and Kidson, 1984). Warm surface waters south of the equator in the Atlantic favor decreased rainfall in the Sahel. If there are also high sea surface temperatures in the eastern tropical Pacific and cold surface waters over the northern Atlantic, then the likelihood of a drought in the Sahel is enhanced. Up to 50% of the interannual variance of Sahel rainfall is related to simultaneous interannual fluctuations in such a sea surface temperature pattern (Folland *et al.*, 1986). To explain what happens in the tropical Atlantic sector, changes in the global atmospheric circulation must be taken into account. Similar factors must influence the Pacific and may be one of the reasons why the frequency of El Niño episodes varies on a decadal time scale.

1.7 Teleconnections

The Southern Oscillation involves interannual variations in the direct thermal circulation of the tropics. It also affects the atmospheric circulation outside the tropics in a more indirect manner. These effects are most pronounced during the mature phase of El Niño when sea surface temperature anomalies cover much of the tropical Pacific as shown in Fig. 1.19. During this stage, an anticyclonic couplet (Fig. 1.17) develops in the upper troposphere of the central Pacific, where the anomalous heating of the atmosphere is at a maximum. This couplet is associated with a strengthening of the Hadley Circulation and with the appearance of upper-level easterly wind anomalies near the equator. On the poleward flanks of the couplet, the subtropical jet streams are intensified and displaced equatorward, especially in the winter hemisphere. During the northern winter of 1982–1983 these conditions contributed to an unusual number of storms that brought damaging high winds to the coast of California, devastating floods to parts of California and the states that border on the Gulf of Mexico, and exceptionally heavy snowfall to the mountainous regions of the western United States (Rasmusson and Wallace, 1983). El Niño of 1982–1983, as noted earlier, was exceptional. In general the correlation between conditions over the western United States and El Niño is not very high: severe winters are possible in the absence of El Niño, and there can be mild winters during warm episodes in the tropical Pacific.

In the Southern Hemisphere the intensification of the subtropical jet stream may be a factor in the exceptionally heavy rainfall that is observed over southern Brazil, Paraguay, and northern Argentina during El Niño (Berlage, 1966; Horel *et al*., 1986).

The intensification and equatorward movement of the subtropical jet during the northern winter months of a warm episode in the tropical Pacific is one aspect of a distinctive pattern of positive and negative geopotential height anomalies over the northern Pacific and North America. Figure 1.26 shows this pattern, which resembles standing waves along a great circle route that starts at the anomalous heat source in the tropical Pacific, proceeds in a northward direction, then recurves eastward and eventually refracts equatorward over the eastern United States. In the vertical these disturbances are equivalent barotropic in structure, in contrast to the baroclinic structure of the variations associated with the Southern Oscillation in the tropics. Surface pressures and temperatures are anomalously low in the negative centers over the northern Pacific Ocean and the southeastern United States, while the opposite conditions are experienced in the positive anomaly over northwestern America (Horel and Wallace, 1981). The pattern in Fig. 1.26 is often referred to as the Pacific North American (PNA) teleconnection pattern. Wallace and Gutzler (1981) defined a PNA Index based on data from the neighborhood of the four black dots in Fig. 1.26 for the months December, January, and February. This index is significantly correlated with Southern Oscillation Indices as is evident in Table 1.1. Though El Niño accounts for approximately 20% of the variance in the centers of action of the PNA pattern, it is nonetheless of importance for long-range forecasting of conditions over northern America (Namias, 1969; Dickson and Namias, 1976).

Despite numerous theoretical studies of the teleconnections from the tropical Pacific to high northern latitudes, a satisfactory explanation for this phenomenon remains elusive. It was first proposed that the pattern in Fig. 1.26 is a stationary train of Rossby waves that emanates from the region of anomalous heating in the tropical Pacific (Hoskins and Karoly, 1981). This explanation is unsatisfactory because the pattern is essentially the same for all El Niño episodes even though the position of the heat source in the tropics changes considerably from event to event. It was then suggested that

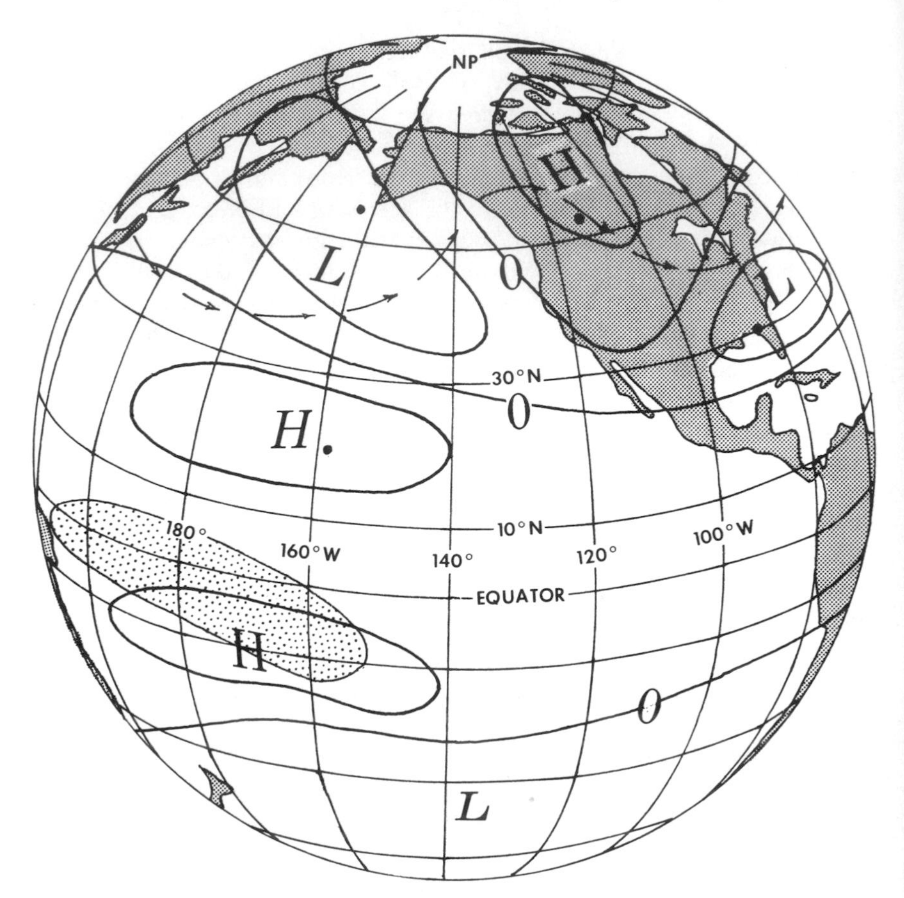

Figure 1.26. A schematic diagram of the Pacific North American (PNA) pattern of middle- and upper-tropospheric geopotential height anomalies during a Northern Hemisphere winter that coincides with El Niño conditions in the tropical Pacific. The arrows depict a midtropospheric streamline as distorted by the anomaly pattern, with pronounced "troughing" over the central Pacific and "ridging" over western Canada. Cloudiness and rainfall are enhanced over the shaded area. The dots indicate the stations used in the time series mentioned in Table 1.1. [From Horel and Wallace (1981).]

the pattern is related to an unstable mode of the midlatitude atmospheric motion. The mode, which depends on longitudinal variations of the zonal winds, can be excited by disturbances from the tropics (Simmons *et al.*, 1983). This thesis seems to be confirmed by experiments with a General Circulation Model of the atmosphere, in which the pattern appears even though climatological sea surface temperatures, without interannual variations, are specified so that the Southern Oscillation is absent from the model (N. C. Lau, 1985). Further experiments in which observed sea surface temperatures are specified to simulate a realistic Southern Oscillation reveal a change in the teleconnection pattern (Kang and Lau, 1987). Apparently there are two qualitatively similar but nonetheless distinct teleconnection patterns.

1.8 Intraseasonal Fluctuations

The dominant intraseasonal variability in the tropics occurs at periods from 30 to 60 days and is associated with a global-scale, zonally oriented circulation cell that propagates eastward. Meridionally it is confined to the tropics—the band of latitudes from 20°N to 20°S approximately—with maximum amplitude along the equator. Its structure, shown schematically in Fig. 1.27, resembles that of the Walker Circulation. Low-level winds converge onto and upper-level winds diverge from a region of rising air and low surface pressure. Fluctuations of the low-level and 250-mbar zonal winds are therefore out of phase at a fixed location. The upper-level winds have a prominent zonal wave number one structure throughout most of the tropics but there are considerable longitudinal variations. The speed of eastward propagation, 10 m/sec approximately, is greater over the eastern tropical Pacific and Atlantic Oceans than it is over the Indian Ocean and western Pacific. The intensity of the cell also varies considerably with longitude. It has a strong effect on cloudiness over the Indian Ocean and western Pacific but has a much weaker effect on the convective zones over the Amazon and Congo River basins. Its influence on cloudiness over the cold eastern tropical Pacific, a region of subsidence, is negligible. Figure 1.14 shows the oscillations over the eastern equatorial Pacific only during El Niño of 1982–1983 when the large convective zone of the western Pacific had moved eastward. This figure also shows that the oscillations occur throughout the annual cycle but exhibit large temporal irregularities in their movement and amplitude (Madden and Julian, 1971, 1972; Weickmann *et al.*, 1985; Lau and Chan, 1986; Krishnamurti *et al.*, 1973; Knutson *et al.*, 1986).

The 30- to 60-day oscillations strongly modulate the Indian monsoon convection during the summer of the Northern Hemisphere. This convection occurs predominantly in east–west cloud zones such as that in Fig. 1.28. These zones migrate northward from the equator at a speed between 1 and 2 m/sec. Over continental Asia (15°N to 30°N approximately) a cloud zone lasts for at most a month before it disappears. This coincides with a break in the monsoon conditions before the arrival of a new northward-propagating convective zone. The repeated northward migrations of the cloud zones, which are illustrated in Fig. 1.29, are initiated near the equator

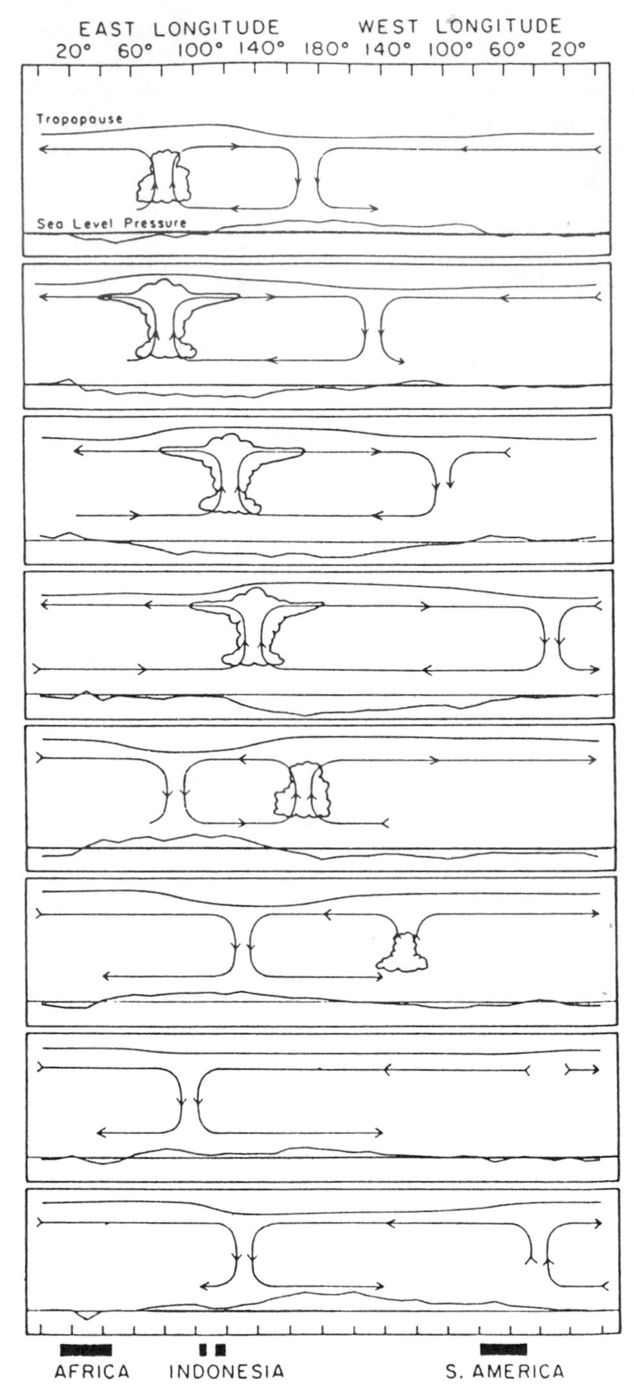

Figure 1.27. Schematic diagram of the longitude–height structure of the atmospheric circulation at different phases of the 30- to 60-day oscillation. The height of the tropopause is shown at the top and the sea level pressure disturbance at the bottom of each chart. Clouds indicate regions of enhanced convection. The interval between successive pictures is approximately 6 days. [From Madden and Julian (1972).]

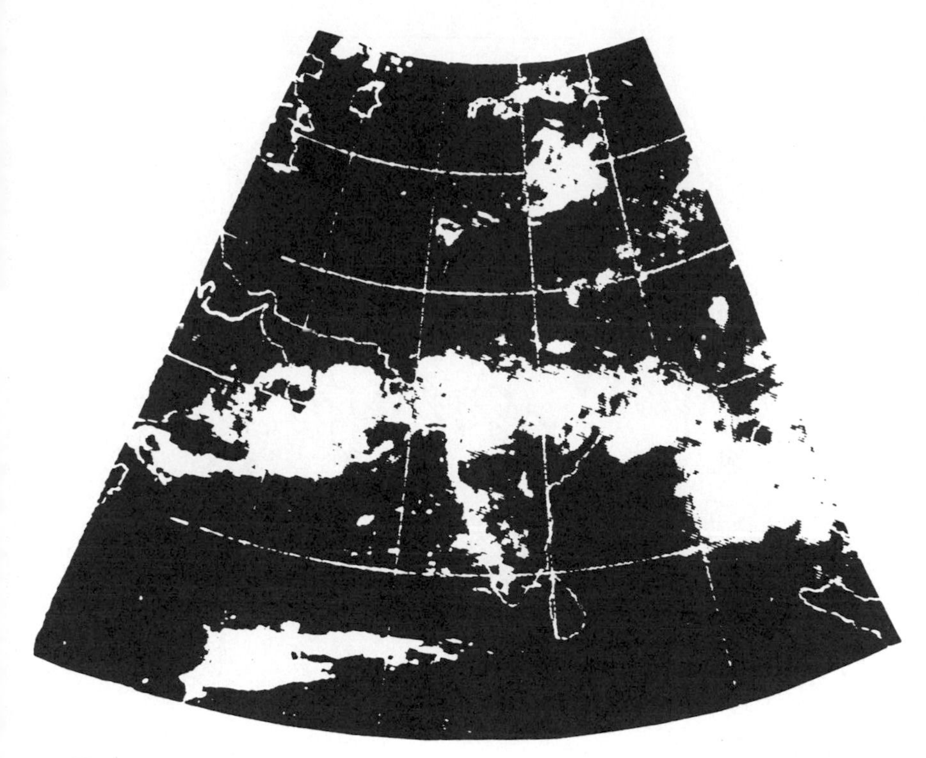

Figure 1.28. A satellite photograph that shows the Intertropical Convergence Zone over India on 8 July 1973. [Reprinted by permission from Gadgil *et al.* (1984). *Nature* (*London*) **312**, 141–143, Copyright ©1984 Macmillan Journals Limited.]

when the eastward-traveling convection cell of Fig. 1.27 arrives in the Indian Ocean sector (Yasunari, 1980; Sikka and Gadgil, 1980; Krishnamurti and Subrahmanyam, 1982).

There have been numerous studies of the 30- to 60-day fluctuations, but many of their features remain unexplained. Madden and Julian (1971) suggested that the oscillations might be equatorial Kelvin waves. However, the speed predicted by theory is greater than the observed speed even when spatial and temporal variations of the mean background flow and dissipative processes are taken into account (Chang, 1977; Lim and Chang, 1983). A further problem is the phase relation between the zonal winds and the geopotential heights; it is inconsistent with that which characterizes Kelvin waves (Murakami *et al.*, 1986). The 30- to 60-day oscillation is not simply a free Kelvin wave but is influenced by interactions between its surface wind fluctuations and the release of latent heat. In the models of Neelin *et al.* (1987) and Emanuel (1987), the anomalous easterly winds east of the region

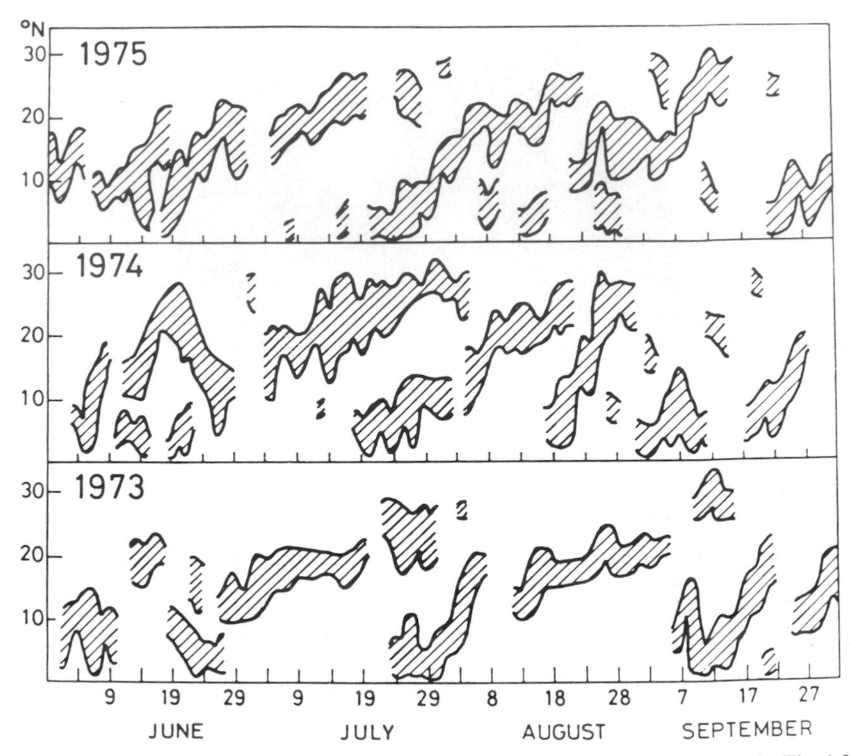

Figure 1.29. Northward movements of bands of cloudiness similar to the one in Fig. 1.28, along the meridian 90°E, during the monsoon seasons of the years 1973, 1974, and 1975. [Reprinted by permission from Gadgil *et al.* (1984). *Nature (London)* **312**, 141–143. Copyright ©1984 Macmillan Journals Limited.]

of rising moist air and the anomalous westerly winds west of this region are superimposed on the mean easterlies in the tropics so that the strength of the surface zonal winds increases east of and decreases west of the heating. Evaporation, which depends on wind speed, therefore increases to the east and decreases to the west. Under certain assumptions this causes the heating anomaly to be strengthened to the east and to be weakened to the west. The resultant eastward propagation is at a speed in reasonable agreement with that which is observed. This theory, however, does not explain the huge zonal scale of the phenomenon because the smallest scales have the largest growth rate. Furthermore, the oscillation appears in General Circulation Models that lack this evaporation–wind feedback.

In General Circulation Models the waves propagate strictly eastward when the lower boundary is idealized and has no geography (Hayashi and Sumi, 1986), but in models with realistic topography and land–ocean contrasts the waves have northward phase propagation over the Indian Ocean (Lau and Lau, 1986; Hayashi and Golder, 1986). This suggests that physical processes related to differences between the land and oceans—surface hydrology, for example (Webster, 1983)—affect the meridional propagation of the oscillation.

At present there is no evidence that the 30- to 60-day oscillations involve interactions between the ocean and atmosphere; the oscillations do not appear to depend on sea surface temperature variations with a similar time

Figure 1.30. A satellite photograph of a pair of cyclones on opposite sides of the equator along 160°E on 18 May 1986. Such pairs are associated with intense bursts of westerly wind anomalies. (Photograph provided by L. Miller.)

scale. However, wind fluctuations over the western equatorial Pacific, associated with the 30- to 60-day oscillations, could affect unstable ocean–atmosphere interactions that lead to El Niño (Chapter 6). Other high-frequency fluctuations that influence the tropical circulation, and hence the Southern Oscillation, include cold surges that stream southeastward off the Asian continent during the northern winter and that excite major outbreaks of precipitation over southeastern Asia (Chang and Lau, 1982).

Cyclone pairs that straddle the equator in the western Pacific are associated with bursts of strong westerly winds that last for a week or two. Figure 1.30 shows such a pair. There is no explanation for this phenomenon but the winds, which are confined to a few degrees latitude of the equator, can force an energetic oceanic response that affects even the eastern equatorial Pacific (Keen, 1982; Luther *et al.*, 1983; Harrison, 1987; Miller *et al.*, 1988).

Notes

1. Eguiguren (1894) first attempted a chronology of El Niño episodes in the distant past on the basis of anecdotal reports of heavy rainfall in Peru and of disruptions of the anchoveta fishery and marine bird life. Hamilton and Garcia (1986) recently surveyed the available historical evidence and concluded that "major" El Niño episodes, sufficiently intense to be recorded as being associated with heavy rains and damaging floods, occurred in 1541, 1578, 1614, 1624, 1652, 1701, 1720, 1728, 1747, 1763, 1770, 1791, 1804, 1814, and 1928.

A continuous record of the Southern Oscillation exists from 1841 onwards when routine pressure measurements were started, first in Madras (India) and subsequently at other stations in the Pacific. Quinn *et al.* (1978) used this information to compile the following list of "strong" El Niño episodes: 1845, 1864, 1877, 1878, 1884, 1891, 1899, 1911, 1918, 1925, 1926, 1941, 1957, 1958, and 1972. Figure 1.2 provides a measure of major and strong El Niño episodes.

Absent from the list of major episodes is the year 1532, when Francisco Pizarro and the Spanish conquistadores marched from Tumbes on the coast to Cajamarca, where the Inca Emperor Atahualpa was captured. Sears (1895) speculated that this celebrated march could only have occurred during a year of abundant rainfall, presumably because the probable route involved a 200-km trek across the coastal desert. Rainfall in an *año de abundancia* would have facilitated the march but it apparently would have been possible in a normal year (Cohen, 1968). Hamilton and Garcia (1986) evaluate the available evidence and conclude that 1532 was not a year of heavy rainfall.

A chronology of El Niño episodes over much longer periods is possible by using biological and geological data. Cores taken from the Quelccaya ice cap in Peru indicate that annual snow accumulation is suppressed during intense El Niño episodes (Thompson *et al.*, 1984), which helps in establishing the dates of major El Niño events as early as 1500 years before the present. A history of El Niño events in the distant past may also be available from anomalies in the annual growth rings of coral skeletons from the Galapagos Islands. Cadmium concentration in the coral shells is reduced during El Niño, presumably because of a change in the chemistry of the ocean when upwelling is reduced (Shen and Boyle, 1984). Yet another means for dating El Niño episodes is afforded by the widths of rings from trees in extratropical regions. Lough and Fritts (1985) have established that these widths are correlated with Southern Oscillation indices for the period 1853 to 1961 and can thus provide information about earlier years. This information may not be entirely reliable, however, because the teleconnections from low to high latitudes during El Niño vary considerably from episode to episode, and because the teleconnections are strongest in winter, whereas the width of rings depends primarily on summer weather.

2. Trenberth and Shea (1987) and Deser and Wallace (1987) choose to define El Niño in terms of sea surface temperature variations and the Southern Oscillation in terms of surface pressure fluctuations. Since these indices are imperfectly correlated, they find that El Niño does not always coincide with a low surface pressure difference across the tropical Pacific.

3. Measurements of outgoing long-wave radiation, though useful, are not entirely reliable indicators of convection, especially outside the tropics, because nonconvective clouds such as cirrostratus in the jet streams can also have cold tops. A subjective analysis that uses both visual and infrared satellite imagery and that takes into account factors such as the size, appearance, and texture of clouds yields a Highly Reflective Cloud Index, which is a more accurate measure of convection (Kilonsky and Ramage, 1976). Conversion of this information to an estimate of rainfall is hampered by insufficient surface rainfall measurements with which to construct a spatially and seasonally varying transfer function that relates cloud cover to areally averaged rainfall.

4. The salient feature of the southwest monsoons is a low-level jet along the highlands of Kenya and Ethiopia (Findlater, 1971). It can be thought of as an atmospheric western boundary current analogous to the Gulf Stream in the Atlantic Ocean (Anderson, 1976). The jet appears south of the equator as early as March and then progresses northwestward while it intensifies. By June it flows along the East African highlands and crosses the coast near Cape Guardafui, the northeastern corner of Africa, where the windspeed can reach 17.5 m/sec at a height of 1 km. The northward progression of the jet causes the monsoons over the Arabian Sea to have an abrupt onset. For example, analyses of wind reports from ships indicate that near 12°N, 54°E the onset of the southwest monsoons is 24 May ± a few days (Fieux and Stommel, 1976).

5. A study of the relation between sea surface temperatures and local cloudiness over the Indian Ocean shows that the probability of about 10 "cloudy" days per month over a certain area is zero if the sea surface temperature in that area is less than 26.5°C. This probability increases with increasing sea surface temperature to a maximum value of 0.65 for the warmest water (Gadgil *et al.*, 1964). Graham and Barnett (1987) confirm this result with data from each of the three tropical oceans but find that an increase in sea surface temperature above 27.5°C has little effect on the intensity of convection. The results demonstrate that a high sea surface temperature is a necessary but not sufficient condition for convection.

6. Analyses of data in terms of Complex Empirical Orthogonal Functions give another perspective on the development of the Southern Oscillation (Barnett, 1985c). This analysis views the oscillation as being quasi-periodic so that the results can be presented in terms of a phase and amplitude. Sea level pressure anomalies are found to propagate eastward across the subtropics of the north and south Pacific Ocean. This result suggests that fluctuations at Tahiti do not precede those at Darwin by a few months but lag by a few seasons. The lag changes as the time scale of the Southern Oscillation changes. Trenberth and Shea (1987) attempt to relate these results from analysis in the frequency domain to results from analysis in the time domain.

7. High sea surface temperatures along the normally cold coast of southwestern Africa have a profound effect on the ecology of that region. The paper by Shannon *et al.* (1986) discusses this issue.

Chapter 2 Oceanic Variability in the Tropics

2.1 Introduction

The appearance of unusually warm surface waters in the eastern tropical Pacific Ocean during El Niño is one of the most prominent aspects of this phenomenon. It is also the most important feature of El Niño because sea surface temperature is the only oceanic parameter that significantly affects the atmosphere. A change in sea surface temperature T can sometimes be caused by the flux of heat Q across the ocean surface:

$$\rho c_{\mathrm{p}} D T_t = Q \tag{2.1}$$

In this equation T is assumed to be the temperature of a mixed surface layer of depth D, t is time, ρ is the mean ocean density, and c_{p} is the heat capacity of water. If D is determined by the local fluxes of heat and momentum across the ocean surface, then Eq. (2.1) can serve as the basis for local models for the prediction of sea surface temperature. Such models, which neglect horizontal variations and hence advection, are reasonably successful at reproducing sea surface temperature variations outside the tropics (Kraus, 1977). There are also parts of the tropics where Eq. (2.1) is valid, but in those regions the depth D of the mixed surface layer is not determined locally, it is influenced by wind conditions over the entire ocean basin. To understand why this occurs, it is necessary to take into account that, in low latitudes, the lower boundary of the mixed surface layer is a zone of extremely large vertical temperature gradients. This zone, known as the thermocline, separates the warm surface waters from the deeper cold water. Spatial variations in the depth of the thermocline are associated with large horizontal density gradients that reflect the dynamical response of the ocean to the surface winds. For example, the thermocline is deep in the western and shallow in the eastern tropical Atlantic and Pacific Oceans because the westward trade winds drive the warm surface waters westward and expose cold water to the surface in the east. The pressure gradient associated with the zonal slope of the thermocline in the equatorial plane balances the windstress. When the winds relax, the thermocline rises in the west and deepens in the east. If the heat flux across the ocean surface remains unchanged, and if entrainment of cold water across the lower boundary of the mixed layer is small, then Eq. (2.1) implies that sea surface temperatures in the west will rise as the value of D decreases. In reality sea surface temperatures in the western equatorial Pacific fall during El Niño when the thermocline in that region shoals because the flux of heat into the ocean decreases (Meyers *et al.*, 1986). The equation with which to calculate the change in sea surface temperature is Eq. (2.1) but the depth D of the thermocline is not determined locally. The principal cause of changes in the depth of the tropical thermocline is a horizontal redistribution of warm surface waters in response to changes in the large-scale winds (Wyrtki, 1975; Merle, 1980a and b; Duing and Leetmaa, 1980; Stevenson and Niiler, 1983; Molinari *et al.*, 1985).

Advection strongly influences sea surface temperatures over much of the tropical Pacific, and especially in the eastern part of the basin, so that Eq. (2.1) fails in those regions. During El Niño, for example, the warming of the eastern tropical Pacific is caused by the eastward advection of warm surface waters associated with the elevation of the thermocline in the west and its deepening in the east. This warming by advection competes with cooling caused by upwelling. The prevailing trade winds drive poleward Ekman drift and hence induce equatorial upwelling that leads to a tongue of cold surface waters along the equator. In Fig. 1.10 this tongue is pronounced during the summer and autumn of the Northern Hemisphere when the easterly winds along the equator are intense, and is almost absent in the spring when the winds are weak. Alongshore winds similarly induce upwelling along the coast of South America by driving offshore Ekman drift. Since the vertical velocity component vanishes at the ocean surface, mixing processes determine the extent to which upwelling affects sea surface temperatures. The upwelling zones are narrow but currents advect the cold surface waters over considerable distances. Surface waters are warmer in the western than eastern tropical Pacific so that bands of high and low sea surface temperatures in Fig. 1.10 are indicative of eastward and westward surface currents, respectively.

The heat flux across the ocean surface, advection, upwelling, and mixing processes all influence sea surface temperatures in the tropical oceans. A change in the balance between these processes causes sea surface temperature variations. On interannual and seasonal time scales such a change in the balance reflects profound changes in the circulation of the entire tropical Pacific Ocean. The circulation is determined primarily by the surface winds. Hence, to understand El Niño and the seasonal cycle—these two phenomena have much in common—it is necessary to understand how the ocean responds to variable winds.

Dramatic basinwide changes in the oceanic circulation occur not only in the Pacific but also in the Atlantic and Indian Oceans. The tropical Atlantic has a counterpart to El Niño; such an event occurred in 1984 when sea surface temperatures off southwestern Africa were exceptionally high. In the Indian Ocean the seasonal reversal of the monsoon winds causes large changes in the sea surface temperatures off eastern Africa and causes various currents, most spectacularly the intense Somali Current, to reverse direction seasonally.[1] There is ample evidence that basinwide changes in the circulations are possible seasonally and interannually in each of the three tropical oceans. From studies of the similarities and differences between these oceans, oceanographers are gaining an understanding of how changes in the windstress patterns and changes in the basin geometries affect the oceanic response. The tropical Pacific and Atlantic Oceans have similarities because both are forced by the trade winds; they have differences because the wind fluctuations are not identical and because the dimensions and geometries of the two basins are vastly different. The dimensions of the Atlantic and Indian Oceans are approximately the same but the monsoons over the Indian Ocean have little in common with the trade winds. The Atlantic and Indian Oceans, in addition to being of interest in their own right, shed light on El Niño.

Any discussion of oceanic variability is handicapped by the severe lack of data. Sea level measurements at islands form an important data set but there are practically no islands in the eastern half of the tropical Pacific Ocean. Regular measurements from commercial vessels leave large parts of the tropical Pacific unsampled, especially in the Southern Hemisphere. Instrumented moorings are expensive and very few can be maintained for extended periods. The rich spectrum of oceanic variability complicates matters further. Fluctuations with a period near 3 weeks are very energetic in the central and eastern equatorial Pacific and, near 110°W for example, can affect the thermal structure as much as seasonal variations do (Hayes *et al.*, 1982). Measurements from research and commercial vessels alias these fluctuations. Interannual variability has such a large amplitude that the definitions of mean conditions and of the climatological seasonal cycle depend heavily on the period over which the data were collected.

The paucity of the available measurements precludes an accurate quantitative description of the oceanic circulation on the basis of measurements only. To answer questions about the mass and heat budgets of the tropical oceans, questions about the role that upwelling and downwelling play in closing the circulation for example, it is necessary to use the available data together with a dynamical model of the ocean. Initial efforts to estimate equatorial upwelling (Wyrtki, 1981; Enfield, 1986; Bryden and Brady, 1985) and the meridional heat transport (Wunsch, 1980, 1984; Roemmich, 1983) combined the data with relatively simple models. The models usually calculate surface Ekman drift from the winds and determine geostrophic currents from density sections by research vessels. Reasonable assumptions about vertically integrated transports determine absolute velocities. These approaches, because the models are simple, make full use of neither the equations of motion nor the information available in the surface winds, which do far more than drive Ekman drift. To exploit all the available information a General Circulation Model of the ocean is required. Ideally the model should be capable of realistic simulations of the oceanic circulation and should assimilate the available data to mitigate its deficiencies. Such a model has not been developed yet but a reasonably realistic model, which uses the measurements as initial conditions, is available.[2] Although further improvements in the model are necessary, the results from simulations of the seasonal cycles of the tropical Atlantic and Pacific will be used in the following discussion of oceanic conditions.

2.2 Mean Conditions

Sea surface temperatures are much lower in the eastern than western tropical Pacific Ocean except for a band of warm surface waters that extends across the width of the ocean basin just north of the equator. This sea surface temperature pattern, shown in Fig. 1.10, is strongly influenced by advection. Westward currents transport cold water that wells up along the American coast. Eastward flow coincides with the band of warm surface waters to the north of the equator. Figure 2.1 depicts these alternating bands of eastward- and westward-flowing currents that characterize the oceanic circulation in the tropical Pacific and Atlantic Oceans even though the prevailing winds are westward. The eastward surface current, which is between 3°N and 10°N approximately, is known as the North Equatorial Countercurrent—it flows counter to the prevailing trade winds—and typically has surface speeds of 50 cm/sec. This narrow current is sandwiched between two broad westward surface currents, the North and South Equa-

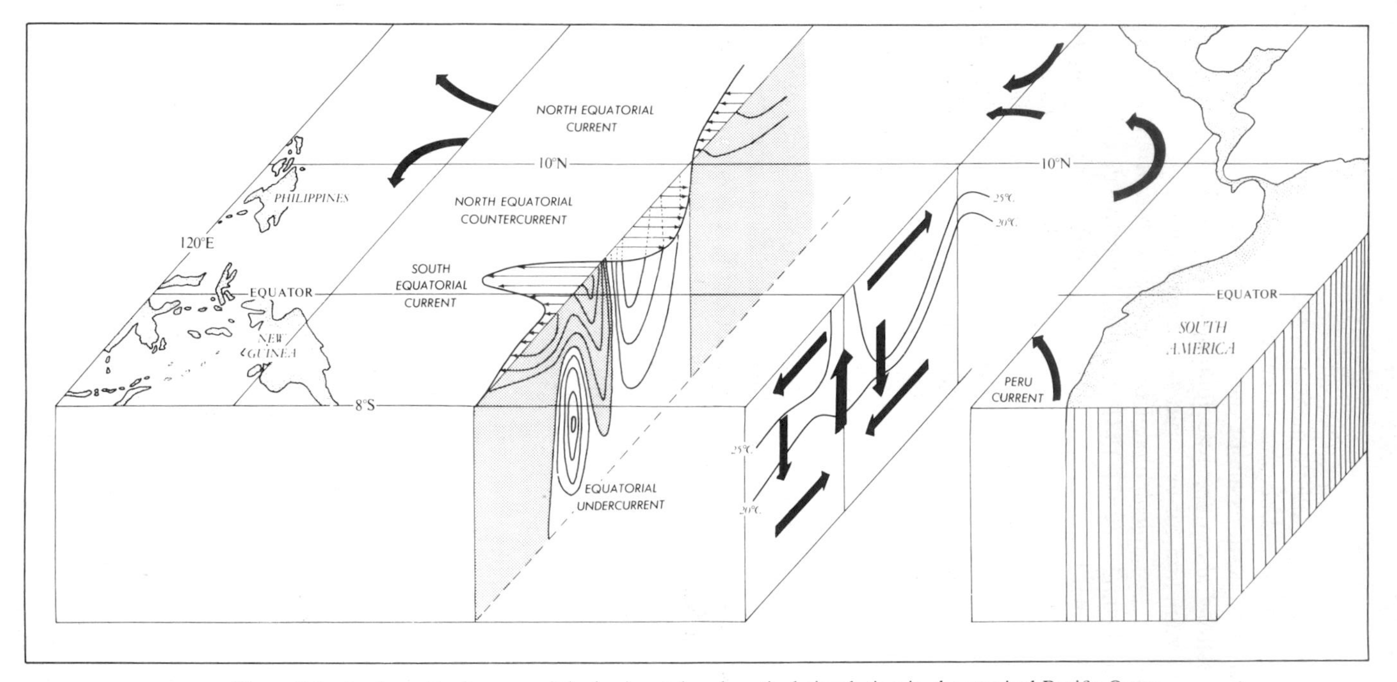

Figure 2.1. A schematic diagram of the horizontal and vertical circulation in the tropical Pacific Ocean.

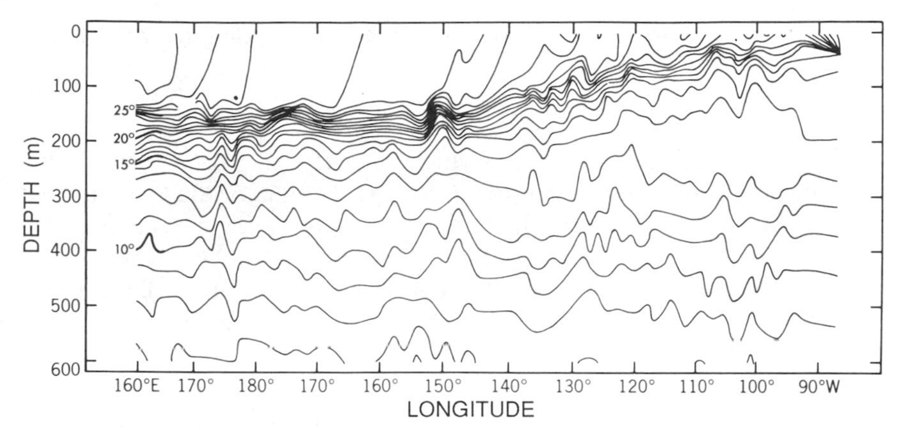

Figure 2.2. Temperature (°C) as a function of depth and longitude along the equator in the Pacific Ocean as measured in 1963 by Colin *et al.* (1971).

torial Currents, to the north of 10°N and to the south of 3°N, respectively. The northern current is relatively weak—its speeds are generally less than 20 cm/sec—but the southern one has an intense core between the equator and 3°N in which speeds can exceed 100 cm/sec. An eastward South Equatorial Countercurrent in the neighborhood of 9°S is usually present west of the date line in the Pacific but it becomes increasingly sporadic in an eastward direction (Merle *et al.*, 1969). Data that describe this current are scant. It is unclear whether a similar current is present in the Atlantic Ocean.

The sea surface temperature patterns reflect subsurface thermal gradients. Along the equator, for example, the decrease in sea surface temperature from west to east is associated with a shoaling of the isotherms as shown in Fig. 2.2. A striking feature of this figure is the region of very high thermal gradients, the thermocline, that separates the warm surface waters from the colder water at depth. The thermocline has a depth of about 150 m in the western equatorial Pacific; it surfaces in the east. This slope implies an eastward pressure force in the upper ocean if a hydrostatic balance is assumed and if pressure gradients in the deep ocean are assumed to be negligible. Under these assumptions, surfaces of constant pressure, including the sea surface slope downward from west to east. Sea level is estimated to be 50 cm higher in the western than eastern equatorial Pacific. This zonal slope and the associated eastward pressure force are not confined to the immediate neighborhood of the equator but are present throughout much of the tropics. Far from the equator and below the directly wind-driven surface layer this pressure force p_x induces equatorward geostrophic motion v in both hemispheres according to the equation

$$-fv + \frac{1}{\rho}p_x = 0 \tag{2.2}$$

Here f is the Coriolis parameter, which changes sign when the equator is crossed, and ρ is the density of the ocean. At the equator the Coriolis force vanishes ($f = 0$) and fails to balance the pressure force. It seems plausible that the fluid that converges onto the equator will, in part, rise to the surface to sustain the divergent Ekman drift and will, in part, accelerate eastward, down the pressure gradient, thus giving rise to an eastward current in the equatorial thermocline. Such a current is indeed observed below the wind-driven westward surface layers and is known as the Equatorial Undercurrent.[3] This swift jet—it can attain speeds in excess of 150 cm/sec—has its core in the thermocline and extends vertically over a depth of approximately 150 m. Although its width is only of the order of 300 km, the current is continuous over a longitudinal distance of more than 10,000 km in the Pacific and 4000 km in the Atlantic Ocean. The Equatorial Undercurrent is effectively a ribbon that marks the location of the equator. Its seasonal appearance in the Indian Ocean—it is present only during the northeast monsoons when the winds near the equator have a westward component—and its disappearance from the Pacific Ocean towards the end of 1982 when westerly winds replaced the trade winds corroborate the thesis that westward winds, which maintain an eastward pressure force, drive the Equatorial Undercurrent.

There are occasions when an eastward equatorial undercurrent appears in the far eastern equatorial Atlantic and Pacific Oceans, where the prevailing winds have a westerly component. This undercurrent is below and not in the thermocline because, at the depth of the thermocline, which slopes downward to the east, the pressure force is westward and opposes an eastward current. The deeper undercurrent appears to be an inertial overshoot of the Equatorial Undercurrent to the west (Wacongne, 1988).

Below the thermocline, to a depth of more than a thousand meters, the zonal flow changes direction repeatedly as shown in Fig. 4.3 (Luyten and Swallow, 1976; Leetmaa and Spain, 1981; Eriksen, 1982). The discussion of these deep equatorial jets, which are far weaker than the Equatorial Undercurrent in the thermocline, and for which there is as yet no explanation, is continued in Section 4.4.

The slope of the isotherms in a meridional plane indicates a geostrophic balance for the various zonal currents in the tropics. In Fig. 2.3 the eastward North Equatorial Countercurrent has isotherms that slope steeply from the deep trough of the thermocline near 3°N to its ridge near 10°N. Farther north, in the region of the westward North Equatorial Current, isotherms slope downward. Direct measurements of these off-equatorial currents come primarily from satellite-tracked drifter buoys and from analyses of ship drift data (Wyrtki *et al.*, 1981; Hansen and Paul, 1987; Richardson and McKee, 1984; Richardson and Walsh, 1986; Richardson and Reverdin, 1987). Indirectly the currents are inferred from density measurements by making the geostrophic approximations. Corrections are necessary to account for Ekman drift near the surface. In Fig. 2.3 the shallow isotherms are elevated but the deeper ones are depressed within a few hundred kilometers of the equator. Such a spreading of the equatorial thermocline is indicative of a geostrophic balance for both the westward surface flow and the deeper eastward Equatorial Undercurrent (Lukas and Firing, 1983). This is remarkable because, right at the equator, the Coriolis parameter vanishes. It appears that the terms $-fu$ and $1/\rho p_y$ can dominate in the meridional momentum equation, even close to the equator. This is not always the case, however, because frequently the equatorial currents are ageostrophic. This happens whenever cross-equatorial winds drive cross-equatorial currents and maintain a north–south pressure force. Exploitation of the geostrophic equations is impractical even when the zonal flow at the equator is geostrophic because small errors in measurements of the latitudinal density gradients imply large errors for the inferred zonal currents. The only reliable estimates of currents at the equator come from direct measurement,[4] examples of which appear in Figs. 2.14 and 2.17.

Salinity gradients in the ocean contribute to density gradients but generally less so than thermal gradients. In Fig. 2.3, for example, the structure of the vertically integrated density (the dynamic topography) between 3°N and 10°N reflects the thermal and not the salinity field. In the western equtorial Atlantic, density at a fixed depth in the upper ocean increases in an eastward direction even though the effect of salinity gradients is in the opposite sense. The sources (evaporation) and sinks (precipitation) for salinity in the tropics are at the ocean surface so salinity can, to some extent, be viewed as a tracer of subsurface motion. Thus the salinity distribution in Fig. 2.3 suggests equatorward flow at the depth of the thermocline in the Southern Hemisphere. In the Atlantic Ocean the salinity field is characterized by a core of very high values, in excess of 36 parts per thousand, that practically coincides with the core of the Equatorial Undercurrent. The origin of this saline water is believed to be the southwestern side of the tropical Atlantic Ocean, where evaporation presumably is high. Studies that trace the saline water across the Atlantic Ocean suggest that the Equatorial Undercurrent bifurcates in the Gulf of Guinea. One branch continues southward as a coastal undercurrent along southwestern Africa and the other branch becomes a westward undercurrent along the northern coast of the Gulf (Voituriez, 1983; Bubnov and Yegorikhin, 1980).

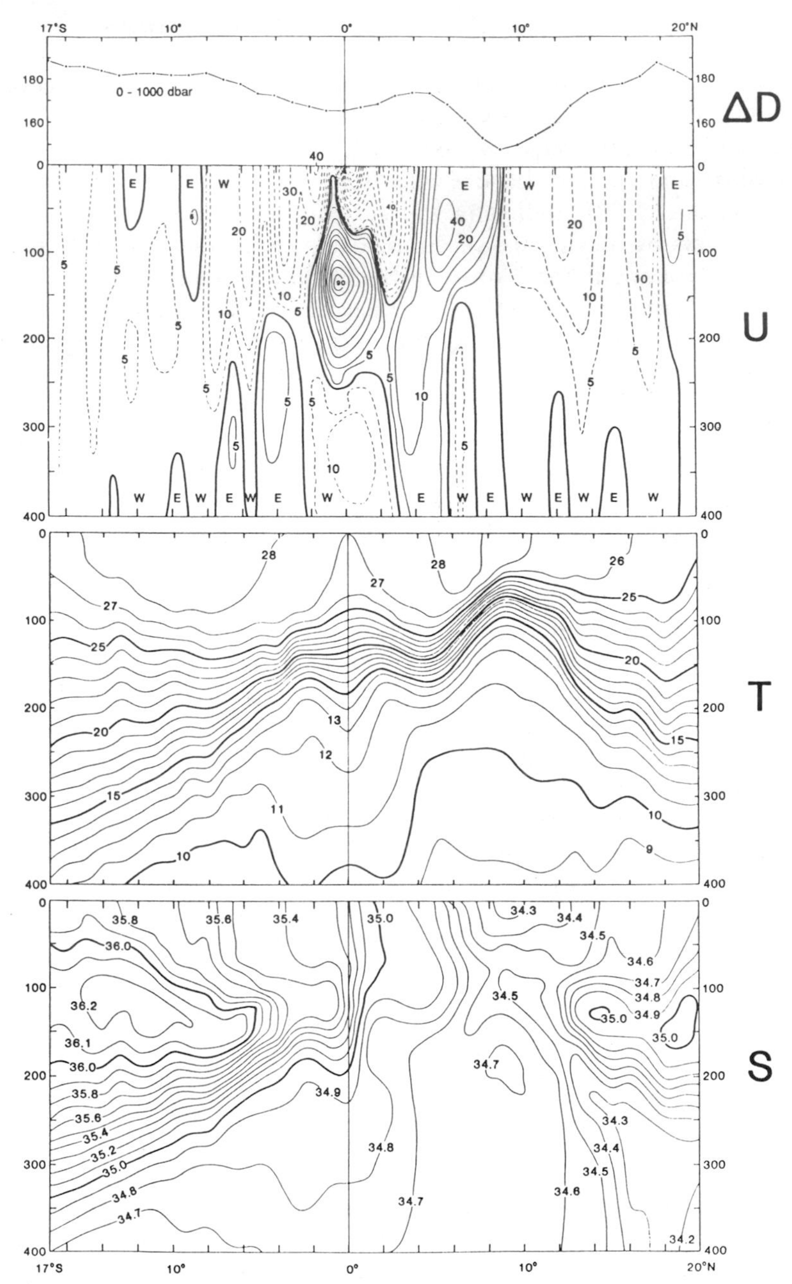

Figure 2.3. Mean distribution, as a function of depth (in meters) and latitude, of dynamic height (ΔD) relative to 1000 dbar in dynamic centimeters, of zonal geostrophic current (U, cm/sec), of temperature (T, °C), and of salinity (S, %) between Hawaii and Tahiti on the basis of measurements during 1979 and 1980. From Wyrtki and Kilonsky (1984). The dynamic height, determined by integrating the density vertically from 1000 dbar to the surface, gives sea level if horizontal pressure gradients at 1000 m are assumed to be zero. The letters E and W denote eastward and westward, respectively. Dashed contours indicate westward currents.

The meridional motion in the tropical Atlantic and Pacific Oceans is poleward Ekman drift in the surface layers and equatorward geostrophic motion at the depth of the thermocline. This circulation implies equatorial upwelling, which can be estimated from the divergence of the horizontal currents as measured with satellite-tracked drifter buoys. Such measurements show that in addition to intense equatorial upwelling there is strong downwelling in the adjacent region of the North Equatorial Countercurrent (Hansen and Paul, 1987). Studies of chemical tracer distributions near the equator (Fine *et al.*, 1983) indicate that the vertical velocity component has a large amplitude only in the upper 150 m of the tropical oceans.[5] This explains why the sea surface temperature is unaffected by equatorial upwelling in regions where the thermocline is deep, in the western Pacific and Atlantic, for example.

To determine the roles that upwelling and downwelling play in closing the oceanic circulation it is necessary to turn to a General Circulation Model of the ocean as explained in Section 2.1. The picture of the circulation that emerges from the model agrees with that shown schematically in Fig. 2.1. The upwelling near the equator and downwelling in adjacent regions are intimately related to the longitudinal variations in the transports of the horizontal currents shown in Fig. 2.4. The transport of the eastward North Equatorial Countercurrent decreases in a downstream direction because the loss of fluid to downwelling and to Ekman drift across 10°N exceeds the gain from Ekman drift across 3°N. The downwelling is into the thermocline, where the motion is equatorward, towards the Equatorial Undercurrent. There is similar equatorward flow in the thermocline in the Southern Hemisphere. Equatorial upwelling transfers the convergent fluid to the surface layers. This diminishes the transport of the Equatorial Undercurrent in a downstream direction—in the Pacific the transport decreases only in the eastern half of the basin—and augments that of the westward South Equatorial Current, which has poleward Ekman drift in its surface layers. The upwelling is most intense in the central Pacific and western Atlantic but is strongly confined to the upper ocean (see Note 5) and attenuates rapidly with increasing depth below 60 m as shown in Fig. 2.5. Figures 2.4 and 2.5 are from a simulation of the seasonal cycle of the tropical Pacific Ocean with a General Circulation Model (see Note 2) that is forced with the climatological seasonal winds as calculated by Hellerman and Rosenstein (1983). Because of the considerable interannual variability of the tropical Pacific, these winds may not be realized in any specific year and may not represent the seasonal cycle as calculated for a period different from and shorter than that used by Hellerman and Rosenstein. This factor, in addition to model deficiencies, can cause conditions observed over a specific short period to differ from those in Figs. 2.4 and 2.5.

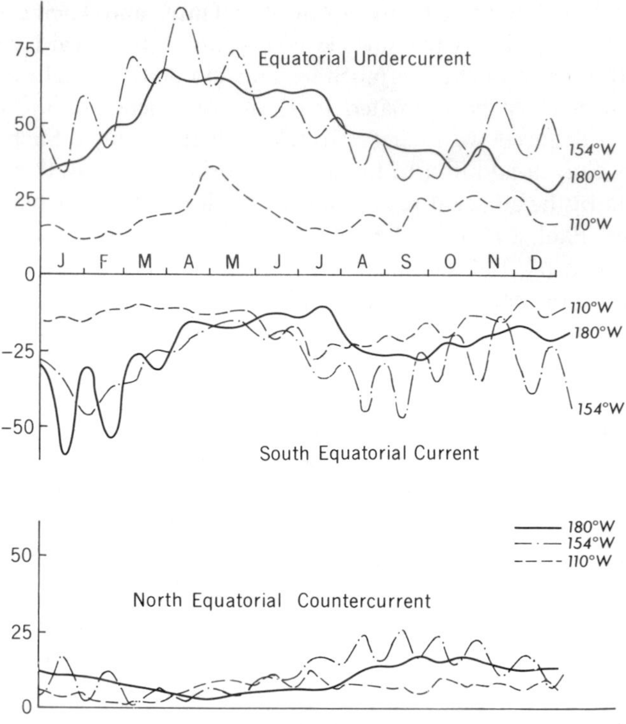

Figure 2.4. Transport (in 10^6 m^3/sec) of the zonal currents across 180°W, 155°W, and 110°W as a function of time in a simulation of the seasonal cycle of the tropical Pacific Ocean with a General Circulation Model (see Note 2). The South Equatorial Current is defined to be the westward flow between 10°S and 6°N; the North Equatorial Countercurrent is the eastward flow between 3°N and 10°N; and the Equatorial Undercurrent is the eastward flow between 3°N and 3°S. To calculate transports, vertical integrals were limited to the upper 317 m of the ocean. Fluctuations with a period of approximately one month are attributable to instabilities of the currents, as discussed in Section 2.7. [From Philander *et al.* (1987).]

The downstream decrease in the transports of the two eastward currents causes both to peter out before they reach the eastern boundaries of the Atlantic and Pacific Oceans.[6] The transports of the westward currents, on the other hand, increase in a downstream direction. In the case of the South Equatorial Current in the Atlantic Ocean this increase continues right to the coast of Brazil, where the fluid is absorbed by the northwestward North Brazil Current[7] as shown schematically in Fig. 2.6. This current in turn feeds the Equatorial Undercurrent and the North Equatorial Countercurrent (Flagg *et al.*, 1986; Metcalf and Stalcup, 1967; Cochrane *et al.*, 1979). The tropical Pacific does not have a counterpart to the North Brazil Current. Information about the circulation in the far western tropical

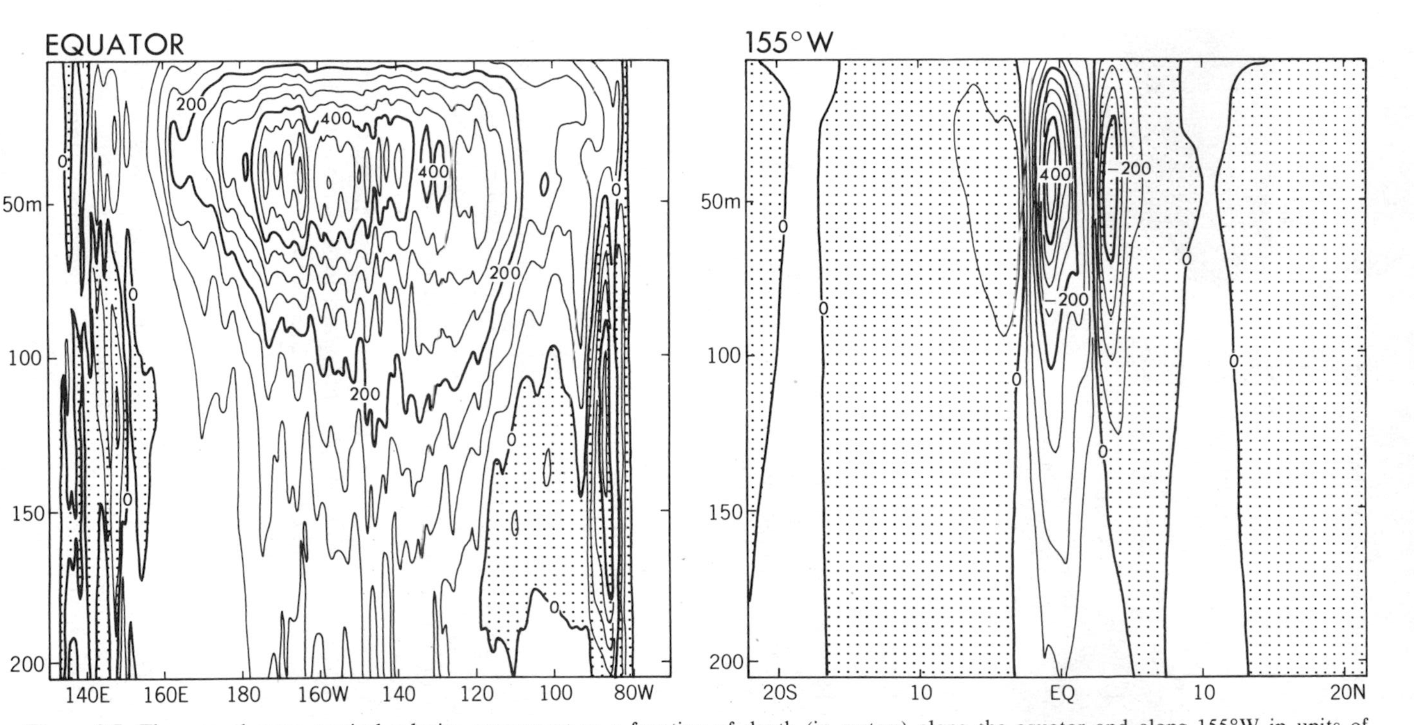

Figure 2.5. The annual mean vertical velocity component as a function of depth (in meters) along the equator and along 155°W in units of centimeter per day in a simulation of the seasonal cycle of the tropical Pacific Ocean with a General Circulation Model (see Note 2). A comparison of this figure with Fig. 2.24 shows that high-frequency fluctuations of the vertical velocity component are comparable in amplitude to its mean value. [From Philander *et al.* (1987).]

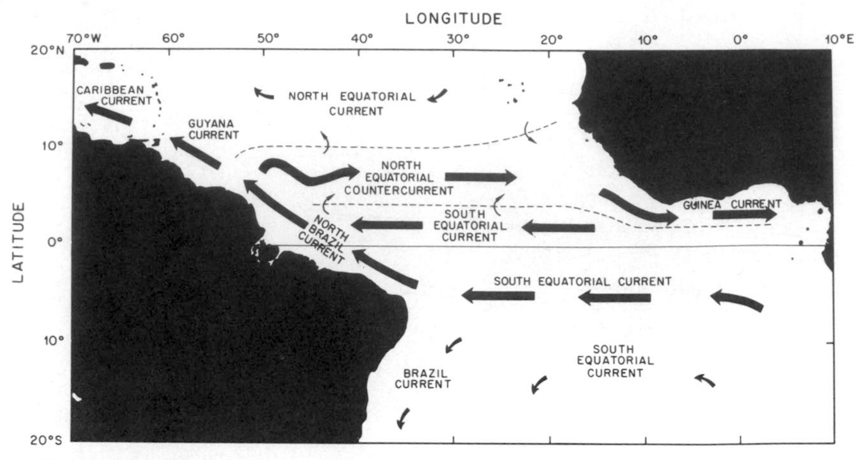

Figure 2.6. A schematic map that shows the major currents of the tropical Atlantic between July and September when the North Equatorial Countercurrent flows swiftly eastward into the Guinea Current in the Gulf of Guinea. From January to May the North Equatorial Countercurrent disappears and the surface flow is westward everywhere in the western tropical Atlantic. [From Richardson and Walsh (1986).]

Pacific is sketchy but it appears that, to the west of the date line, water from the North and South Equatorial Currents converges onto the equatorial region to feed the eastward currents (Lindstrom *et al.*, 1987). Of particular importance is a northwestward subsurface current along the coast of New Guinea in the Southern Hemisphere (Lukas, 1987) and the southward-flowing Mindanao Current along the eastern coast of the Philippines (Cannon, 1966; Nitani, 1972). In the far western equatorial Pacific there appears to be a loss of warm surface waters to the Indian Ocean but the volume is difficult to estimate (Fine, 1985).

The westward North and South Equatorial Currents, in addition to their equatorward loss of mass, also have a poleward loss of mass so that they form part of the subtropical ocean gyres in the Atlantic and Pacific Oceans and are linked to the Gulf Stream and Kuroshio Current, for example.

The connection between the tropical and extratropical circulations is important because the oceanic circulation effects a transport of heat from the equatorial zone, which gains heat across the surface throughout the year, to higher latitudes, where the ocean, on the average, loses heat. Figure 2.7 shows one estimate of the annual mean heat flux across the ocean surface. This flux Q, which also influences sea surface temperatures, can be written

$$Q = QS - QL - QE - QH \tag{2.3}$$

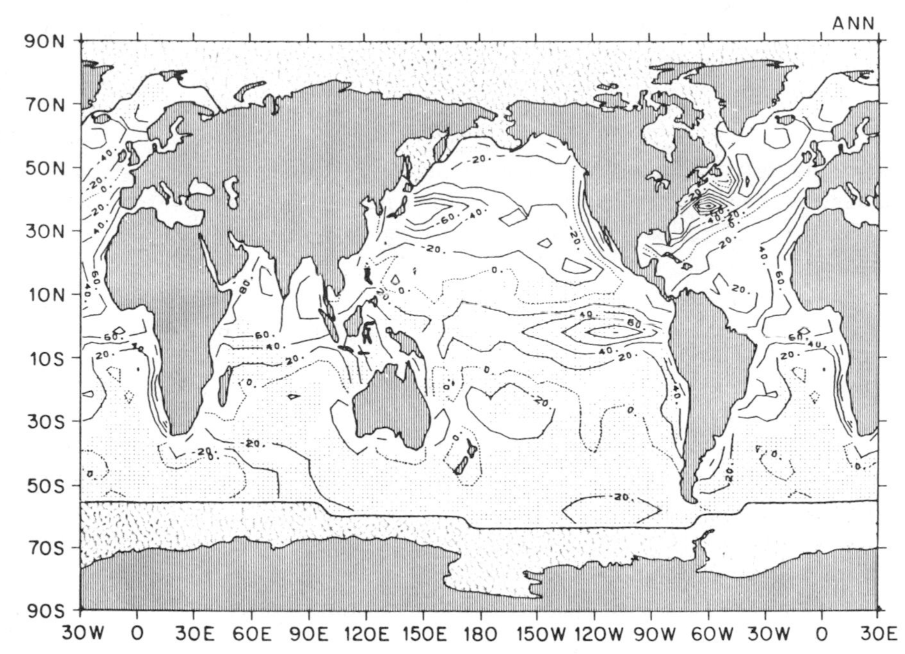

Figure 2.7. The annual mean heat flux across the ocean surface in watts per square meter. Positive values indicate a flux into the ocean. In stippled areas there are fewer than 500 observations in an area of 5° longitude by 5° latitude. [From Esbensen and Kushnir (1981).]

where QS is the short-wave solar radiation, QL is the long-wave radiation back to space, QE is the latent heat flux, and QH is the sensible heat flux. These terms are usually calculated from bulk aerodynamic formulae. Data for calibration are inadequate so it is difficult to determine the various coefficients and functions that appear in the formulae (Blanc, 1985). For a given set of formulae, inadequate data with which to calculate the fluxes create further problems. (Presently the data are relatively abundant only along shipping lanes, but satellite measurements should improve matters in the near future.) For these reasons estimates of Q by different investigators (Hsuing, 1985; Hastenrath and Lamb, 1977; Molinari and Hansen, 1987) differ significantly, often by 20 W/m^2 or more. All the estimates indicate that, in the tropics, two terms dominate the flux of heat across the ocean surface: QE, the loss of latent heat, and QS, the solar radiation into the ocean. There is agreement that, near the equator, the flux of heat is into the ocean throughout the year. In higher latitudes the ocean gains heat during the summer but loses even more during the winter so that there is a net loss in the course of the year. Oceanic currents transport heat to higher latitudes to maintain a balance.

The heat budget of the ocean can be studied by integrating the equation for the conservation of heat

$$T_t + (uT)_x + (vT)_y + (wT)_z = \text{Diffusion} \tag{2.4}$$

over the volume under consideration. Here T denotes temperature, u, v, and w are the velocity components in the eastward (x), northward (y), and upward (z) directions, respectively, and t denotes time. If the volume of interest extends across the full zonal width and depth of the ocean, and if Eq. (2.4) is also integrated in time to obtain annual mean values, then the divergence of the meridional heat transport HT is given by

$$(HT)_y = Q \tag{2.5}$$

provided horizontal diffusion is ignored. This equation can be integrated southward, from a high northern latitude where HT vanishes, to obtain the meridional heat transport of the ocean once the flux across its surface is known. The results of such calculations (Fig. 2.8), give an indication of the extent to which different estimates for Q affect the inferred oceanic heat transport. The figure also shows the heat transport as calculated with models.

In Fig. 2.8 the three oceans have remarkably different meridional heat transports. In the Pacific the heat gained across the ocean surface near the equator is transported poleward, towards both hemispheres. In the Atlantic it augments a northward transport from high southern towards high northern latitudes, and in the Indian Ocean the heat is transported southward (Bryan, 1982; Hastenrath, 1980, 1982; Stommel, 1980). The Pacific has a poleward heat transport because warm surface waters move poleward while colder waters return equatorward, primarily at the depth of the thermocline. The very deep meridional cell below the thermocline—northward abyssal motion and overlying deep southward flow—appears not to interact strongly with the shallower cell that accomplishes most of the meridional heat transport in the Pacific (Roemmich, 1987). Matters are very different in the Atlantic Ocean, where the zonally integrated transport from the surface to depths of approximately 1500 m is northward in both hemispheres. Below that depth it is southward. The deep water crosses the equator relatively unmodified. The northward flow in the surface layers has a large upward velocity component near the equator, where the water is heated so that the northward heat transport increases there. In the far northern Atlantic the surface waters sink to feed the deep southward flow. Much of the cross-equatorial flow of warm surface waters is in the North Brazil Current (Roemmich, 1983; Sarmiento, 1986; Philander and Pacanowski, 1986b). In the Indian Ocean there is a southward transport of heat because warm

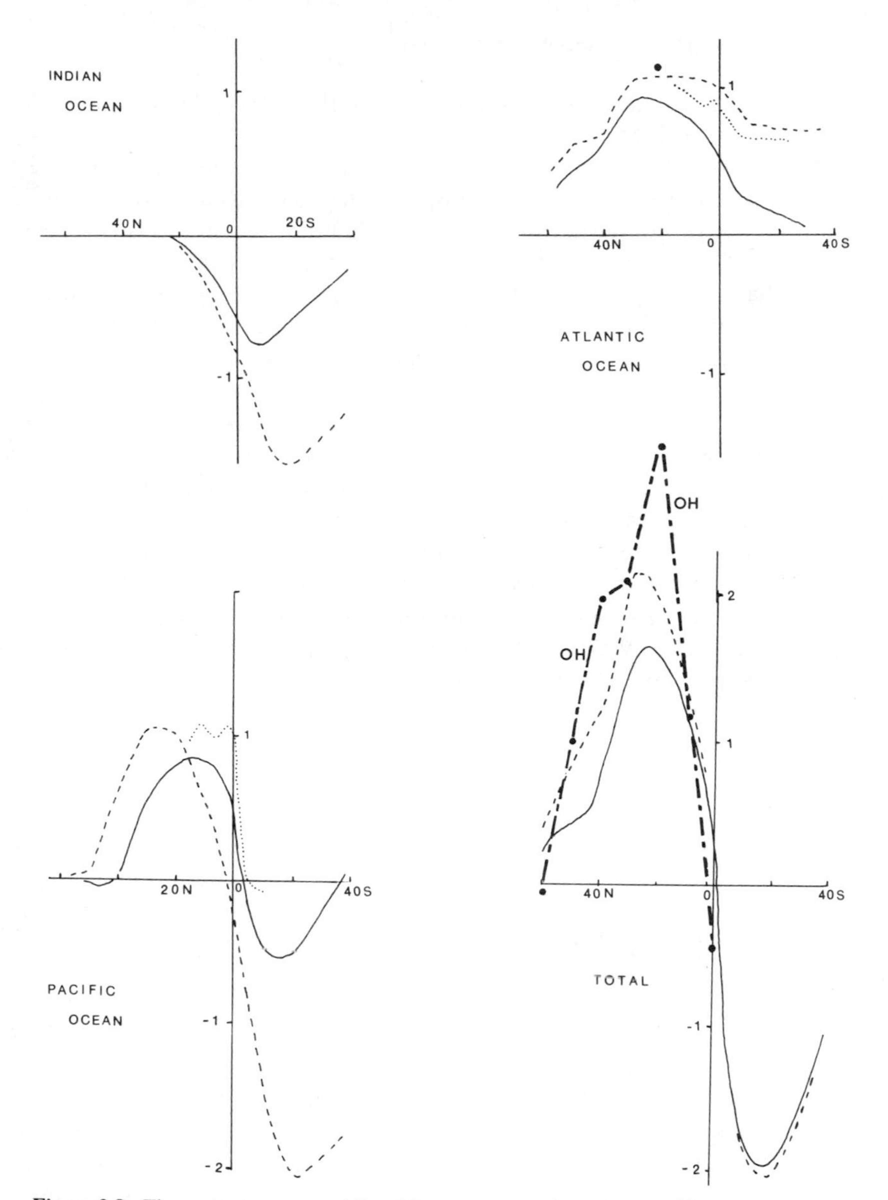

Figure 2.8. The annual mean meridional heat transport (in units of 10^{15} W) in each of the oceans and in all the oceans combined. Two of the estimates, by Hsuing (1985) (shown by the solid line) and Hastenrath (1982) (shown by the dashed line), use data for the heat flux across the ocean surface plus Eq. (2.5). The dotted line shows results from a simulation of the seasonal cycle with a General Circulation Model (see Note 2). The dot near 20°N in the Atlantic is an estimate based on a temperature section along that latitude plus a simple dynamical model of the ocean (Bryden and Hall, 1980). The OH curve in the panel for the total transport (-·-·-·) is that of Oort and von der Haar (1976), who use satellite data to determine the net heating of a band of latitudes and then subtract out the divergence of the atmospheric heat transport (estimated from meteorological data) to infer the oceanic heat transport. The error bars for the various estimates are so large that the differences between the estimates are not statistically significant.

surface waters tend to flow southward while the colder, deeper waters flow northward (Fu, 1986).

If the heat flux across the ocean surface is specified then Eq. (2.5) determines the meridional heat transport. The flux across the ocean surface depends on the sea surface temperature, which in turn depends on oceanic processes. Of particular importance are the global-scale meridional circulations that are superimposed on the intense shallow cells that involve upwelling at the equator and downwelling in adjacent latitudes. Models suggest that the large-scale meridional circulations, and hence the heat transports, are different in the three oceans because conditions in high latitudes are different. (The only high northern latitudes where deep water is formed—where there is deep convection that carries surface waters to the ocean floor—are in the Atlantic Ocean.) These results imply that conditions in high latitudes can influence the tropics. The time scales on which this happens have not been determined yet and it is unclear whether El Niño is affected by oceanic conditions outside the tropics. It has been proposed that variations in the heat exported from the tropical Pacific to higher latitudes influence El Niño (Wyrtki, 1985; Zebiak and Cane, 1987) so it is of concern that different estimates of the heat transport can be significantly different.

Thus far, in studies of El Niño, it has been assumed that the upper tropical oceans can be considered in isolation from the subtropics and higher latitudes, and also from the deep ocean well below the thermocline, If changes in the meridional heat transport of the ocean affect El Niño, then this assumption is questionable.

2.3 The Seasonal Cycle

Seasonally, sea surface temperatures in the eastern tropical Pacific are at a maximum during the northern spring when the southeast trades are relaxed and are at a minimum during the northern summer and autumn when these winds are intense. The change in sea surface temperatures is but one aspect of the oceanic response to the seasonally varying winds. During the period of weak southeast trades, equatorial upwelling is minimal, the surface currents in the tropical Pacific are weak—the North Equatorial Countercurrent can disappear completely from the surface layers of the central Pacific—but the transport and speed of the Equatorial Undercurrent attain maxima as shown in Fig. 2.4. Figure 2.25 shows how the surface currents at the equator surge eastward at this time. This surge is apparently accompanied by only a modest change in the zonal slope of the thermocline in the central Pacific. This means that the eastward pressure force, which is associated with the slope of the thermocline, and which is usually balanced by the westward windstress, is temporarily unbalanced in March and April when the southeast trades relax. The pressure force therefore accelerates the equatorial currents eastward. [During some years westerly winds appear along the equator and directly drive eastward equatorial jets (Halpern, 1987).] The eastward advection of warm surface waters contributes to the high sea surface temperatures in the eastern tropical Pacific during the northern spring.

The intensification of the southeast trades and the northward retreat of the northeast trades, from May onwards usually, increase the equatorial upwelling and the downwelling a few hundred kilometers farther north. The warm water along the equator is now displaced, mostly northward, and contributes to the formation of a trough in the thermocline near 3°N. To the north of 8°N the thermocline rises because the Ekman drift there is divergent. The steepening of the slope of the thermocline between 3°N and 10°N is associated with the intensification of the North Equatorial Countercurrent shown in Fig. 2.4. The westward South Equatorial Current also strengthens and acquires a swift core just north of the equator. The latitudinal shear near 3°N becomes so large that instabilities result and waves, described in Section 2.7, appear.

In Fig. 2.9 the seasonal movements of the thermocline near 3°N and 10°N are out of phase. The thermocline has a pronounced trough near 3°N and a prominent ridge near 10°N when the North Equatorial Countercurrent is most intense, usually in September. In the northern spring the thermocline is elevated near 3°N and depressed near 10°N. These seasonal changes suggest a flux of warm surface waters, and presumably of heat, back and forth across 8°N. Simulations of the seasonal cycle with a General Circulation Model, shown in Fig. 2.10, confirm that this is the case. The model also shows that there is an interesting difference between the two phases of the seasonal cycle. The warm surface waters flow northward across 8°N as Ekman drift in the central and eastern Pacific but do not flow back in the same manner. During the northern winter, when the ITCZ is farthest south, the northeast trades near 8°N and the Ekman drift across that latitude are at a seasonal maximum in the central and eastern Pacific. This northward flow of warm surface waters contributes significantly to the northward heat transport. During the northern summer, when the ITCZ is near 9°N and the winds there are weak, the northward Ekman drift across that latitude is small. The southward transport of heat is now associated primarily with the equatorward-flowing Mindanao Current off the Philippines. The western tropical Pacific gains warm surface waters from the two westward-flowing currents, the North and South Equatorial Currents, which are heated by the flux across the ocean surface as they flow westward. Warm surface waters are advected away from the west by the eastward

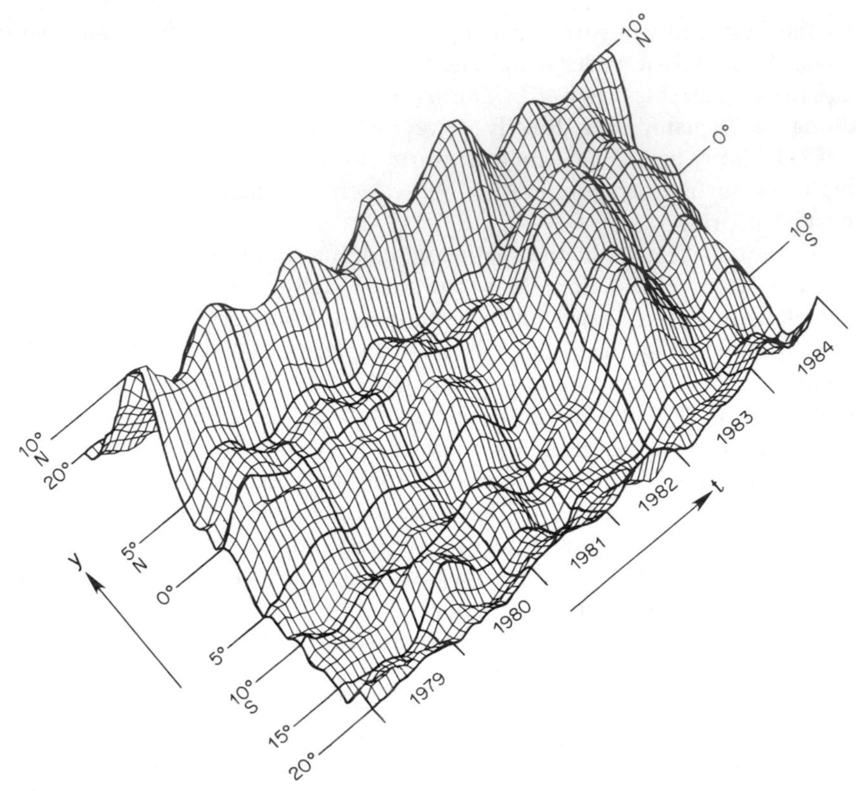

Figure 2.9. The depth of the 20°C isotherm, as a function of latitude and time, along ship tracks that run from approximately 20°S, 180°W to 20°N, 150°W and that cross the equator between 170°W and 160°W. Seasonally, vertical excursions along 10°N and 3°N are out of phase. In late 1982 the elevation of isotherms near 10°N was at a maximum as was the eastward transport of the North Equatorial Countercurrent. [From Kessler and Taft (1987).]

surge along the equator during the northern spring, and by the North Equatorial Countercurrent. In summary, there is an interesting asymmetry between the two phases of the seasonal cycle. When the thermocline rises near 3°N and sinks near 10°N, Ekman drift carries warm surface waters directly from one region to the other. The warm surface waters, once north of 10°N, flow westward and some of it ultimately joins the equatorward current off the Philippines. Warm water surges eastward along the equator in March and April and, when the southeast trades intensify, is displaced northward, towards 3°N, where the thermocline deepens. The intensifying North Equatorial Countercurrent continues to advect warm water from the west.

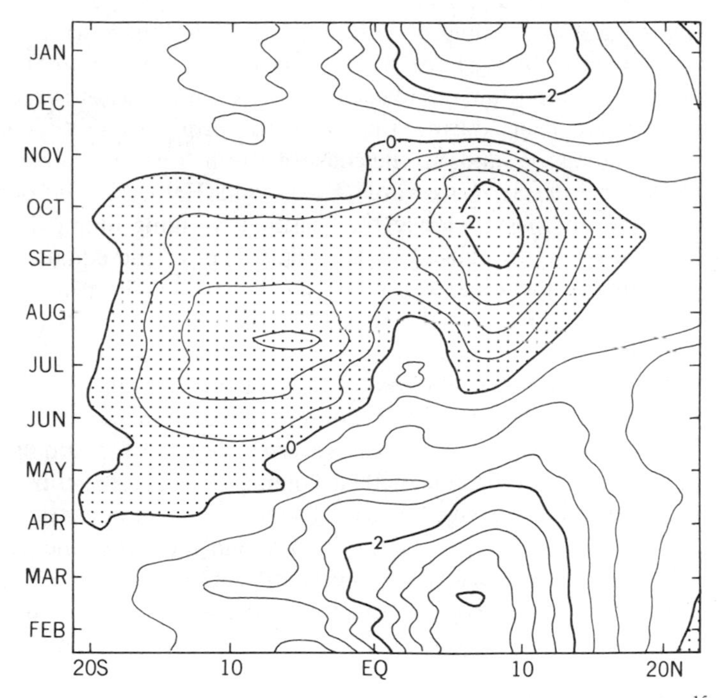

Figure 2.10. The zonally integrated meridional heat transport (in units of 10^{15} W) as a function of month and latitude in a simulation of the seasonal cycles of the tropical Pacific Ocean with a General Circulation Model. The transport is southward in shaded areas. [From Philander *et al.* (1987).]

South of the equator the meridional heat transport involves an asymmetry similar to that north of the equator. The southward transport of heat in the Pacific, across 10°S, say, in Fig. 2.10, is attributable primarily to southward Ekman drift when the southeast trades are intense during the northern summer. During the northern winter, when the southeast trades are relaxed and the southward Ekman drift across 10°S is weak, a current off the coast of New Guinea advects warm water equatorward.

The southeast trades along the western coasts of South America and the northeast trades along the coasts of northern and central America induce offshore Ekman drift, coastal upwelling, and low sea surface temperatures. The winds also drive equatorward surface coastal currents and poleward undercurrents (Allen, 1980). The dynamics of the coastal undercurrents is similar to that of the Equatorial Undercurrent in that the driving force is a pressure force maintained by the wind in a direction opposite to that of the wind. Seasonal changes in the intensity of the coastal upwelling and in sea surface temperatures are highly correlated and in phase with seasonal changes in the local alongshore winds, but factors other than the local winds can also be important. One such factor is the eastward surge of the Equatorial Undercurrent during the northern spring when it penetrates beyond the Galápagos Islands to the coast of South America (Lukas, 1986). The warm water advected eastward (Hayes and Halpern, 1984) spreads poleward along the coasts. Hence the seasonal warming along the South American coast is in part caused by the adjustment of the equatorial zone to the weakening of the zonal winds. Along the coast of Peru this effect is secondary to that induced by the relaxation of the local winds (Bigg and Gill, 1986). This is true on seasonal but not on interannual time scales.

The trade winds prevail to the east of the date line but the winds over the far western Pacific form part of the monsoon circulation so that the zonal component of the surface windstress reverses direction seasonally. This gives rise to equatorial currents with a complex vertical structure in and above the thermocline in the western Pacific. For example, during the period of westward winds, in the northern summer usually, the eastward Equatorial Undercurrent below the westward surface flow can have two distinct cores, one in the deep thermocline and one at shallower depths in the thermally mixed layer. When the winds are eastward, during the early months of the calendar year, a westward equatorial jet is often sandwiched between eastward flow at the surface and eastward flow in the thermocline (Hisard *et al.*, 1070; Burkov and Ovchinikov, 1960; Bubnov *et al.*, 1982). In the eastern equatorial Indian Ocean, where the winds also reverse direction, the currents can have a similarly complicated vertical structure (Taft, 1967).

2.4 Interannual Variability

The response of the tropical Pacific Ocean to interannual wind variations has many similarities with the response to seasonal wind variations. For example, sea surface temperatures in the eastern Pacific are high when the southeast trades are relaxed and are low when these winds are intense, on both seasonal and interannual time scales. A zonal redistribution of warm surface waters is involved in both cases but the magnitude of the interannual redistribution is far larger. In Fig. 2.11 the November 1981 and March 1982 sections along 85°W are reasonably representative of the cold and warm seasons, respectively. The deepening of the thermocline during the subsequent months, as El Niño developed, was far larger than the seasonal deepening. Such an interannual deepening of the thermocline in the eastern tropical Pacific Ocean is usually accompanied by a shoaling in the west (Hickey, 1975). This can be inferred from the sea level changes in Fig. 2.12

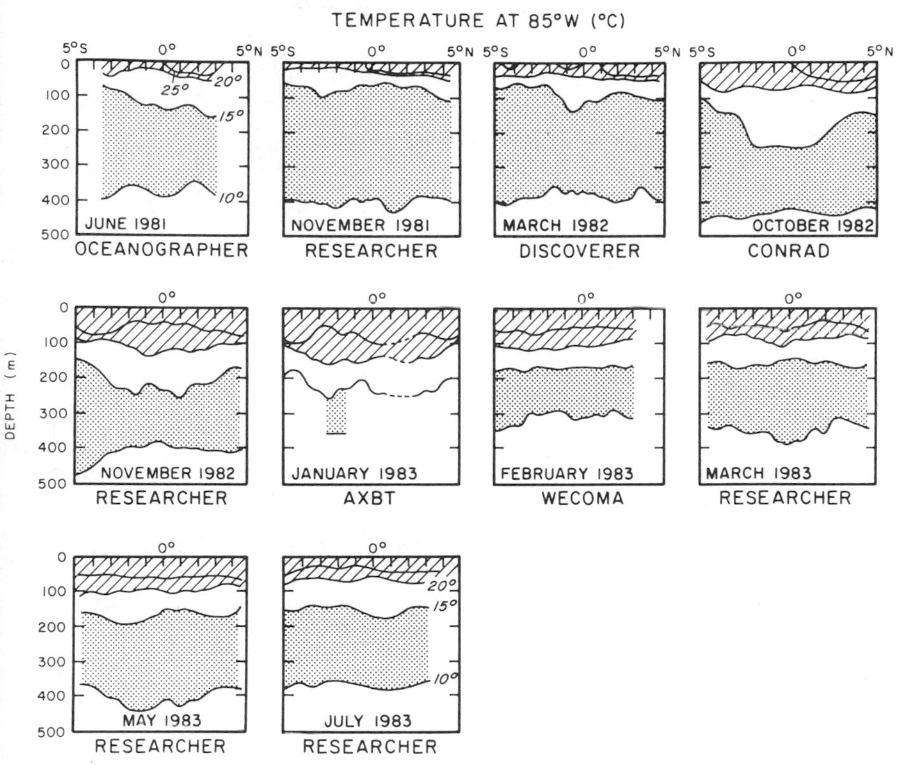

Figure 2.11. Changes in the thermal structure of the eastern tropical Pacific during El Niño of 1982–1983. (Figure provided by A. Leetmaa.)

because sea level provides a measure of the depth of the thermocline. (A deepening of the thermocline causes an increase in the average temperature of the water column and hence an expansion of the column so that sea level increases. Thermocline depth and sea level are therefore positively correlated.)

The interannual zonal redistribution of warm surface waters in the tropical Pacific involves interannual variations in the zonal currents. Figure 2.13 shows these variations, which are inferred from sea level measurements and geostrophic calculations. Estimates based on subsurface temperature measurements are consistent with those based on sea level measurements (Meyers and Donguy, 1984; Kessler and Taft, 1987). The changes in the surface currents indicate that during El Niño the warm eastward North Equatorial Countercurrent intensifies while the cold westward South Equatorial Current is relaxed. During La Niña the situation is reversed. (Note the interesting difference, as yet unexplained, between the interannual and

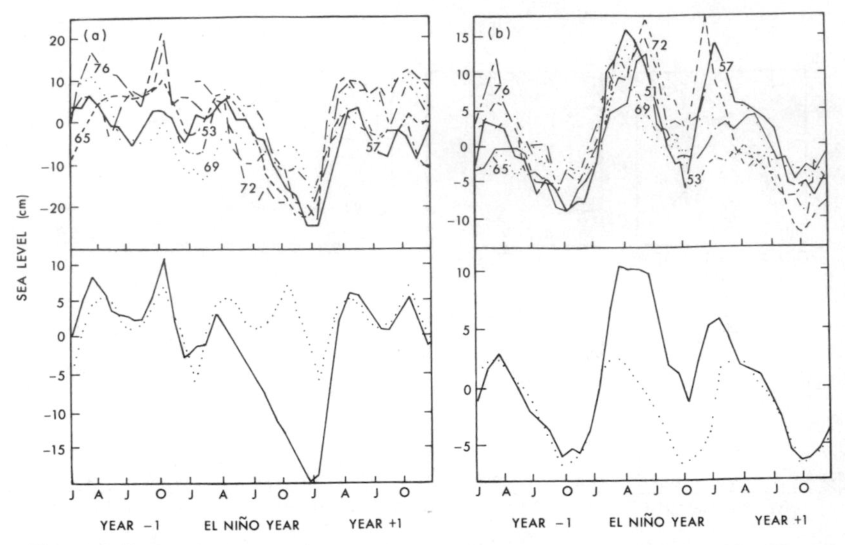

Figure 2.12. Sea level at (a) Truk (152°E, 7°N) and (b) Callao (79°W, 12°S). The upper figures show the strong similarity between variations during different El Niño years between 1953 and 1976. The continuous lines in the lower figures are the averages of the upper curves. The dotted lines in the lower figures are the mean curves for La Niña years. In the eastern tropical Pacific, El Niño is an enhancement of the warm phase of the annual cycle. [Adapted from a paper by Meyers (1979b).]

seasonal variations in the intensity of the North Equatorial Countercurrent. Seasonally this current is strong when the southeast trades are intense; interannually it is strong when these winds are weak.) The intensified currents transport warm surface waters eastward during El Niño but transport cold surface waters westward during La Niña. It clearly is not simply a matter of the same water sloshing back and forth across the Pacific.

Because of a lack of measurements it is impossible to determine exactly how the surface waters of the tropical Pacific are redistributed interannually and how this redistribution is related to changes in the surface winds. Figure 2.12 raises several questions that cannot be answered at present. Sea level is seen to rise in the east before it falls in the west during El Niño. Does that happen because the initial wind changes are confined to the eastern side of the ocean basin? Is the second rise in sea level in the east, towards the end of year (0), a local coastal phenomenon? Or is it a consequence of the preceding fall in sea level in the west?

It is fortunate that extensive measurements are available for El Niño of 1982–1983. (It is also important to keep in mind that El Niño of 1982–1983 developed in an atypical manner, as mentioned in Section 1.5.) Most El

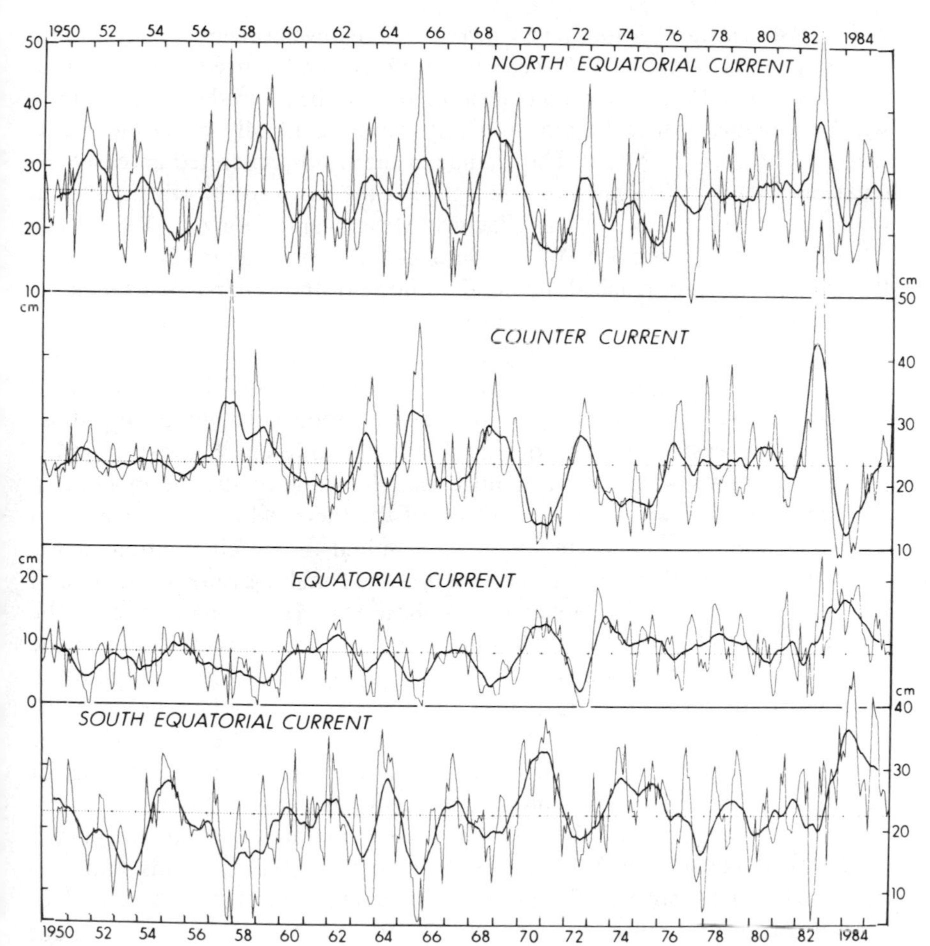

Figure 2.13. Time series of sea level differences across the zonal currents in the central Pacific between 1950 and 1985. The thin line gives the monthly averages and the heavy line is a 12-month running mean. The sea level differences, which provide a measure of the geostrophic zonal currents, are across the troughs and ridges of the upper curve in Fig. 2.3. For the North Equatorial Current the difference is between 10°N and 20°N, for the North Equatorial Countercurrent between 3°N and 10°N, for the southern part of the South Equatorial Current between the equator and 15°S, and for the northern part of the latter current between the equator and 3°N. [From Wyrtki (1974), who later extended the time series to 1985.]

Niño episodes start off the coast of Peru, where developments precede those in the west. El Niño of 1982–1983 was different and started in the western Pacific in May 1982 with a relaxation of the southeast trades. The westerly wind anomalies, which became westerly winds, gradually penetrated eastward as shown in Fig. 1.20. The change in the winds generated an eastward equatorial jet in the ocean and strengthened the eastward North Equatorial Countercurrent in the western Pacific (Meyers and Donguy, 1984). The equatorial jet reached 159°W by October 1982 (Firing *et al.*, 1983). Although the jet never reached the neighborhood of the Galápagos Islands the westward surface flow in the eastern equatorial Pacific was much weaker than normal and was sporadically eastward as is evident in Fig. 2.14.

Seasonally the equatorial currents surge eastward in March and April and contribute to the warming of the eastern tropical Pacific during those months. The eastward surge during 1982 was different because it redistributed so much warm water zonally that the slope of the thermocline in the central Pacific was altered. The rise of the thermocline in the west and its fall in the east, shown in Fig. 2.15, resulted in the elimination of the zonal slope of the thermocline. This caused the zonal pressure gradient, and hence the Equatorial Undercurrent, to disappear. It first happened in the central Pacific (Firing *et al.*, 1983) and occurred a few months later in the eastern Pacific, as shown in Fig. 2.14. This disappearance of the Equatorial Undercurrent is to be contrasted with its seasonal strengthening when the easterly winds relax. Chapters 3 and 4 present theoretical evidence that the crucial difference between the seasonal and interannual weakening of the winds is the longer duration of the relaxation during El Niño.

The first phase of El Niño of 1982–1983 terminated towards the end of 1982 when northerly winds rapidly replaced the westerly winds that had prevailed over the western Pacific. The abrupt change in the west excited an eastward-traveling pulse that elevated the thermocline in the east without affecting the sea surface temperature. In the west El Niño had effectively come to an end and sea level started to rise as shown in Fig. 2.15. In the east the pulse just mentioned was associated with a brief intensification of westward currents and a fall in sea level, but then westerly winds appeared in the east and generated an eastward equatorial oceanic jet there. In Figs. 2.14 and 2.15 this is seen to happen in late April 1983 at 95°W. The second peak in sea level in the east in Fig. 2.15 is therefore attributable to local winds. By June 1983 the southeast trades started to be reestablished and cold La Niña conditions developed rapidly.

Interannual variability in the tropical Pacific is characterized by a zonal redistribution of heat in the upper ocean. There probably is a meridional redistribution too but very little is known about it. Wyrtki (1985), on the basis of sea level measurements at islands and along the coast of the

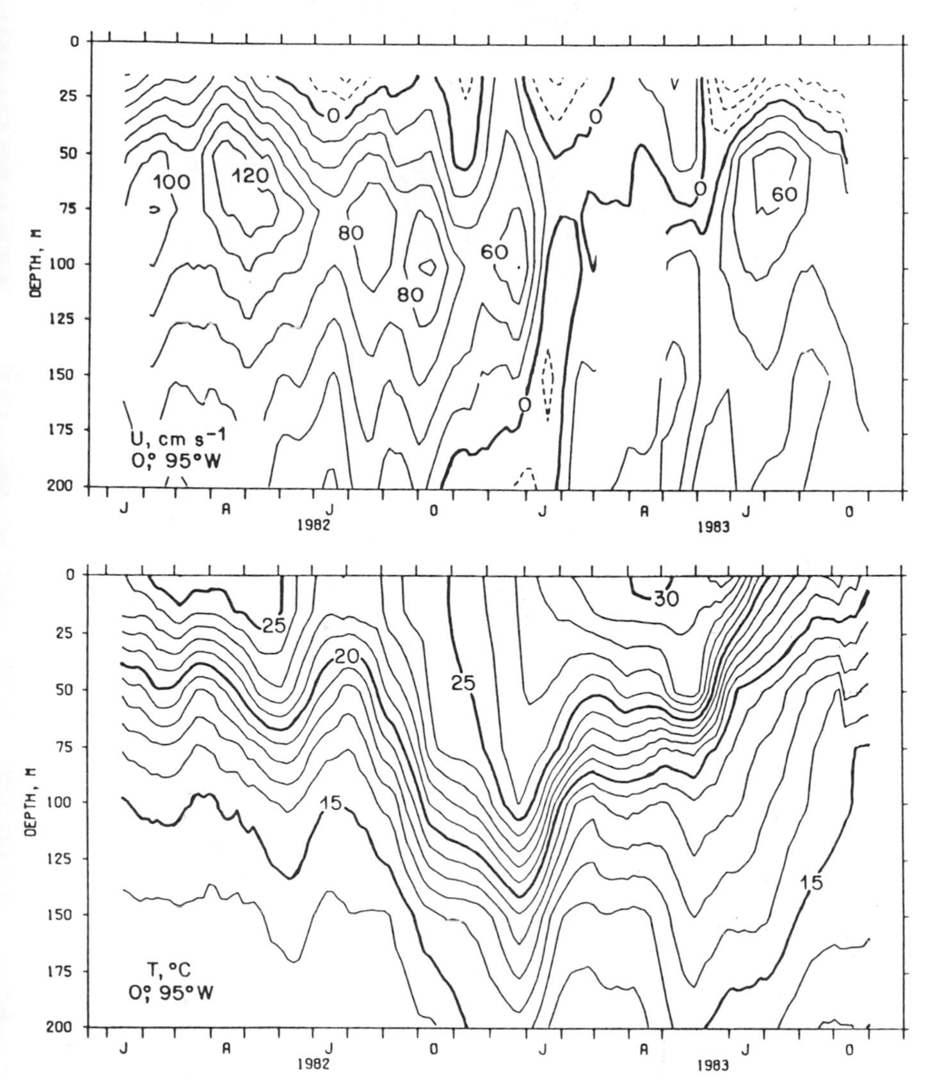

Figure 2.14. Changes in the zonal current (centimeters per second; dashed lines indicate westward flow) and changes in the temperature (°C) on the equator in the eastern equatorial Pacific at 95°W during El Niño of 1982–1983. As isotherms deepened towards the end of 1982, the Equatorial Undercurrent decelerated until it disappeared altogether. The shoaling of the thermocline in early 1983 was associated with wind changes over the western equatorial Pacific; it did not immediately lead to the termination of El Niño in the east. [From Halpern (1987).]

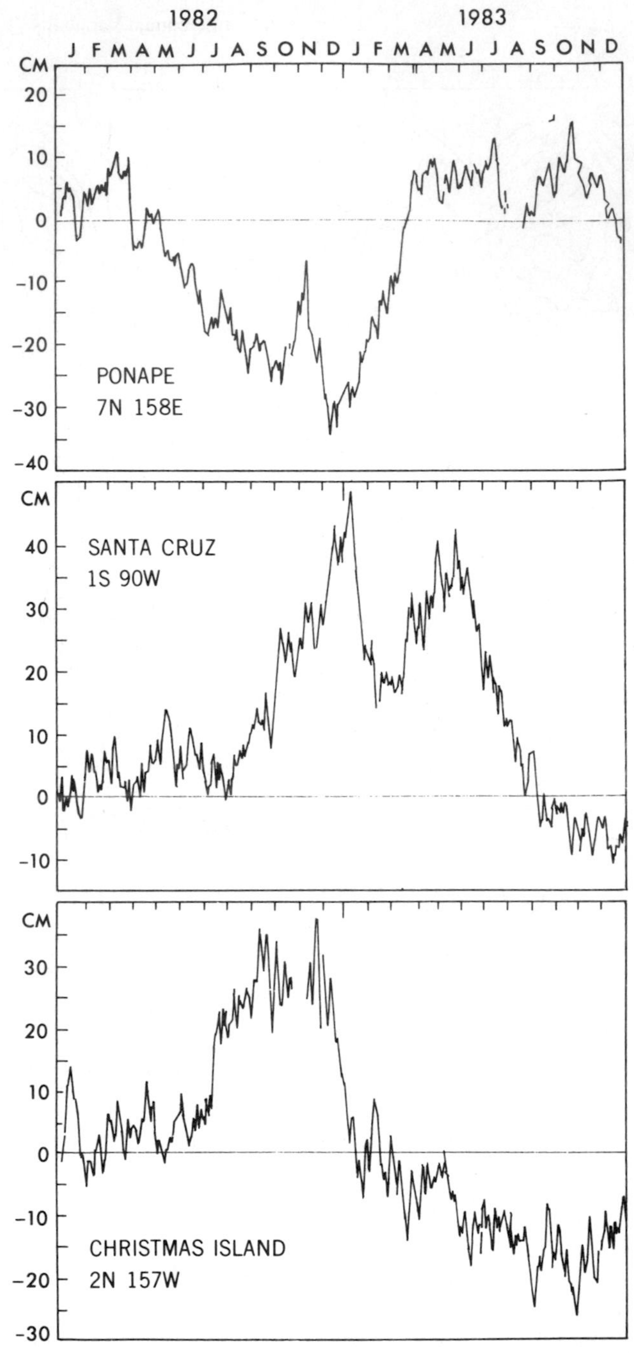

Figure 2.15. Sea level variations in the western, central, and eastern tropical Pacific Ocean during El Niño of 1982–1983. Changes in sea level reflect changes in the depth of the thermocline. (Compare the sea level change at the Galápagos with the thermal changes in Fig. 2.14.) [From Wyrtki (1984a).]

Americas, has concluded that the amount of warm surface waters in the tropical Pacific decreased during 1982–1983. A simulation of that event with a realistic General Circulation Model (see Note 2) corroborates this result and indicates that, between 12°N and 12°S approximately, the averaged depth of the thermocline decreased for two reasons: a decrease in the heat gained across the surface because of increased evaporation associated with higher sea surface temperatures, and an increase in the heat exported northward. Seasonally the northward heat transport occurs during the northern winter when the ITCZ is close to the equator. In early 1983 the ITCZ moved farther south than usual and lingered near the equator longer than usual. The large Ekman drift across 10°N and neighboring latitudes during this period contributed to the increase in northward heat transport. Presumably the equatorial Pacific recovers the heat lost during El Niño when cold La Niña conditions return. The lower sea surface temperatures could mean less evaporation and a greater heat flux across the ocean surface. The poleward displacement of the ITCZ during La Niña could mean less Ekman drift across approximately 10°N and hence a smaller northward transport of heat out of the equatorial zone. Interannual variations in the heat budget need further investigation because they could be a factor that influences the time interval between El Niño episodes. A particularly important aspect of the heat budget concerns the difference between El Niño and La Niña, which is similar to the differences between the warm and cold phase of the seasonal cycle. During El Niño, warm water is advected eastward to the south of approximately 10°N, but during La Niña, the intensified westward current along the equator is cold and does not advect warm water back west. It appears that the warm water that accumulates in the east during El Niño first flows poleward as Ekman drift and then is advected westward. (Poleward-propagating coastal disturbances do not appear to be important in the heat budget.) Analyses of thermal data indicate the presence of westward-moving depressions of the thermocline between about 10°N and 20°N during the years preceding El Niño of 1982–1983 (White *et al.*, 1985; Pazan *et al.*, 1986). Upon arrival in the far west, these depressions could increase the equatorward transport of warm surface waters in that region. Whether the same happens south of the equator is unclear. The relative importance, during La Niña, of increased heat flux across the ocean surface and increased equatorward transport of warm surface waters in the far western Pacific in restoring a deep thermocline to the equatorial zone is not known.

The warm water that surges eastward in the tropical Pacific gives rise to disturbances that propagate poleward along the coasts of the Americas in both hemispheres. Sea level fluctuations as far north as San Francisco and as far south as Valparaiso are correlated with those in the eastern tropical

Pacific (Enfield and Allen, 1980). These disturbances have some of the properties of coastal Kelvin waves but certain aspects of the coastal response indicate that other mechanisms are also involved. Warming attributed to Kelvin waves should have its largest amplitude at the coast, but off Peru the invasion of warm water affects the region closest to land last. [Fish get trapped in shrinking pockets of cold surface waters near the coast (Barber and Chavez, 1986).] Along the coast of California, sea level and sea surface temperatures increase during El Niño in part because of changes in the atmospheric conditions over the northern Pacific. These changes affect the oceanic gyre of the northern Pacific and in particular the California Currents (Simpson, 1984).

2.5 The Atlantic Ocean

The tropical Atlantic and Pacific Oceans, although forced by seasonally varying winds that are very similar, respond in remarkably different ways. In the central equatorial Pacific the zonal slope of the thermocline hardly changes with the seasons. In the equatorial Atlantic, on the other hand, the zonal slope of the thermocline varies practically in phase with the winds (Katz and collaborators, 1977; Lass *et al.*, 1983). In Fig. 2.16 the thermocline is relatively horizontal during the first half of the year when the winds

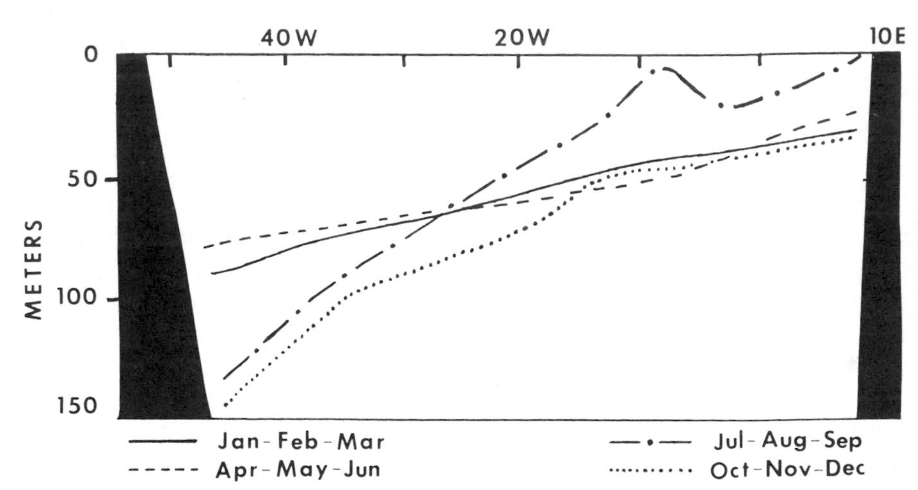

Figure 2.16. The depth (in meters) of the 23°C isotherm in the equatorial plane of the Atlantic Ocean in different seasons. The slope of the isotherm is small when the winds near the equator are weak, during the first half of the year, and is large when the winds are intense, during the northern summer and autumn. [From Merle (1980a).]

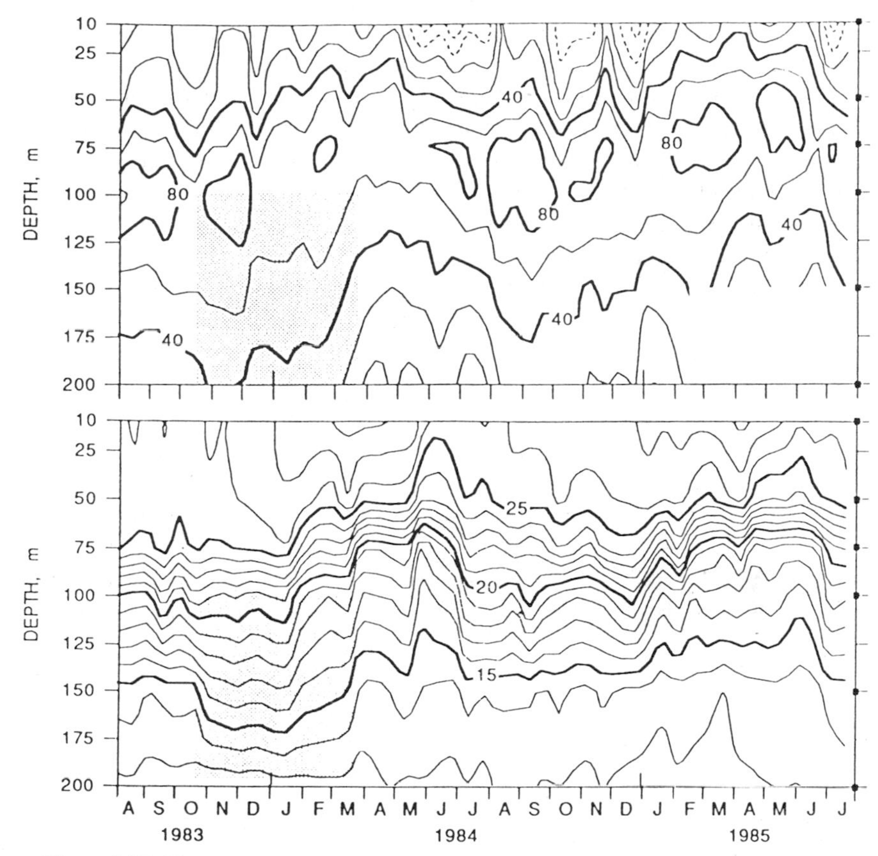

Figure 2.17. The zonal velocity component (centimeters per second) and temperature (°C) as measured from instrumented moorings on the equator at 28°W in the western Atlantic Ocean. The depths of the instruments are indicated by the dots along the right-hand edge of the figure. Dashed lines correspond to westward motion. Unlike the situation depicted in Fig. 2.14, the maximum speed of the Equatorial Undercurrent in the Atlantic is remarkably insensitive to changes in the thermal structure. [From Weisberg and Colin (1986).]

along the equator are weak, and it has a steep slope when the westward winds are intense during the second half of the year. In the central equatorial Pacific, comparable changes in the slope occur not seasonally but interannually. It happened in late 1982 and was associated with a weakening, and at one stage with a disappearance, of the Equatorial Undercurrent. It is therefore surprising that, as shown in Fig. 2.17, the seasonal changes in the slope of the thermocline in the Atlantic are associated with only modest variations in the maximum speed of the Equatorial Undercurrent. The puzzle is not completely solved yet but Wacongne (1988) points to two factors that help explain what happens in the Atlantic. One is the attenuation, with depth, of the amplitude of the seasonal variations in the zonal pressure gradient (Weisberg and Weingartner, 1986) and the other is the complex vertical structure of the changes in the currents. At a fixed depth in the lower part of the Atlantic Undercurrent in Fig. 2.17 there is indeed eastward acceleration as the westward winds intensify and the eastward pressure force increases. The core of the Equatorial Undercurrent moves downward because the wind intensifies the westward surface current, which penetrates farther down. Chapter 3 explores how the dimensions of an ocean basin affect its response to given winds and explains how the different sizes of the equatorial Atlantic and Pacific Oceans contribute to their different seasonal variations.

Off the equator, the western tropical Atlantic and central tropical Pacific respond similarly to the seasonally varying winds. The surface currents are weak and westward when the southeast trades are relaxed during the northern spring. Both the westward South Equatorial Current and eastward North Equatorial Countercurrent intensify when the winds strengthen, usually in May. Figure 2.18 depicts these seasonal changes. The geostrophic balance of the North Equatorial Countercurrent is reflected in a deepening

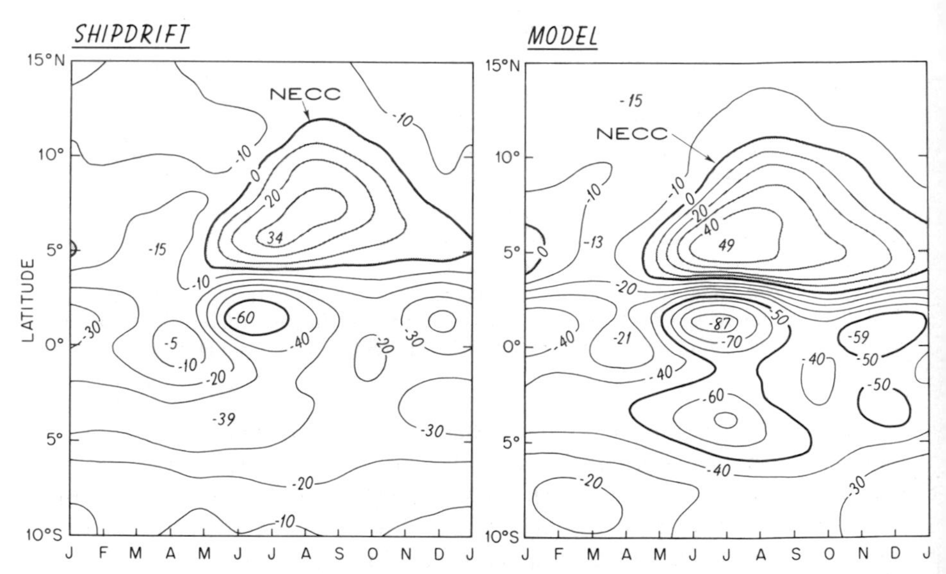

Figure 2.18. The seasonal disappearance of the North Equatorial Countercurrent from the western tropical Atlantic. The data, which have been averaged over the band of longitudes 23°W to 33°W, are from shipdrift records and from a simulation of the seasonal cycle with a General Circulation Model. The units are centimeters per second and motion is eastward in shaded areas. [From Richardson and Philander (1986).]

of the thermocline near 3°N and its elevation near 10°N. This north–south slope of the thermocline decreases when the winds start to relax, towards the end of the calendar year. Seasonal vertical movements of the thermocline to the north and south of 8°N are therefore out of phase (Katz, 1981; Garzoli and Katz, 1983; Merle and Arnault, 1985). This suggests that, seasonally, warm surface waters move back and forth across 8°N. In the Pacific, where the same happens, the northward flow is in the form of Ekman drift but the southward flow is in a current in the far western side of the basin, The Atlantic is interestingly different.

The North Brazil Current flows continuously along the coast into the Gulf of Mexico, between November and April approximately, when the southeast trades are weak. During those months, this coastal current and Ekman drift transport warm surface waters across 8°N so that the thermocline shoals to the south of 8°N while it deepens farther north. The heat transport across 8°N is large at this time, as is evident in Fig. 2.19.

The intensification of the southeast trades in May affects the North Brazil Current dramatically. It abruptly veers offshore between 5°N and 10°N to feed the North Equatorial Countercurrent and it persists in this mode through October[8] (Bruce and Kerling, 1984). The heat transport across 8°N is now at a minimum, and in Fig. 2.19 it is even zero in September. This interruption of the northward flow of warm water results in a deepening of the thermocline south of 8°N. To maintain the relatively steady northward heat transport across 15°N in Fig. 2.19, the region between 8°N and 15°N is drained of warm water so that the thermocline rises. That region is replenished once the ITCZ moves equatorward and the North Brazil Current flows continuously along the coast. At that time the region to the south of 8°N is drained of warm water. The latitudinal bands to the north and south of 8°N act like capacitors that are out of phase.

In the eastern tropical Atlantic the meridional redistribution of surface waters contributes to a curious phenomenon along the northern coast (near 5°N) of the Gulf of Guinea. Although the winds along that coast have almost no seasonal variation, the sea surface temperatures have the enormous fluctuations shown in Fig. 2.20 (Bakun, 1978; Verstraete and Picaut, 1983; Voituriez, 1983). The prevailing southwesterly winds along the coast are relatively steady and should induce coastal upwelling throughout the year. However, cold waters appear at the surface only during the northern summer because the thermocline at the coast is shallow at that time. The vertical movements of the thermocline form part of the basinwide response of the Atlantic to the seasonally varying winds. The response to the intense southeast trades during the northern summer includes a strong North Equatorial Countercurrent that penetrates deep into the Gulf of Guinea as shown in Fig. 2.6. The associated geostrophic slope of the thermocline

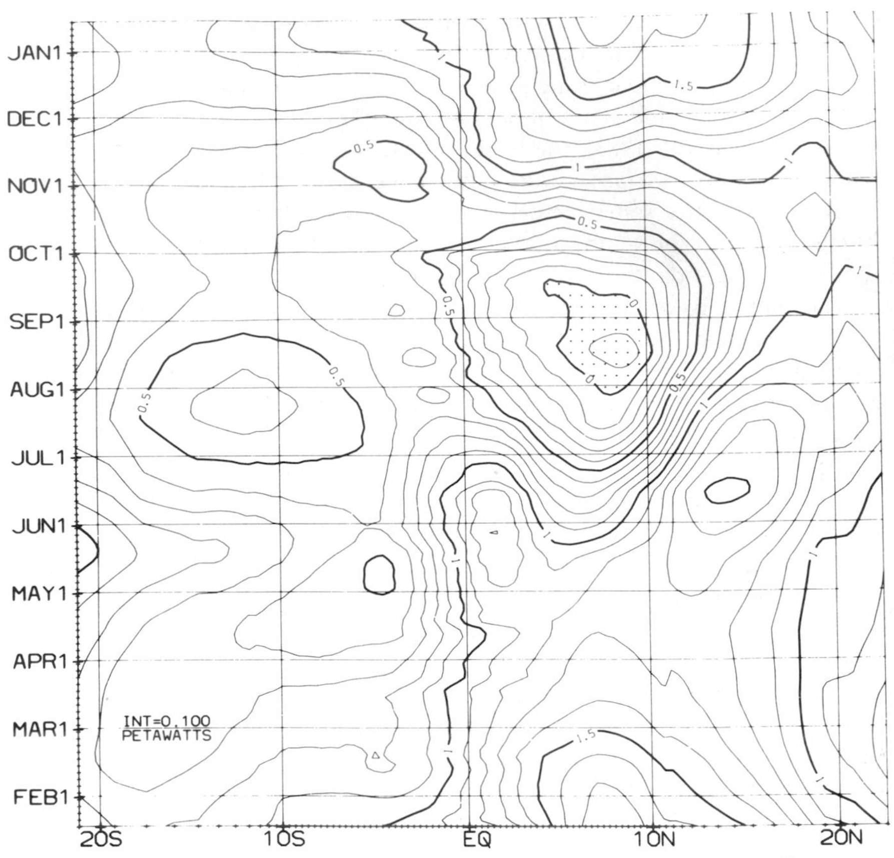

Figure 2.19. The zonally integrated meridional heat transport (in units of 10^{15} W) as a function of month and latitude in a simulation of the seasonal cycle of the tropical Atlantic Ocean with a General Circulation Model. The transport is southward in shaded areas. [From Philander and Pacanowski (1986a).]

elevates isotherms along the coast near 5°N and thus facilitates the coastal upwelling of cold water. In Fig. 2.21 the North Equatorial Countercurrent is most intense during the season of low temperatures at the coast.

Both a zonal and meridional redistribution of surface waters influence sea surface temperatures in the Gulf of Guinea. In Fig. 2.16, a deepening of the 23°C isotherm in the west during the northern summer coincides with a shoaling in the east. The deepening in the west is primarily in response to the intensification of the local westward winds. The strengthening of the winds can be sudden. The transients that are excited persist longest in the Gulf of Guinea and contribute not only to the major upwelling season

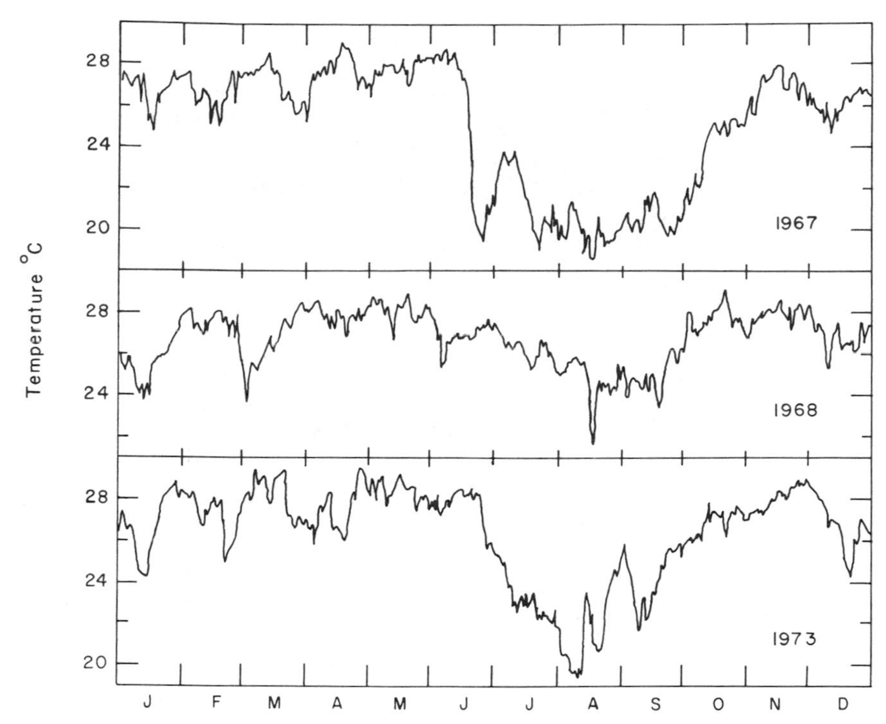

Figure 2.20. Sea surface temperature variations in the Gulf of Guinea at Tema, Ghana (5°37′N, 0°0′E), on the African coast when the amplitude of the seasonal cycle was exceptionally large (1967 and 1973) and exceptionally small (1968). [From Houghton (1976).]

during the northern summer but also to a minor upwelling season in December and January. Because of the latter event, the seasonal cycle in the Gulf of Guinea has a semiannual harmonic. It is more pronounced along the equator in Fig. 2.21 than along the coast in Fig. 2.20.

In 1984 the tropical Atlantic experienced a phenomenon similar to El Niño (Section 1.6). An intensification of the North Equatorial Countercurrent and the appearance of an unusual eastward current south of the equator contributed to a warming of the southeastern tropical Atlantic (Hisard and Henin, 1987). The zonal pressure gradient along the equator disappeared for a while (Katz *et al.*, 1986); however, the intensity of the Equatorial Undercurrent was almost unaffected (Weisberg and Colin, 1986). Another intriguing aspect of the Atlantic episode concerns the upwelling along the northern coast of the Gulf of Guinea. It remained normal during 1984 even though upwelling along the southwestern coast of Africa was weakened significantly (Shannon *et al.*, 1986). Further analysis of the data from 1984 will be of considerable interest because information about earlier

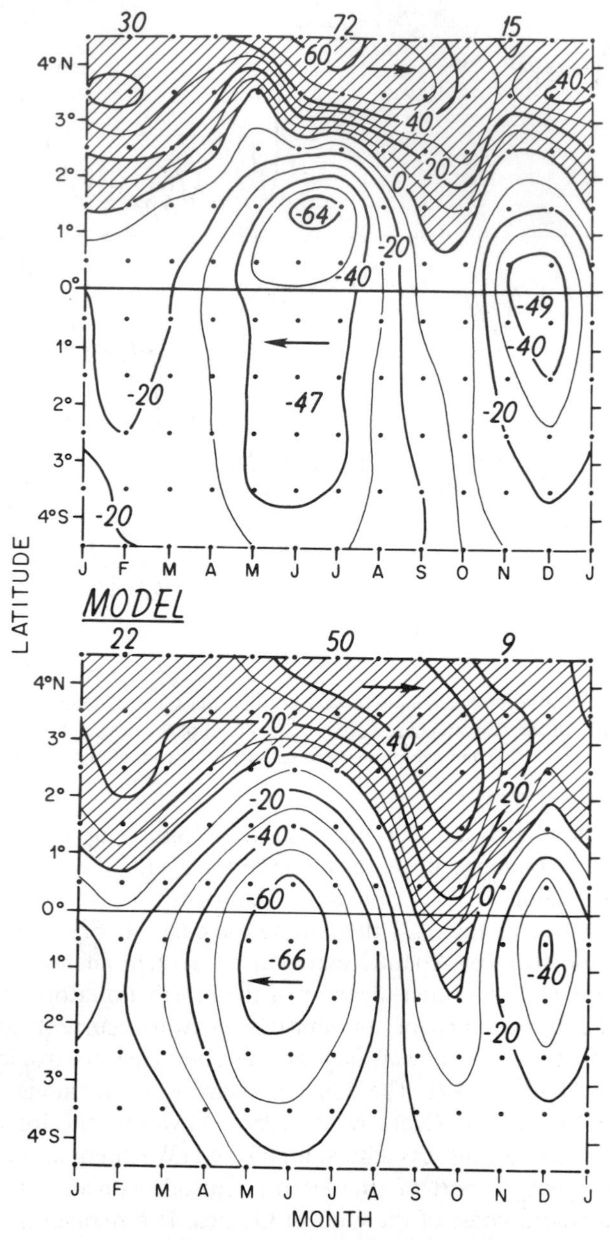

Figure 2.21. Seasonal changes in the zonal component of the surface currents in the eastern tropical Atlantic (data are averaged over the band of meridians 0° to 10°W) as a function of month and latitude as determined from shipdrift data and as calculated with a General Circulation Model. The units are centimeters per second and motion is eastward in shaded areas. [From Richardson and Philander (1986).]

events is scant. Figure 1.23 gives an indication of the frequency of warm events. Figure 2.20 illustrates the amplitude of interannual variations in sea surface temperature along the northern coast of the Gulf of Guinea.

2.6 The Indian Ocean

The monsoons over the Indian Ocean drive a circulation that has features in common with the circulations of the tropical Pacific and Atlantic. The Indian Ocean has two gyres that share the westward South Equatorial Current, a permanent surface current south of approximately 10°S. The anticlockwise southern gyre is relatively steady and involves the Agulhas Current along the southeastern coast of Africa, and the eastward Antarctic Circumpolar Current. The northern gyre is highly variable. During the southwest monsoons it is clockwise and includes the northwestward Somali Current, which feeds a broad eastward surface current that occupies the region north of 10°S. During the northeast monsoons this eastward flow contracts to the narrow band of latitudes just south of the equator in Fig. 2.22. The surface flow north of the equator is now westward, towards the Somali Current, which has reversed direction and flows southwestward. An eastward Equatorial Undercurrent appears late in the northeast monsoon season. Little is known about the gyres of the Indian Ocean, but two regions, the equatorial zone and the region off the eastern coast of Africa, have been studied extensively.

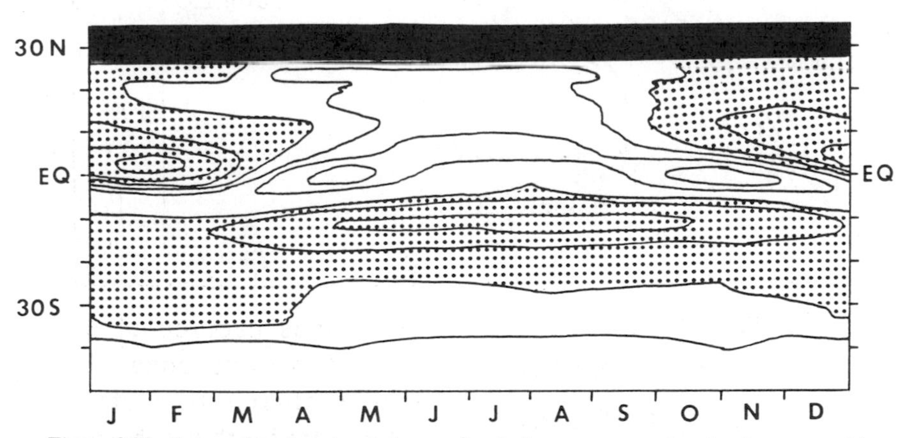

Figure 2.22. Seasonal variations of the zonal velocity component (motion is westward in shaded areas) in the central Indian Ocean as a function of latitude. The contour interval is 10 cm/sec. [From Levitus (1984).]

During the southwest monsoons, the Somali Current has two branches. The southern one appears as early as May, crosses the equator, and turns offshore near 4°N (Leetmaa, 1972). The onset of strong winds in June intensifies this current—surface speeds can reach 3.7 m/sec (Duing *et al.*, 1980)—and also drives a northern branch, which is part of the clockwise gyre between 4°N and 10°N in Fig. 2.23. Wedges of cold surface waters that extend eastward from the coast near 4°N and 10°N indicate where upwelling is intense (Brown *et al.*, 1980; Bruce, 1979; Swallow *et al.*, 1983).

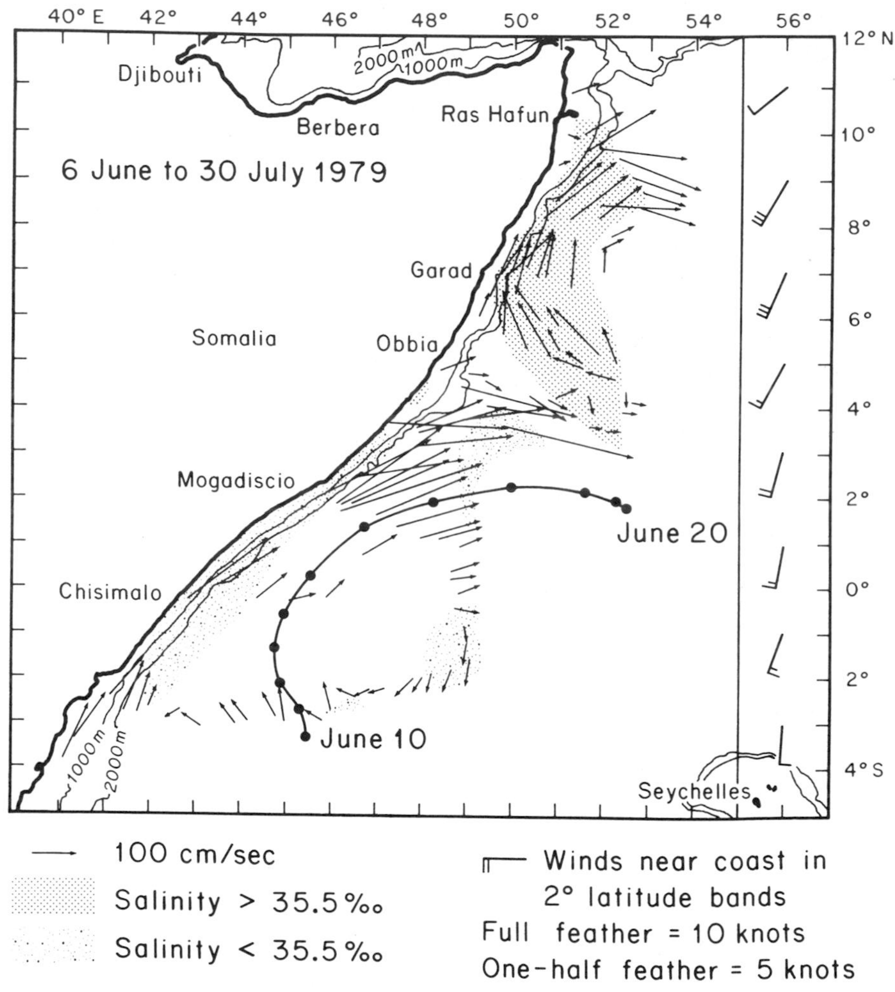

Figure 2.23. A composite view of surface current vectors, surface salinities, a drifter trajectory, and surface winds during June and July 1979. [From Duing *et al.* (1980).]

Towards the end of the southwest monsoons, in August and September, the southern wedge of cold water propagates northward until it merges with that near 10°N. By this time the Somali Current flows as one continuous current along the African coast (Evans and Brown, 1981; Schott, 1983; Knox and Anderson, 1985). The reason for the initial complex structure of the currents near 4°N is not obvious. The windstress has a maximum near 10°N, where the low-level atmospheric jet mentioned in Section 1.2 crosses the equator, but there is nothing distinctive about the winds near 4°N. It is unclear why the cold wedge near 4°N starts to drift northward in August. By November the northeast monsoons prevail and drive a southwestward Somali Current in the surface layers. Below the surface the clockwise gyre between 4°N and 10°N persists into February at least (Bruce, 1979). The vertical structure of the flow further south is complex and, near the equator, includes a northward coastal undercurrent just below the surface layers (Schott, 1987; Schott and Quadfasel, 1980; Leetmaa *et al.*, 1980).

The most prominent features of the oceanic circulation near the equator are intense equatorial surface jets that appear twice a year and are evident in Figs. 2.22 and 3.1 (Wyrtki, 1973b). These currents are in response to strong westerly winds that prevail over the central Indian Ocean during April and May and again during November and December. The response penetrates to considerable depths; a semiannual signal in the currents is evident well below the thermocline (Luyten and Roemmich, 1982). The relation between the surface winds and the various equatorial currents, including the transient Equatorial Undercurrent, which appears towards the end of the northeast monsoons (Swallow, 1967), sheds considerable light on how tropical oceans respond to variable winds (Chapter 3).

2.7 Instabilities

When the southeast trades intensify, sea surface temperatures near the equator start to fall. Surface waters remain warm to the north of 3°N so that a temperature front appears near 3°N. The latitudinal shear of the currents in that region becomes so large that instabilities result. The associated eddy motion transports heat equatorward so that the rate at which sea surface temperatures near the equator decrease also declines.

The instabilities near 3°N cause the thermal front to have westward-propagating undulations with a wavelength of the order of 1000 km and a period between 3 and 4 weeks (Legeckis, 1977; Hansen and Paul, 1984; Philander *et al.*, 1985). Similar waves appear in the western equatorial Atlantic (Weisberg, 1984). The structure of the waves is essentially that shown in Fig. 2.24. Motion is convergent in the cold crests and is divergent

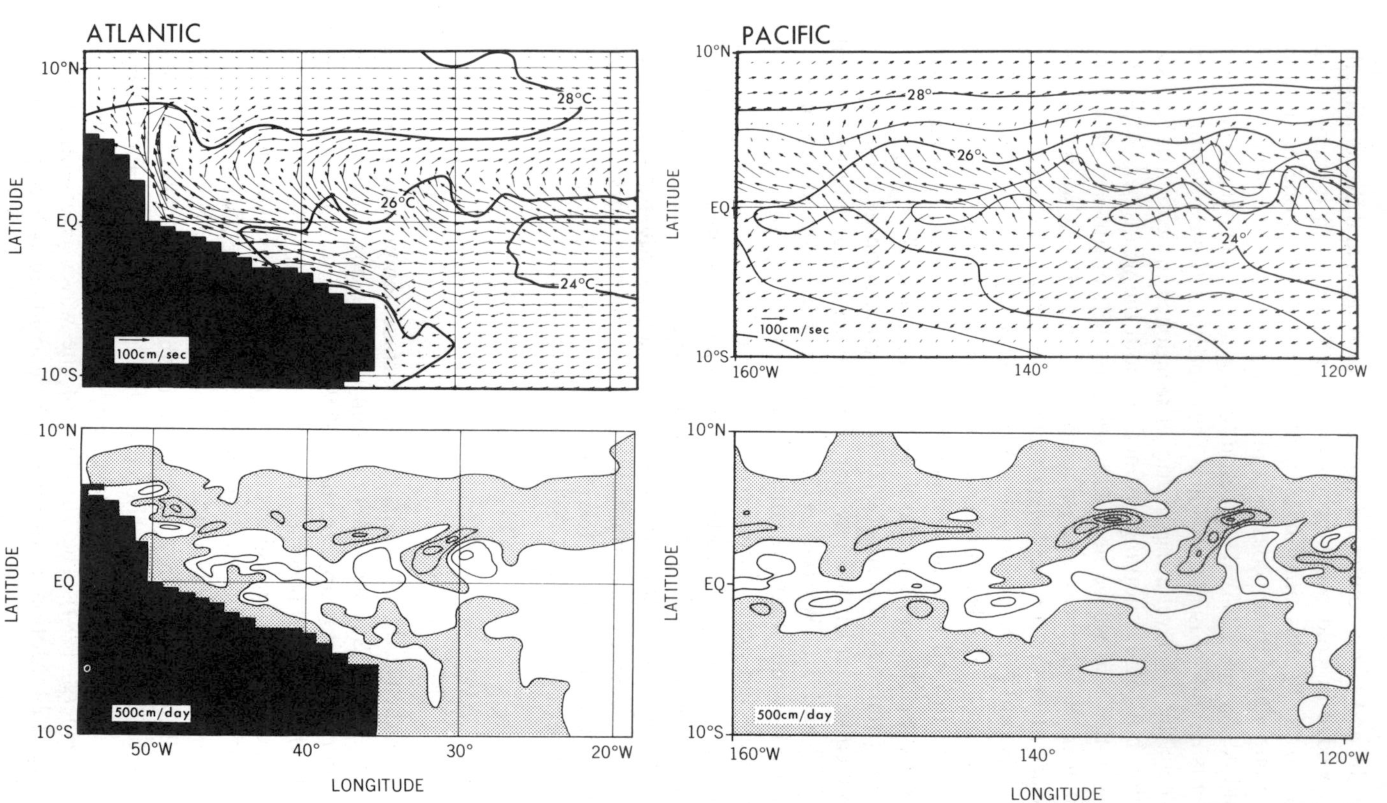

Figure 2.24. (Top) Instabilities of the equatorial currents as manifested in the instantaneous surface currents, superimposed on surface isotherms, and (bottom) the vertical velocity component at a depth of 60 m. The results are from a simulation of the seasonal cycles of the tropical Atlantic and Pacific Oceans (see Note 2). Motion is downward in shaded areas and the contour interval for the vertical velocity component is 500 cm/day. [From Philander *et al.* (1986).]

in the warm troughs of the waves. The vertical velocity fluctuations are comparable in magnitude to the annual mean vertical velocity component. Meridional velocity fluctuations can be ± 80 cm/sec as shown in Fig. 2.25. These large amplitudes are confined to the surface layers of the ocean and attenuate rapidly with increasing depth below the thermocline. The waves are spatially inhomogeneous and nonstationary in time so that their period and wavelength vary with longitude and time. They are absent from the eastern Atlantic and western Pacific. In the Atlantic they appear when the southeast trades intensify, in May usually, and persist for a few months only so that there are at most three or four wave crests. In the Pacific they are present for more prolonged periods and can be remarkably regular as is evident in Fig. 2.25. This figure also shows that they tend to disappear when the southeast trades relax, seasonally in March and April and interannually during El Niño.

Stability analyses (Section 3.10) indicate that the large latitudinal shear of the surface currents, especially that of the westward jet just north of the equator, is the principal cause of instabilities and waves. This hypothesis explains why the waves fail to appear at certain times and in certain regions —they are absent when and where the necessary shear for instabilities is absent—and explains the observed period and wavelength as that of the waves with the most rapid growth rate. Reynolds' stress calculations on the basis of current measurements with surface drifters (Hansen and Paul, 1984) and with meters on moorings (Weisberg, 1984) confirm that the waves draw on the kinetic energy of the mean flow. The upwelling in warm troughs and downwelling in cold crests, shown in Fig. 2.23, suggest that the waves also draw on the potential energy of the mean flow. The local equatorward heat flux effected by the waves is estimated to be equivalent to a heating of the equatorial mixed layer of approximately 180 W/m^2 (Hansen and Paul, 1984; Bryden and Brady, 1985; Halpern, 1987). The contribution that the waves make to the heat transport across a circle of latitude, 3°N say, is nonetheless small because the waves are important only in a confined region, the surface layers of the eastern equatorial Pacific.

Despite the magnitude of the associated velocity fluctuations, the waves are surprisingly linear. The undulations of the isotherms in Fig. 2.23 are modest in comparison with the tortuous nonlinear meanders of the Gulf Stream, for example. Whereas the equatorial waves have energy in a relatively narrow band of frequencies (near $2\pi/1$ month) and wave numbers (near $2\pi/1000$ km), the instabilities of the Gulf Stream cascade energy over a very broad band of frequencies and wave numbers. This implies that the contribution of instabilities to the spectrum of oceanic variability is considerable in the neighborhood of the Gulf Stream but not in the vicinity of the equatorial currents. This has fortuitous implications for the simula-

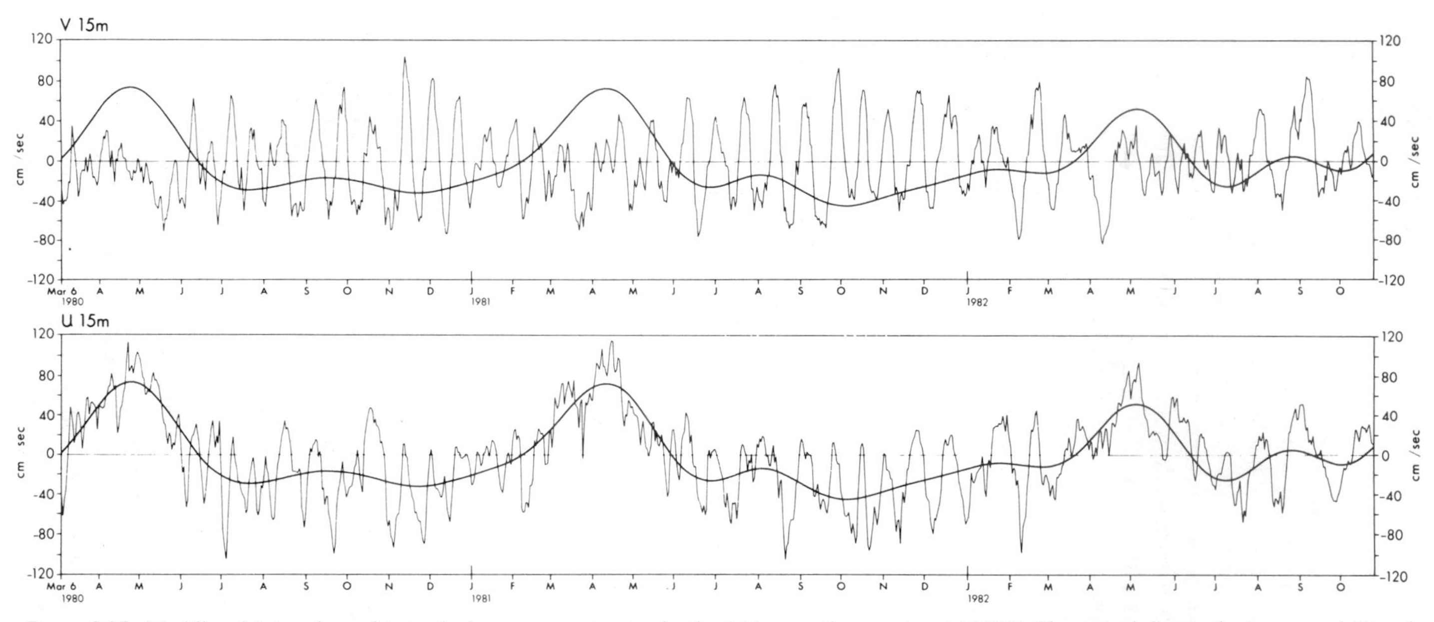

Figure 2.25. Meridional (v) and zonal (u) velocity components at a depth of 15 m on the equator at 110°W. The smooth line is the low-passed filtered zonal current. The energetic fluctuations with a period near 3 weeks almost disappear when the "mean" current is eastward during the spring of the Northern Hemisphere. The fluctuations also disappear during El Niño. [From Halpern (1987).]

tion and predictability of oceanic variability in the tropics, which is discussed in Chapter 4.

Two factors contribute to the linearity of the instabilities near the equator. One is the availability of free waves at the period of the instability that can propagate downward into the deep ocean, thus draining energy from the surface layers. (Such waves are unavailable in higher latitudes.) Equatorially trapped waves with the spatial and temporal scales of the surface instabilities are indeed present below the thermocline (Weisberg *et al.*, 1979). Another mechanism that limits the amplitude of the equatorial instabilities is turbulent mixing and dissipation, which are discussed next.

2.8 Mixing Processes

If mixing processes were entirely absent, then sea surface temperatures would be unaffected by equatorial and coastal upwelling, which determine many of the distinctive features of the sea surface temperature patterns in Fig. 1.10. Mixing processes also affect the intensity of the equatorial currents. The energetics of the Equatorial Undercurrent, for example, is to some extent a balance between the work done by the zonal pressure gradient and turbulent dissipation (Crawford and Osborn, 1981). Small-scale turbulence clearly has a significant effect on the large-scale oceanic circulation.

The turbulent quantities that are of most interest, the rate of kinetic energy dissipation and the vertical turbulent fluxes of heat and mass, cannot be measured directly but can be inferred from measurements of vertical temperature gradients and horizontal velocity components. Such measurements, by means of free-falling instruments that are tracked very accurately, reveal that turbulent dissipation is far higher near the equator than in extraequatorial latitudes, and is high even below the core of the Equatorial Undercurrent (Gregg, 1976; Gregg and Sanford, 1980; Gargett and Osborn, 1981). One reason for this is the large vertical shear of the currents near the equator. These results have been used to devise parameterizations for the turbulent mixing of momentum and heat in terms of the Richardson number of the motion (Moum and Caldwell, 1985; Peters *et al.*, 1987). When used in numerical models, these parameterizations lead to simulations that are more realistic than those with constant values for coefficients of eddy viscosity and diffussivity (Pacanowski and Philander, 1981). The parameterizations are nonetheless inadequate and need to be improved.

Measurements in the central equatorial Pacific towards the end of 1984 indicate that the turbulent dissipation rates have a pronounced diurnal cycle and are much higher at night than during the day. Measurements of intense mixing in the surface layers of the ocean at night, when a loss of heat to the atmosphere destabilizes the upper ocean, have been made in many locations. What is special about the measurements in the central equatorial Pacific is the intensity of the turbulence and its presence in the stratified region below the mixed layer. A diurnal cycle is present in the dissipation rates even when it is absent from the Richardson number. This suggests that, in addition to the vertical shear, other factors influence the turbulence. It is probable that the convective instability near the surface at night excites downward-propagating internal waves that break at depth; further studies are necessary.

The measurements of turbulence near the equator were made primarily in the central Pacific. The western and eastern equatorial Pacific are very different because the former has a very deep, thermally homogeneous surface layer and the latter has none (Fig. 2.2). Temperatures are uniform in the upper part of the western equatorial Pacific but this is not indicative of strong mixing because the currents have considerable vertical shear and the salinity appears to have vertical gradients near the surface (Lindstrom *et al.*, 1987). A zero heat flux across the ocean surface seems necessary to explain the absence of thermal stratification in the upper western equatorial Pacific Ocean. Measurements that provide information about mixing processes in this region are unavailable.

The region below the core of the Equatorial Undercurrent, where the vertical shear is large, has the very small vertical temperature gradients evident in Figs. 2.2 and 2.3. This deep, mixed layer, known as the thermostad, extends from approximately 4°N to 4°S. The steep slopes of the 14°C and 15°C isotherms at the northern and southern boundaries of the thermostad in Fig. 2.3 imply deep eastward geostrophic currents that are distinct from the Equatorial Undercurrent and that are well below the thermocline. These currents are present in both the Atlantic and Pacific Oceans (Tsuchiya, 1975; Hisard *et al.*, 1976).

The high salinity core of the Equatorial Undercurrent loses salt in a downstream direction primarily because of upwelling and vertical mixing into the surface layers and into the deep thermostad. The salinity core tends not to become diffuse in a latitudinal direction. This indicates a balance between horizontal mixing processes and convergent motion at the depth of the core. Katz *et al.* (1979) use such observations to infer that an appropriate value for a lateral eddy diffusion coefficient is approximately 2×10^7 cm^2/sec. Fahrbach *et al.* (1986), on the basis of experiments with clusters of drifting buoys in the Atlantic Equatorial Undercurrent, arrive at a slightly lower value but the difference is not statistically significant.

Notes

1. Although it has been known since at least the time of the early Roman Empire that the winds over the Indian Ocean reverse seasonally, the first written description of the reversal of the current now known as the Somali Current appeared in the ninth century in a book by the famous geographer Ibn Khordazbeh, who noted that "the sea flows during the summer months to the northeast" and "during the winter months to the southwest" (Aleem, 1967; Warren, 1965).

2. The results in Figs. 2.4, 2.5, 2.10, and 2.24 are from the General Circulation Model described in Section 4.9. The model is reasonably realistic—Figs. 2.18 and 2.21 compare observed and simulated currents—but is not perfect and the results depend on the parameterization of mixing processes. The model uses the Richardson number scheme mentioned in Section 2.8 (Pacanowski and Philander, 1981). This parameterization leads to flaws such as the absence, in the model, of a deep thermally homogeneous layer in the western equatorial Pacific. Apparently improved parameterizations of mixing processes should take into account not only the Richardson number dependence but also processes such as convection and internal wave breaking. Recent measurements by Lindstrom *et al.* (1987) imply that incorrect specification of the heat flux across the ocean surface also contributes to the unrealistic thermal structure in the western equatorial Pacific. Apparently the ocean gains very little heat across its surface in that region.

3. The original discovery of the Equatorial Undercurrent in the Atlantic Ocean in 1886 by Buchanan (1886, 1888) fell into oblivion until the accidental rediscovery of the current in the central Pacific Ocean in 1951 by biologists who were fishing for tuna from the research vessel *Hugh M. Smith* of the U.S. Fish and Wildlife Service. They observed that the winds and surface currents on the equator were westward so that the ship drifted in that direction. The longline fishing gear, however, drifted to the east! The first detailed survey of the current was by Cromwell *et al.* (1954). After Cromwell's death in a plane crash, Knauss and King (1958) proposed that the current be named the Cromwell Current. Its Atlantic counterpart is sometimes referred to as the Lomonosov Current because the first modern measurements of the current in that ocean were made from on board the *R. V. Mikhael Lomonosov* (Voight, 1961). The oceanographic practice of not naming currents after persons has prevailed and the term Equatorial Undercurrent is now the accepted one. McPhaden (1986) reviews its history.

4. Density measurements and geostrophic calculations provide information about off-equatorial currents, but within a few hundred kilometers of the equator currents have to be measured directly. To obtain long time series, instruments are attached to wires that are moored to the ocean floor at one end and that are tied to floats at the other end. This arrangement poses a formidable engineering challenge if the floats are on the ocean surface (where they move with the waves) and near the equator (where the vertical shear of the currents is enormous). Only since 1979 has it been possible to measure the equatorial currents in this manner (Halpern, 1987).

5. Estimates of the vertical velocity component, based on the distribution of bomb radiocarbon, by Broecker *et al.* (1978) and Wunsch (1984) are apparently for different depths and hence differ because this variable has a strong depth dependence.

6. Some of the North Equatorial Countercurrent water joins the westward North Equatorial Current by flowing counterclockwise around an elevation of the thermocline near 10°N in the eastern Atlantic and Pacific Oceans, where these thermocline features are known as the Dakar and Costa Rica Domes, respectively. In the Atlantic, that portion of the current that is south of 5°N penetrates into the Gulf of Guinea, where it is sometimes referred to as the Guinea Current.

7. The coast of Brazil juts farthest eastward near 8°S. The westward South Equatorial Current bifurcates near this cape. The water south of approximately 8°S feeds the southward-flowing Brazilian Coastal Current. The water north of approximately 5°S feeds the northwestward-flowing North Brazil Current (Molinari, 1983). Between May and November the latter current turns offshore near 5°N and is distinct from the Guiana Current, which flows northwestward farther along the coast off Guiana. Between approximately December and April the two currents form one continuous coastal current. They become distinct when the southeast trades suddenly intensify, usually in May. Their merger occurs over a more prolonged period. Figure 2.6 depicts these currents.

8. The North Brazil Current veers offshore near 5°N only between May and November. However, an anticyclonic eddy continues to be evident in that region near the coast for several more months.

Chapter 3 Oceanic Adjustment: I

3.1 Introduction

Twice a year, at the time of the equinoxes, westerly winds abruptly start to blow over the central Indian Ocean. The response of the ocean is dramatic and curious. An intense narrow eastward jet, with speeds comparable to that of the Gulf Stream, appears in the surface layers of the ocean within weeks after the onset of the winds (Wyrtki, 1973b). The jet is only a few hundred kilometers wide and clearly marks the location of the equator even though there is nothing exceptional about the structure of the winds in the neighborhood of the equator. At first the jet accelerates—during October 1973 as shown in Fig. 3.1, for example—but this acceleration stops abruptly after a few weeks, although the winds continue to provide eastward momentum to the ocean. Subsequently the jet decelerates and reverses direction. The winds, however, never reverse direction. The trajectories of buoys that drifted in the current in late 1979 indicate that the reversal of the jet first occurred in the east near Sumatra. In Fig. 3.2, buoy 1804 is seen to start traveling westward at a time when buoys 1090 and 1803 farther to the west are still traveling eastward at a high speed. The latter two buoys start to drift westward at points increasingly farther from the coast of Sumatra. It is as if a signal propagating westward at an approximate speed of 55 cm/sec caused the buoys to reverse direction. Presumably the arrival of this signal at Gan (Fig. 3.1) causes the deceleration of the jet observed there.

The events described raise a number of questions. Why does a jet appear at the equator and what determines its width? What causes it to accelerate initially and to reverse direction subsequently? Why does the reversal start

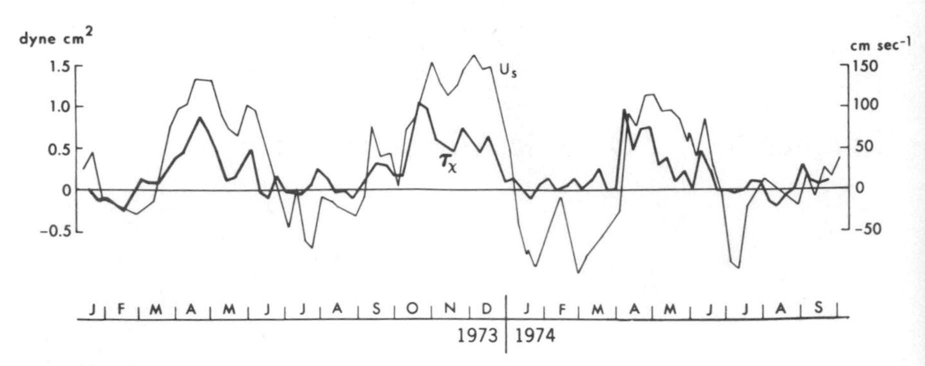

Figure 3.1. The zonal velocity component at the surface and the zonal component of the windstress, as measured in the central Indian Ocean on the equator near Gan (70°E). [From Knox (1976).]

in the east? How long would it have taken the ocean to reach a state of equilibrium had the winds remained constant after their sudden onset? In other words, how long is the "memory" of the ocean?

Examples of the rapid response of tropical oceans to large-scale changes in the surface winds are numerous. Chapter 2 describes basinwide seasonal and interannual changes in the circulations of each of the three tropical oceans. In the subtropics and midlatitudes, on the other hand, comparable variability involving the Gulf Stream and Antarctic Circumpolar Current, for example, appears only on far longer time scales. This suggests that the time it takes the ocean to adjust to a change in the winds increases with increasing latitude. The memory of the ocean is far longer in high than in low latitudes. What physical processes determine the memory?

The measurements in Figs. 3.1 and 3.2 describe variations in the Indian Ocean but an understanding of these variations can shed light on a host of other phenomena, including El Niño. It is important to know how long the memory of the ocean is because it is this memory that presumably permits anomalous oceanic and atmospheric conditions associated with the Southern Oscillation to persist for several seasons. (The time it takes the atmosphere to return to a state of radiative equilibrium after a change in its heating is of the order of a month, far shorter than the time scale of the Southern Oscillation.) Of interest is how long it takes the ocean to reach a state of dynamic equilibrium. For example, how long would it take the ocean to become motionless should the winds suddenly stop blowing? In the final state of rest there are no horizontal density gradients but there can be vertical gradients. The time to establish the vertical gradients, the stratification of the ocean, is believed to be of the order of decades. This is much longer than the time for the changes in the oceanic circulation

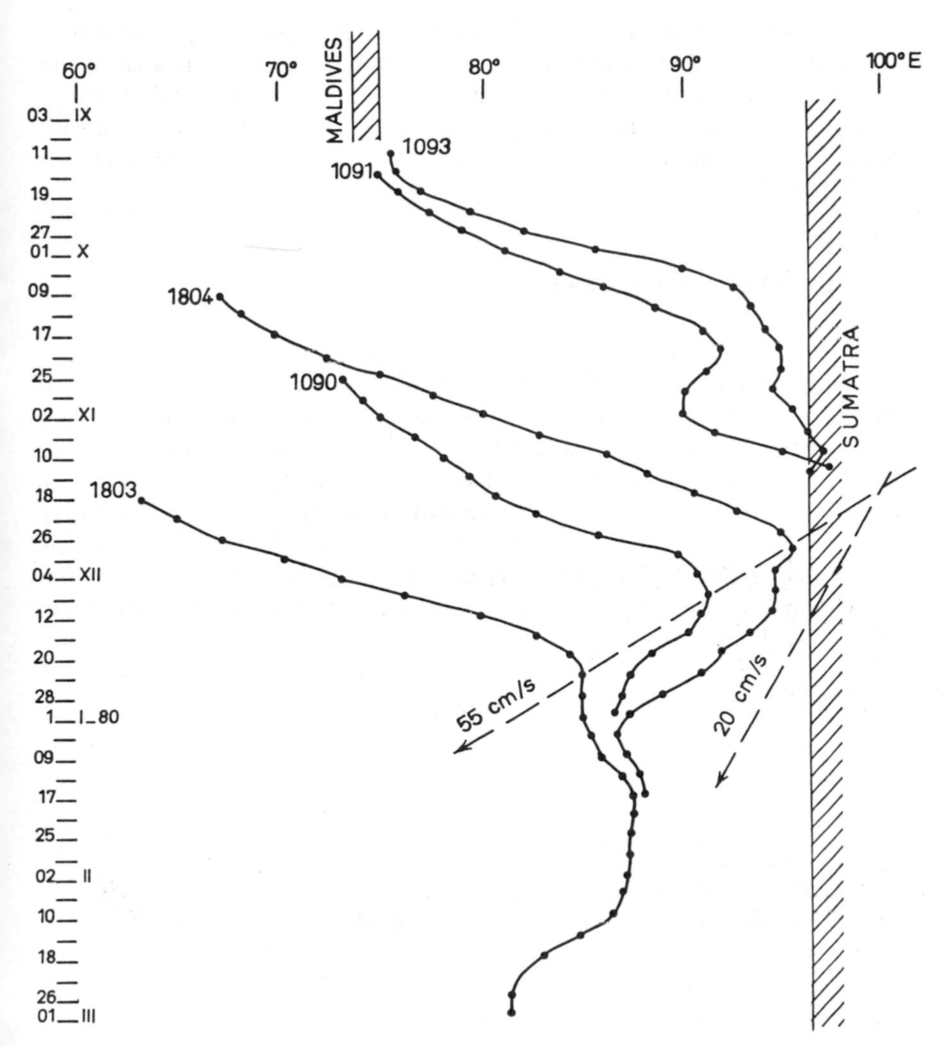

Figure 3.2. The movement of buoys that drifted with the surface currents along the equator in the Indian Ocean between 3 September 1979 and 1 March 1980. The first two buoys washed ashore on Sumatra. [From Gonella *et al.* (1981).]

between El Niño and La Niña. In studies of the oceanic adjustment it is therefore assumed that the vertical stratification of the ocean is given—the processes that maintain the thermocline are not considered—and attention is focused on the manner in which the winds generate horizontal density gradients and currents on the relatively short time scale of the Southern Oscillation.

3.2 The Shallow-Water Model

An appealingly simple model of the tropical oceans can provide answers to the questions raised by Figs. 3.1 and 3.2. In the model, the interface between two immiscible layers of fluid, each of constant density, simulates the sharp and shallow tropical thermocline that separates the warm surface waters from the cold waters of the deep ocean. The upper layer has density ρ_1, has a mean depth H, and is bounded above by a rigid lid. The lower layer has density ρ_2 and is infinitely deep so that it must be motionless for the kinetic energy to be finite. (In reality the ocean is 4000 m deep and the thermocline is at a depth of approximately 100 m.) Linear hydrostatic motion in the upper layer is driven by the windstress τ that acts as a body force. This motion is associated with a displacement η of the interface and is described by the shallow-water equations[1]

$$u_t - fv + g'\eta_x = \tau^x/H \tag{3.1a}$$

$$v_t + fu + g'\eta_y = \tau^y/H \tag{3.1b}$$

$$g'\eta_t + c^2(u_x + v_y) = 0 \tag{3.1c}$$

The Cartesian coordinate system, which is fixed in the rotating earth, is shown in Fig. 3.3. The velocity components in the eastward (x) and northward (y) directions are u and v, respectively, while t measures time. The equator is at $y = 0$. Effects caused by the rotation and curvature of the earth enter through the Coriolis parameter

$$f = \beta y, \qquad \text{where } \beta = 2\Omega/a \tag{3.2}$$

Here Ω denotes the rate of rotation of the earth and a its radius. The gravitational acceleration g, because of the stratification, is effectively reduced to

$$g' = \frac{\rho_2 - \rho_1}{\rho_1} g \tag{3.3}$$

The gravity wave speed is

$$c = (g'H)^{1/2} \tag{3.4}$$

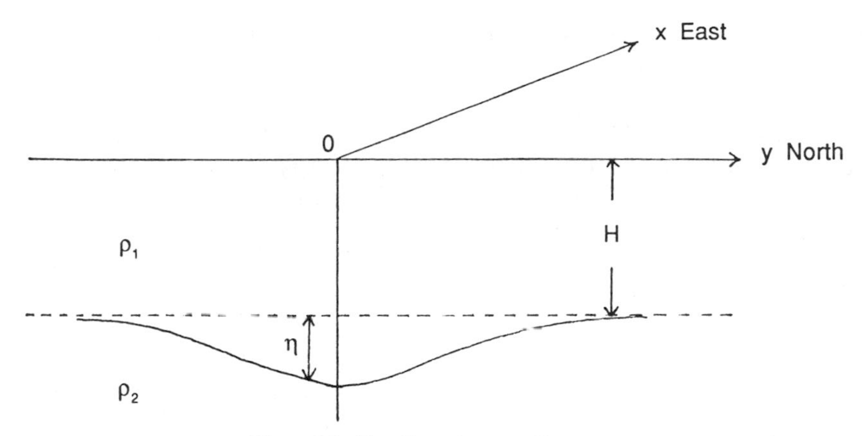

Figure 3.3. The Cartesian coordinate system.

This is sometimes written

$$c = (gh)^{1/2} \tag{3.5}$$

where h is known as the equivalent depth. Reasonable numerical values are

$$\frac{\rho_2 - \rho_1}{\rho_1} = 0.002; \qquad H = 100 \text{ m}; \qquad h = 20 \text{ cm}; \qquad c = 1.4 \text{ m/sec} \tag{3.6}$$

Equations (3.1) can be reduced to a single equation for the northward velocity component v:

$$(v_{xx} + v_{yy})_t + \beta v_x - c^{-2} v_{ttt} - \frac{f^2}{c^2} v_t = F \tag{3.7}$$

where

$$F = \frac{f^2}{c^2}\left(\frac{\tau^x}{H}\right)_t - \frac{1}{c^2}\left(\frac{\tau^y}{H}\right)_{tt} + \frac{1}{H}\frac{\partial}{\partial x}\left(\frac{\partial \tau^x}{\partial y} - \frac{\partial \tau^y}{\partial x}\right)$$

In regions that are far from the equator and that have a latitudinal extent L sufficiently small for the Coriolis parameter to be regarded as a constant (even though its y derivative, β, is also regarded as a nonzero constant), the vorticity equation (3.7) can be simplified provided the following conditions are satisfied: the time scale under consideration must be much longer than the local inertial period and the rotational Froude number of the motion (fL/c) must be at most order one. Under these conditions, Eq. (3.7) can be written in terms of a stream function such that $v = \psi_x$ and $u = -\psi_y$:

$$\left(\psi_{xx} + \psi_{yy} - \frac{f^2}{c^2}\psi\right)_t + \beta\psi_x = \text{curl}_z\,\tau \tag{3.8}$$

A steady solution to the original equations (3.1) can readily be written down without invoking the approximations that lead to Eq. (3.8):

$$\beta v = \text{curl}_z\,\tau \tag{3.9}$$

This equation was first used by Sverdrup (1947) to explain how the curl of the wind drives the surface currents in Fig. 2.1. If the curl of the wind is zero then the solution to Eqs. (3.1) is

$$u = v = 0, \qquad g'\eta_x = \tau^x/H \tag{3.10}$$

This solution, and hence the inviscid model, appears unpromising at first because it predicts that steady, uniform, zonal winds maintain a pressure gradient but do not drive any currents. The model nonetheless deserves attention for its description of the evolution of equilibrium conditions after the sudden onset of the winds. To put it another way, the journey is more important than the destination. An understanding of the fascinating transients that appear before equilibrium is attained greatly facilitates the interpretation of results from more realistic and complex models.

3.3 The Equatorial Jet

Consider the motion induced by the sudden onset of spatially uniform zonal winds that then remain steady. Initially the flow in the interior of the ocean basin, far from coasts, is independent of longitude. Zonal variations become important when the effects of coasts penetrate to the interior of the basin, but until such time set $\partial/\partial x = 0$ in Eq. (3.7):

$$v_{tt} + f^2 v - c^2 v_{yy} = -f\tau^x/H \tag{3.11}$$

At large distances L from the equator, and after a time longer than the local inertial period $2\pi/\beta L$, the second term dominates the left-hand side of this equation so that

$$v = -\tau^x/fH \tag{3.12}$$

This expression for the Ekman drift, which is to the right of the wind in the Northern Hemisphere and to the left in the Southern Hemisphere, is valid provided the distance L from the equator exceeds the local value of the radius of deformation λ^*, where

$$\lambda^* = c/f \tag{3.13}$$

The Ekman drift, which converges on the equator if the winds are eastward, amplifies with decreasing latitude so that downwelling must be intense near the equator. Not only the vertical component of the velocity but also the zonal component is strong because Eq. (3.1a) implies that, in the absence of zonal gradients, the wind accelerates the ocean steadily: $u_t = \tau^x/H$ at $y = 0$. A distinctive equatorial zone clearly exists. Its width can be inferred from a scale analysis of (3.11) and is the distance from the equator where the second and third terms on the left-hand side of this equation have comparable magnitudes. This distance

$$\lambda = (c/\beta)^{1/2} \sim 250 \text{ km} \tag{3.14}$$

is known as the equatorial radius of deformation. The time scale

$$T = (\beta c)^{-1/2} \sim 1.5 \text{ days} \tag{3.15}$$

which determines the relative importance of the first two terms in Eq. (3.11), is the inertial time $1/f$ at the latitude λ. Shortly after the winds start to blow ($0 < t \ll T$) the first term in (3.11) is far larger than the second term, which represents rotational effects. During this period there is nothing distinctive about the neighborhood of the equator because the rotation of the earth is unimportant. It follows that T is the time it takes for a distinctive equatorial zone to form.

Next confine attention to times much longer than T so that the first term in Eq. (3.11) is negligible. Physically, this approximation filters out high frequency inertia-gravity waves that are excited by the sudden onset of the winds. The solution to Eq. (3.11) can then be written as (Yoshida, 1959)

$$v = -\frac{\tau^x}{H}(\beta c)^{-1/2} Q \tag{3.16a}$$

$$u = \frac{\tau^x}{H} t(1 - \xi Q) \tag{3.16b}$$

$$\eta = -\tau^x \frac{t}{c} Q_\xi \tag{3.16c}$$

where

$$\eta = y/\lambda$$

The function Q, the velocity components, and the interface displacements are shown in Fig. 3.4. The solution describes an accelerating equatorial jet whose half-width is twice the radius of deformation, approximately 500 km. The meridional Ekman drift is steady and is maintained by a steady deepening of the thermocline near the equator. The associated increase in the latitudinal density gradient is such that the accelerating equatorial jet is

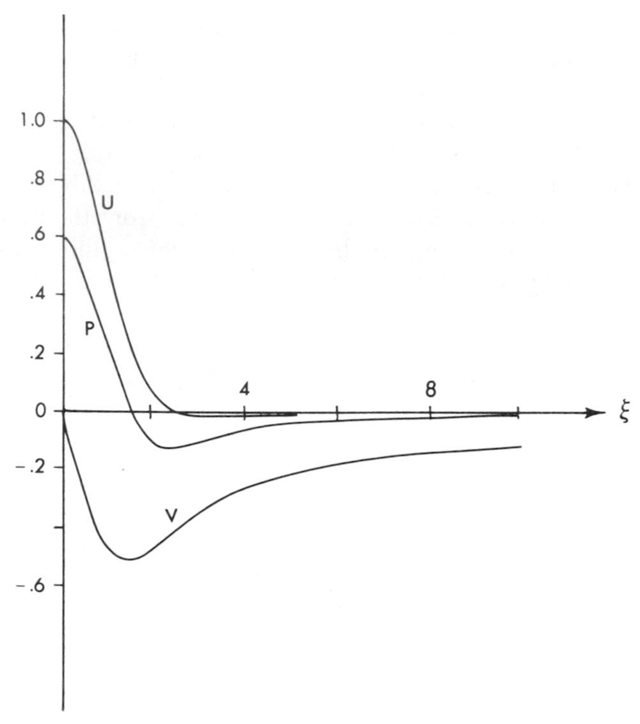

Figure 3.4. The latitudinal structure of the accelerating equatorial jet (U), of the steady meridional flow (V), and of the thermocline displacement (P) in response to spatially uniform eastward winds. The unit for latitude is the equatorial radius of deformation. [From Moore and Philander (1977).]

always in geostrophic balance:

$$fu + g'\eta_y = 0 \tag{3.17}$$

Geostrophic motion is a form of resonance because it persists in the absence of any forcing. Hence, if the winds should suddenly stop blowing after a time t_0, then the acceleration of the jet, the equatorward Ekman drift, and the deepening of the equatorial thermocline will all stop, while a steady geostrophic equatorial jet of intensity $u = \tau^x t_0/H$ at $y = 0$ will persist indefinitely in the absence of dissipation and meridional coasts. Under these conditions the ocean has a memory and records how long the winds had blown. To destroy the equatorial jet generated by winds that had blown eastward for a certain time, it is necessary to blow winds westward for exactly the same time.

Accelerating jets eventually become nonlinear and unstable, but the next complication to account for is the effect of meridional coasts at $x = 0$ and

$x = L$. At these walls the zonal flow u must vanish at all times. This can be accomplished by superimposing on the wind-driven jet just described the free modes of oscillation of the ocean. These are discussed in the next section.

3.4 Waves

The model described by Eqs. (3.1) permits two types of waves: inertia-gravity waves, which have restoring forces owing to the stratification of the ocean and the rotation of the earth, and Rossby waves, which have restoring forces owing to the latitudinal variation of the Coriolis parameter. Consider waves with zonal wavelength $2\pi/k$ and frequency σ:

$$v = V(y)\exp(ikx - i\sigma t) \tag{3.18}$$

Adopt the convention that σ is always positive so that the sign of k determines the direction of zonal phase propagation. (Phase propagation is eastward if $k > 0$ and westward if $k < 0$.) Substitution of (3.18) into (3.8) gives

$$V_{yy} + \frac{\beta^2}{c^2}(Y^2 - y^2)V = 0 \tag{3.19}$$

where

$$Y^2 = \left(\frac{\sigma^2}{c^2} - k^2 - \frac{\beta k}{\sigma}\right)\frac{c^2}{\beta^2}$$

Solutions to Eq. (3.19) are wavelike (oscillatory) in an equatorial zone of width $2Y$ but are exponentially decaying poleward of latitudes $\pm Y$. These latitudes depend on the wave number and frequency of the wave and have a maximum value when

$$k = -\beta/2\sigma \tag{3.20}$$

in which case

$$Y^2 = Y_{\max}^2 = \frac{\sigma^2}{\beta^2} + \frac{c^2}{4\sigma^2} \tag{3.21}$$

In Fig. 3.5, which shows $Y_{\max}$ as a function of frequency, the width of the equatorial waveguide has a minimum value at the frequency

$$\sigma = (\beta c/2)^{1/2} \tag{3.22}$$

In extraequatorial latitudes no waves are possible in a band of frequencies centered on this value. The approximate limits of the band, which separates

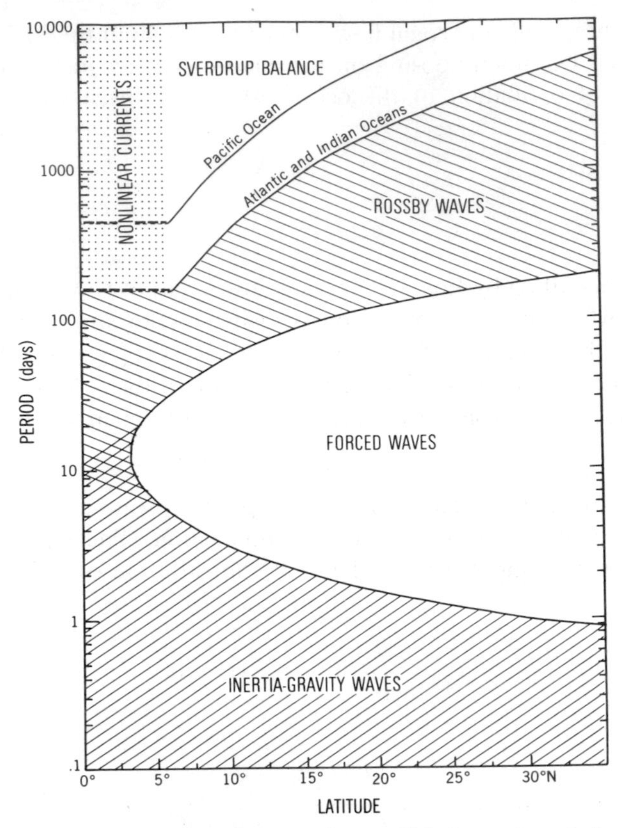

Figure 3.5. Periods, as a function of latitude, at which fluctuating winds excite oceanic inertia-gravity, Rossby, and no free waves—in other words, only forced waves. At very long periods, which depend on the zonal extent of the basin, the response to variable winds is an equilibrium Sverdrup balance. The Indian and Atlantic Oceans are assumed to be 5000 km wide and the Pacific 15,000 km wide.

high-frequency inertia-gravity waves from low-frequency Rossby waves, can be obtained by assuming that the waves far from the equator have such short meridional wavelengths that the value of the Coriolis parameter is practically constant over several wavelengths. Locally the latitudinal dependence of V can then be written $\exp(iny)$. A dispersion relation follows from (3.19):

$$\sigma^2 = f_0^2 + gH(k^2 + n^2 + \beta k/\sigma) \tag{3.23}$$

At high frequencies this expression simplifies to

$$\sigma^2 = f_0^2 + gH(k^2 + n^2) \tag{3.24}$$

which is a dispersion relation for inertia-gravity waves. Their frequency always exceeds the local inertial frequency f_0.

At low frequencies $\sigma \ll f_0$,

$$\sigma = -\beta k/(k^2 + n^2 + f_0^2/c^2) \tag{3.25}$$

The frequency σ is by definition positive so that the zonal wave number k must have a negative value. In other words, all Rossby waves have westward phase propagation. A further restriction on the wave number can be inferred from (3.25):

$$k^2 + \beta k/\sigma < 0 \tag{3.26}$$

This means that Rossby waves are excited in the ocean only if the Fourier components of the forcing function have wave numbers and frequencies that satisfy this condition. The dispersion diagram for Rossby waves shows that although the waves have westward phase speed, their zonal group velocity can be either eastward (for waves with a zonal scale smaller than the radius of deformation λ^*) or westward (for long waves with a horizontal scale that exceeds λ^*) (Fig. 3.6). Waves with zero group velocity have the highest possible frequency

$$\sigma_{\max}^2 = \beta c/2f_0 \tag{3.27}$$

This relation gives the maximum value that the Coriolis parameter can have for waves with frequency σ. In other words, Rossby waves can propagate only equatorward of the latitude where the Coriolis parameter has the value given by Eq. (3.27). The very long waves are nondispersive, have a zonal velocity component that is in approximate geostrophic balance, and have a

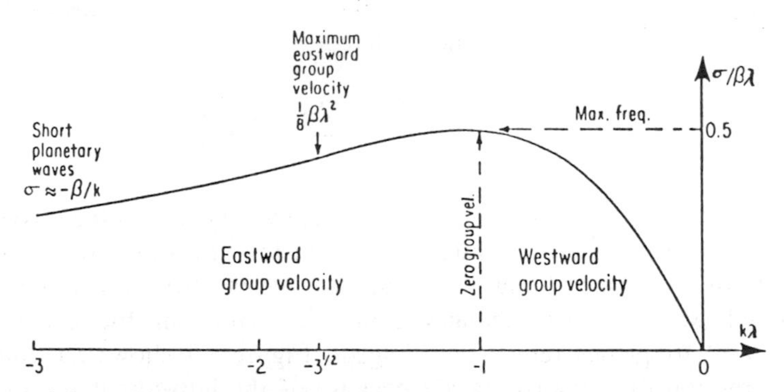

Figure 3.6. A dispersion diagram for Rossby waves that can be written as $\sigma/\beta\lambda = -k\lambda/(1 + (k\lambda)^2)$, where σ is frequency and $\lambda = c/f$ is the radius of deformation. [From Gill (1985).]

speed

$$s = -\beta c^2/f_0^2 \tag{3.28}$$

The short waves, with eastward group velocities, are much slower than the long waves—their maximum group velocity is $s/8$—and are prone to dissipation. These short waves are relatively unimportant in the oceanic adjustment to a change in wind conditions, but the long nondispersive waves are of paramount importance in the adjustment. The long waves reflect off the western boundaries of ocean basins as short Rossby waves that do not propagate far offshore before being dissipated, so that energy tends to accumulate close to western boundaries. The zonal asymmetry of Rossby wave dispersion is of great importance in a number of phenomena to be studied later.

3.4.1 Ray Paths

Consider a packet of Rossby waves with a latitudinal scale sufficiently small for the Coriolis parameter f to be regarded as a local constant. As the packet propagates over a large latitudinal distance, the value of f changes so that the dispersion relation (3.25) can be regarded as a slowly varying function of latitude y. This implies that the group velocity vector of the packet changes gradually with latitude. This vector is the tangent to the ray path so that

$$\frac{dx}{dy} = \frac{\partial \sigma}{\partial n} \bigg/ \frac{\partial \sigma}{\partial k} = -\frac{2\sigma}{\beta}\left(-\frac{\beta k}{\sigma} - \frac{\beta^2 y^2}{c^2}\right)^{1/2} \tag{3.29}$$

The coordinates of the ray path are (x, y), and the dispersion relation (3.25) has been simplified by confining attention to long Rossby waves ($k^2 \gg f^2/c^2$). From (3.29) it follows that

$$y = \left(-\frac{c^2 k}{\sigma\beta}\right)^{1/2} \cos\left(\frac{2\sigma}{c}x + \theta_0\right) \tag{3.30}$$

The constant of integration is θ_0. Figure 3.7 shows these rays for waves that emanate from the eastern boundary of the basin with a period of 0.5 year and with an initial meridional wave number that is zero. (The value of the meridional wave number changes along the path but the zonal wave number and frequency remain unchanged.) Figure 3.7 shows the equatorward refraction of the wave packet because of the latitudinal variation of the Coriolis parameter. This refraction can result in waves that are equatorially trapped.

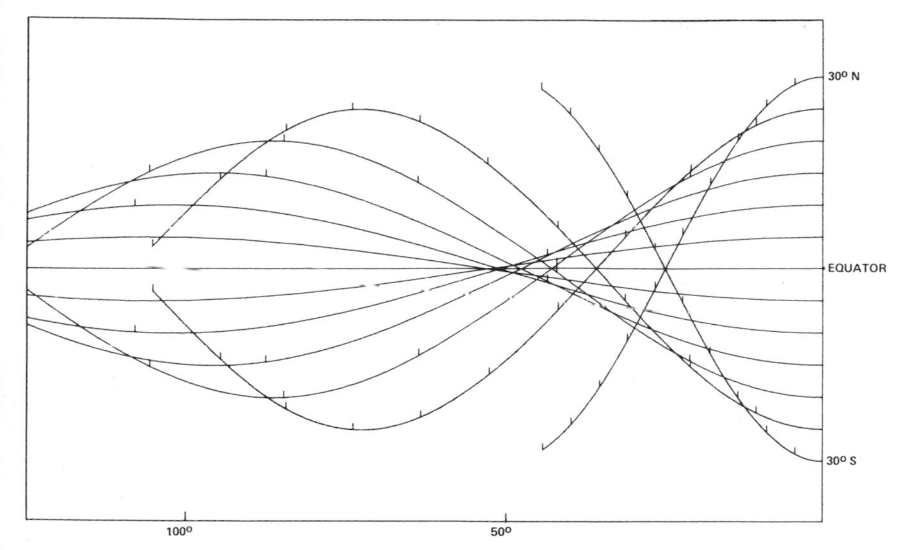

Figure 3.7. Ray paths for Rossby waves with a period of one year that start at the eastern boundary of the ocean basin with a meridional wave number equal to zero. [From Schopf *et al.* (1981).]

The rays in Fig. 3.7 seem to be refracted towards a focal point (caustic) on the equator approximately 50° in longitude from the eastern coast. From Eq. (3.30) it follows that this distance is given by

$$x = (2r + 1)\pi c/4\sigma, \qquad r = 0, 1, 2, \ldots \tag{3.31}$$

(The constant θ_0 is zero for the case under consideration.) After passing through a focus the waves propagate poleward as far as their turning latitude and are then refracted towards another caustic on the equator. These results explain features of certain models but are of limited relevance to the ocean because of the neglect of mean currents.

3.4.2 Equatorially Trapped Waves

The superposition of a wave that propagates towards its northern turning latitude and another wave with the same frequency and wave number that propagates to the south could result in a standing latitudinal mode. Such modes, which span the equator and which are evanescent poleward of the turning latitude, are known as equatorially trapped modes.[2] An arbitrary superposition of waves traveling in opposite directions will in general not result in standing modes. Modes are possible only for certain discrete values of the meridional wave number n that are eigenvalues of Eq. (3.19) subject to the condition that solutions are bounded at large values of $|y|$. These eigenvalues are given by the expression

$$\frac{c}{\beta}\left(\frac{\sigma^2}{c^2} - k^2 - \frac{\beta k}{\sigma}\right) = 2n + 1, \qquad n = 0, 1, 2, \ldots \tag{3.32}$$

This is the dispersion relation for equatorially trapped modes whose latitudinal structure is described by the eigenfunctions of (3.19):

$$V = D_n(y/\lambda) \tag{3.33}$$

The D_n are Hermite functions of order n,

$$D_n = e^{-\xi^2/2}H_n(\xi), \qquad \xi = y/\lambda \tag{3.34}$$

where H_n is the nth Hermite polynomial.[3] The functions are orthogonal so that

$$\int_{-\infty}^{\infty} D_m D_n \, d\xi = 2^n n! \pi^{1/2} \delta_{mn} \tag{3.35}$$

where

$$\delta_{mn} = 1, \text{ if } m = n \qquad \text{and} \qquad \delta_{mn} = 0, \text{ if } m \neq n$$

The zonal velocity component and pressure are given by the expressions

$$u = i(2\beta)^{1/2} \exp(ikx - i\sigma t)\left[\frac{n^{1/2}D_{n-1}}{\sigma + ck} + \frac{(n+1)^{1/2}D_{n+1}}{\sigma - ck}\right]$$

$$\eta = -(2\beta)^{1/2} \exp(ikx - i\sigma t)\left[\frac{n^{1/2}D_{n-1}}{\sigma + ck} - \frac{(n+1)^{1/2}D_{n+1}}{\sigma - ck}\right] \tag{3.36}$$

For an equatorial wave mode with amplitude A the energy density is

$$E = \frac{1}{2}\int_{-\infty}^{\infty} \overline{(u^2 + v^2 + c^2\eta^2/H^2)} \, dy$$

$$= \frac{1}{4}A^2\left[1 + \frac{n+1}{(\sigma - ck)^2} + \frac{n}{(\sigma + ck)^2}\right] \tag{3.37}$$

and the zonal energy flux is

$$F = g' \int_{-\infty}^{\infty} \overline{\eta u} \, dy$$

$$= \frac{1}{4}A^2\left[\frac{n+1}{(\sigma - ck)^2} - \frac{n}{(\sigma + ck)^2}\right] \tag{3.38}$$

The overbar denotes a time average. Since

$$F = Ec_g$$

the energy equation, which is derivable from Eqs. (3.1) in the absence of forcing, can be written

$$E_t + c_g E_x = 0 \tag{3.39}$$

The group velocity c_g ($= \partial\sigma/\partial k$) can be calculated from the dispersion relation (3.32). Equation (3.32) gives two curves for each value of n: one for inertia-gravity waves and the other for Rossby waves, as shown in Fig. 3.8. The odd modes ($n = 1, 3, 5, \ldots$) are symmetrical about the equator and the even modes are antisymmetrical. Figure 3.9 depicts their structure. Since n

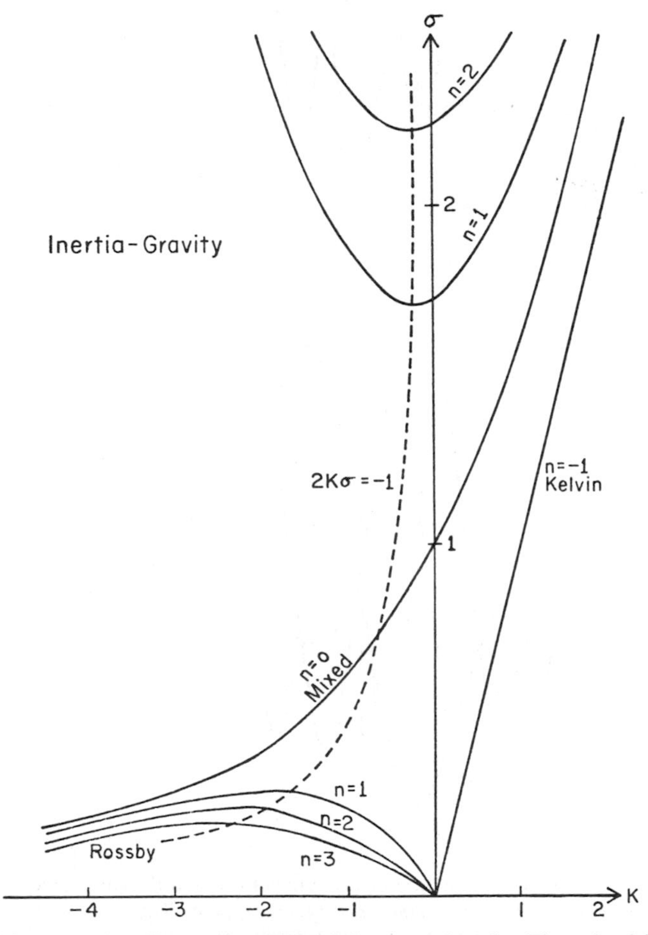

Figure 3.8. Dispersion diagram for equatorially trapped modes. The unit of frequency is $(\beta c)^{1/2}$ and the unit of zonal wave number k is the inverse of the radius of deformation $(c/\beta)^{1/2}$. [From Cane and Sarachik (1976).]

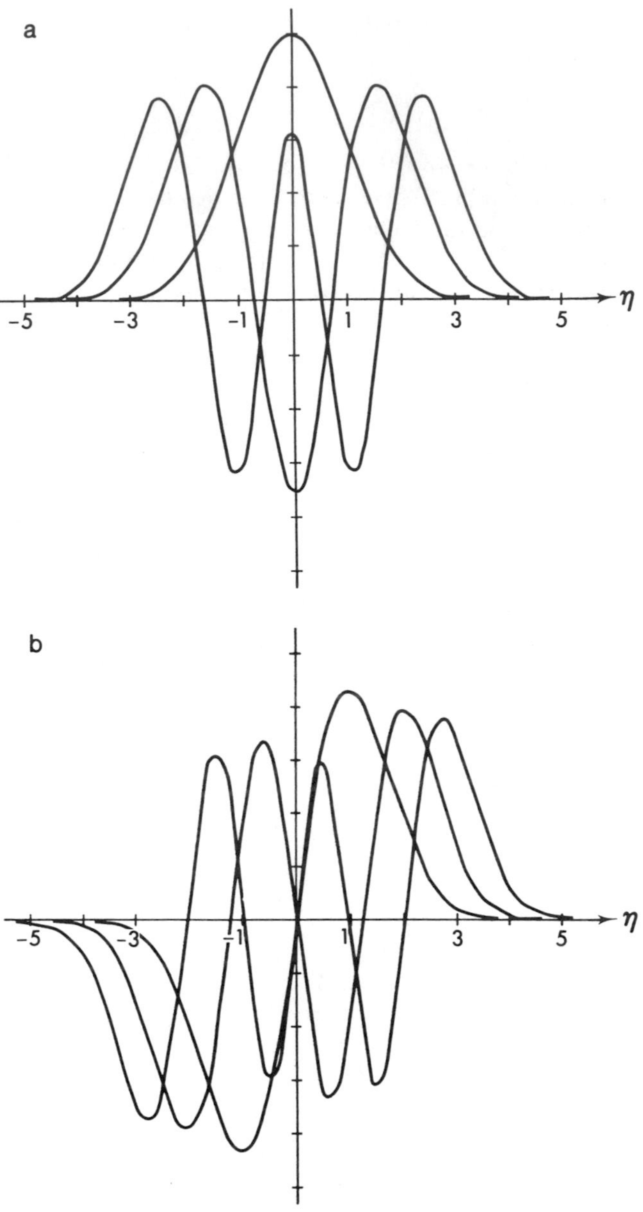

Figure 3.9. The latitudinal structure of (a) symmetrical and (b) antisymmetrical Hermite functions that describe the meridional velocity component. The unit of distance in the northward direction is the equatorial radius of deformation.

corresponds to the number of zeroes that a mode has, it can be regarded as a meridional wave number. As n increases, the turning latitude increases. For large values of n the Hermite functions are essentially sinusoidal except in the neighborhood of and poleward of their turning latitudes. For $n \gg 1$ the dispersion relation (3.23) is a good approximation to (3.32).

The gravest equatorially trapped mode deserves special comment. According to Eq. (3.32) there are two roots for $n = 0$:

$$k = -\sigma/c$$

$$k - \sigma c - \beta/c \tag{3.40}$$

The first root must be discarded because the associated u and η grow exponentially for large values of y even though v is bounded. The other root is known as the Rossby-gravity mode (see Note 2) because it is similar to inertia-gravity waves at high frequencies and similar to Rossby waves at low frequencies. The latitudinal shape of the meridional velocity component is a Gaussian centered on the equator; the zonal flow is antisymmetrical about the equator. Weisberg *et al.* (1979) describe measurements that are consistent with the structure of this wave.

At low frequencies the dispersion relation (3.32) simplifies to

$$\sigma = \frac{-\beta k}{k^2 + \dfrac{2n+1}{\lambda^2}}, \qquad n = 1, 2, 3, \ldots \tag{3.41}$$

from which it follows that equatorially trapped Rossby waves are very similar to the Rossby waves discussed earlier: the slow, short, dispersive waves have eastward group velocities and the fast, long, nondispersive waves have westward group velocities $c/(2n + 1)$. The most rapid Rossby wave ($n = 1$) travels at one-third the speed of long gravity waves. Its structure is shown in Fig. 3.10. The zonal current has a maximum and the thermocline displacement a minimum on the equator.

The long Rossby waves are of paramount importance in the oceanic adjustment to a change in the winds. The inertia-gravity, Rossby-gravity, and short Rossby waves are relatively unimportant and can be filtered from the equations of motion by making the "long wave" approximations. It is necessary that the scale of zonal variations, L, be much larger than the radius of deformation λ. The scale of latitudinal variations can, however, be comparable to λ. It is also necessary that the magnitude of the zonal flow exceed that of the meridional flow by a factor of L/λ at least. This assumption is justified by measurements at the equator that consistently show that at low frequencies—periods longer than a month—zonal velocity fluctuations are far more energetic than meridional velocity fluctuations

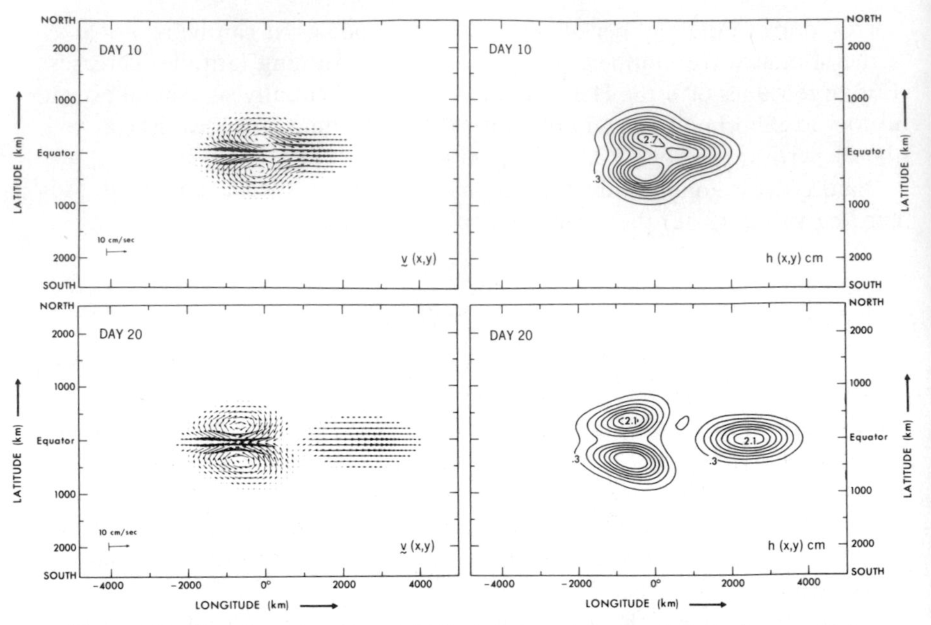

Figure 3.10. The dispersion of an initially bell-shaped thermocline displacement into an eastward-propagating Kelvin wave and westward-propagating Rossby waves. The left-hand panels show the horizontal currents and the right-hand panels the thermocline displacements. [From Philander *et al.* (1984).]

(Knox and Anderson, 1985). Under these conditions the shallow-water equations simplify to

$$u_t - fv + g'\eta_x = 0$$

$$fu + g'\eta_y = 0$$

$$g'\eta_t + c^2(u_x + v_y) = 0 \tag{3.42}$$

These equations yield a vorticity equation

$$\left(v_{yy} - \frac{f^2}{c^2}v\right)_t + \beta v_x = 0 \tag{3.43}$$

so that

$$v = F\left(x + \frac{c}{2n+1}t\right)D_n(y) \tag{3.44}$$

where F is an arbitrary function and D_n is a Hermite function. These long Rossby waves have their zonal flow in geostrophic balance.

To isolate short, low-frequency Rossby waves, which are important primarily in the neighborhood of the western boundaries of ocean basins, it is necessary to assume that the zonal wave number is large ($k \gg 1/\lambda$) and that the frequency is low [$\sigma \ll (\beta c)^{1/2}$]. The equations of motion then simplify to

$$\begin{aligned} -fv + g'\eta_x &= 0 \\ v_t + fu + g'\eta_y &= 0 \\ u_x + v_y &= 0 \end{aligned} \tag{3.45}$$

The meridional flow is in geostrophic balance and the horizontal motion is nondivergent. A single equation for the meridional velocity component readily follows:

$$v_{xt} + \beta v = 0 \tag{3.46}$$

This equation has solutions of the form $(\beta t/x)^{\nu/2} J_\nu(4x\beta t)$, where ν is a constant and J_ν is a Bessel function of order ν.

3.4.3 Kelvin Waves

Equatorial Kelvin waves have no meridional velocity fluctuations so that the equations of motion (3.1) simplify to

$$u_{tt} - c^2 u_{xx} = 0 \tag{3.47a}$$

$$fu_t - c^2 u_{xy} = 0 \tag{3.47b}$$

The first equation implies that

$$u = E(y)F(x \pm ct)$$

where E and F are arbitrary functions. Disturbances propagate nondispersively either eastward or westward with speed c. Equation (3.47b) determines the function E. It is unbounded at large values of y in the case of westward-propagating disturbances, which must therefore be ruled out. However, eastward equatorially trapped Kelvin waves are possible:

$$u = g'\eta/c = e^{-y^2/2\lambda^2} F(x - ct), \qquad v = 0 \tag{3.48}$$

For the case of wave disturbances $F = \exp(ikx - i\sigma t)$, Eq. (3.47a) gives the dispersion relation

$$\sigma = ck \tag{3.49}$$

A disturbance that is symmetrical about the equator will disperse into an eastward-traveling Kelvin wave and a westward-traveling Rossby pulse as shown in Fig. 3.10.

A Kelvin wave packet is nondispersive so that its components are always in phase with each other and can interact nonlinearly[4] in an efficient manner. Nonlinearities modify the equation

$$u_t + cu_x = 0 \tag{3.50}$$

in essentially two ways. Advection increases the phase speed from c to $c + u$. There is an additional change in the phase speed because of the deepening of the thermocline by the wave itself:

$$H \to H + \eta = H + cu/g' \tag{3.51}$$

$$\therefore\ c \to (g'H + cu)^{1/2} \sim c + u/2 \qquad \text{for small } u/c$$

According to this heuristic argument (Ripa, 1982) these two nonlinear corrections amount to the replacement of c by $c + 3u/2$ in Eq. (3.53). The zonal current u is a function of latitude so that its effect has to be averaged in that direction. This procedure gives the following equation for the nonlinear Kelvin wave:

$$u_t + \left(c + \sqrt{\tfrac{3}{2}}\,u\right)u_x = 0 \tag{3.52}$$

Equation (3.54), which can be derived formally by means of a perturbation expansion provided $u/c \ll 1$, has a solution that can be written in parametric form (Ripa, 1982; Boyd, 1980a). Figure 3.11 shows this solution for a disturbance that initially is Gaussian:

$$u(x, t = 0) = A\exp(-x^2/2a^2) \tag{3.53}$$

The leading edge of the disturbance, which introduces eastward currents and deepens the thermocline as it propagates eastward, steepens until it forms a front after a time

$$t = \frac{a}{A}\left(\frac{2e}{3}\right)^{1/2}$$

This singularity can be avoided by permitting not only Kelvin waves but also Rossby or inertia-gravity waves in the interactions. A Kelvin pulse with an amplitude of $+50$ cm/sec and a zonal scale of 5000 km will steepen into a front after approximately 100 days—before it has crossed the Pacific if it were excited in the west. By that time its speed would have increased by almost 30%. If the amplitude were $A = -50$ cm/sec, so that the initial pulse elevates the thermocline, then nonlinearities decrease both the speed and zonal gradients across the pulse. Nonlinear effects such as these have been identified in numerical models and may contribute to discrepancies between observed and predicted Kelvin wave speeds.

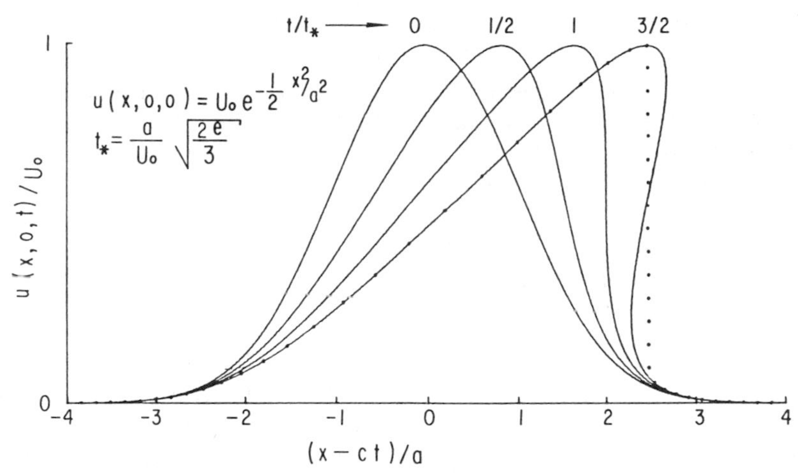

Figure 3.11. The nonlinear evolution of a pulse of Kelvin waves, associated with eastward currents and a depression of the thermocline. Initially, at time $t = 0$, the pulse is a Gaussian. The curves show the zonal structure of the eastward current at different times. The x axis is shifted for each curve in such a way that the departure from the initial curve is a nonlinear effect. The dots show the front at time $t = t^*$. [From Ripa (1982).]

Although observations that show eastward phase propagation along the equator are plentiful, it is difficult to find measurements that unambiguously show the presence of Kelvin waves. This is because the waves are superimposed on other waves and on time-dependent wind-driven currents (which have no dispersion relation). Measurements that filter out some of this variability provide the most persuasive evidence of Kelvin waves. Knox and Halpern (1982), for example, integrate the zonal currents vertically to reveal a pulse that propagated nondispersively from the central to the eastern equatorial Pacific in the northern spring of 1980 (Fig. 3.21). The vertical structure of the pulse changed significantly during its journey, presumably because of the presence of fluctuations not attributable to Kelvin waves. Tide gauges and Inverted Echo Sounders (which detect changes in the travel time of sound pulses through the water column) measure a vertical average of the density. This filter tends to bring Kelvin waves into prominence (Eriksen *et al.*, 1983; Katz, 1987a). As the 1982–1983 El Niño developed, tide-gauge records showed an eastward-traveling signal that appears to have been a first baroclinic mode Kelvin wave (Lukas *et al.*, 1984). Of far greater importance to the development of El Niño, however, was the considerably slower, eastward migration of isotherms on the ocean surface shown in Fig. 1.20. This migration, and more generally the development of all El Niño episodes, depends on far more than oceanic Kelvin waves and involves the unstable ocean–atmosphere interactions described in Chapter 6.

Kelvin waves are possible along the equator and also along coasts. If the coast is north–south then the mathematical description of these waves is complex because of the latitudinal variation of the Coriolis parameter (Moore, 1968). Section 3.4.5 on reflections pursues this matter. The next topic concerns the effect of east–west coasts.

3.4.4 Coasts Parallel to the Equator

Consider a zonal coast at such a high latitude that the value of the Coriolis parameter is essentially a constant f_0 within a radius of deformation c/f_0 of the coast. Explore motion with no meridional velocity component so that Eqs. (3.47) are the governing equations with $f = f_0$. It follows that disturbances propagate nondispersively either eastward or westward along the coast with speed c. In the case of eastward-traveling waves, the amplitude grows exponentially with increasing distance from the coast at $y = L$. In the case of westward-propagating signals,

$$u = \exp[(y - L)f_0/c]f(x - ct) \tag{3.54}$$

The e-folding distance for these coastal waves is the local value of the radius of deformation. Their dispersion relation is

$$\sigma = -ck \tag{3.55}$$

These are coastal Kelvin waves that travel with the coast on their right in the Northern Hemisphere and on their left in the Southern Hemisphere.

If a zonal coast is near the equator then the properties of Kelvin waves along that coast are affected by the latitudinal variations of the Coriolis force. Such a coast will also affect the equatorially trapped waves. This can happen in the eastern tropical Atlantic, in the Gulf of Guinea, which has a coast near 5°N. Equation (3.19), which describes equatorial waves, must now be solved subject to the condition that $v = 0$ at 5°N. This means that the equatorial Kelvin wave is unaffected because it has no meridional velocity component. The dispersion relation for the other waves continues to be Eq. (3.32) but the integers n are replaced by positive eigenvalues μ_n $(n = 0, 1, 2, \dots)$. The eigenfunctions are Parabolic Cylinder Functions. (When the coasts are far from the equator, at $\pm\infty$ then $\mu_n = n$ and the Cylinder Functions are Hermite functions.) The first few eigenvalues that correspond to the conditions $v = 0$ at 5°N and v bounded at $y = -\infty$ are (Cane and Sarachik, 1981)

$$\mu_0 = 0.01, \qquad \mu_1 = 1.1, \qquad \mu_2 = 2.2, \qquad \mu_3 = 3.4, \qquad \mu_4 = 4.8$$

The difference between μ_n and n is a measure of the degree to which the

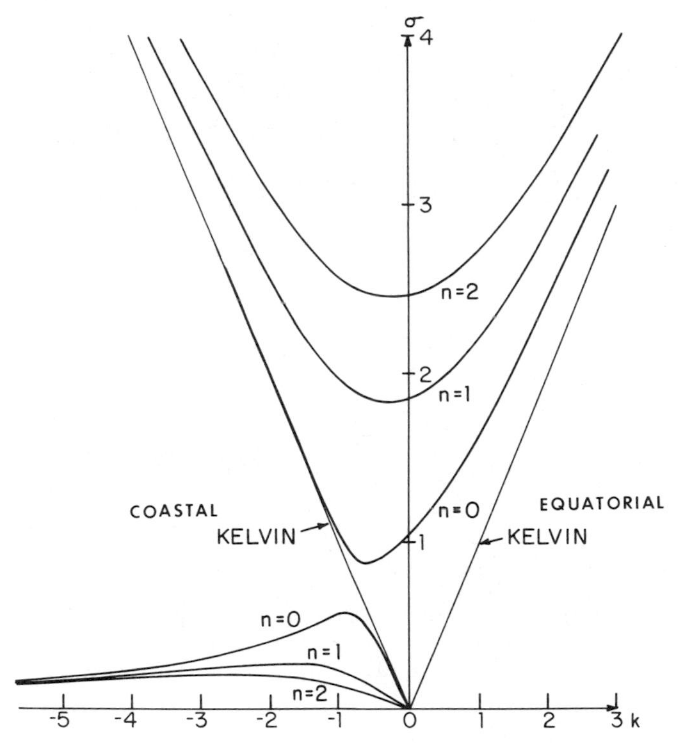

Figure 3.12. The dispersion diagram when a wall is present along a circle of latitude 1.7 radii of deformation north of the equator. This is the approximate location of the northern coast of the Gulf of Guinea. The unit of frequency is $(\beta c)^{1/2}$ and the unit of wave number k is $(\beta/c)^{1/2}$. The $n = 0$ Rossby-gravity curve of Fig. 3.8, which would have intersected the line for coastal Kelvin waves, now becomes two curves, for inertia-gravity-Kelvin and Rossby-Kelvin modes. [From Cane and Sarachik (1979).]

east–west coast affects the equatorial waves. The gravest mode appears to be little affected but the dispersion diagram (Fig. 3.12) indicates otherwise. Instead of a Rossby-gravity and coastal Kelvin wave there are Rossby-Kelvin and inertia-Kelvin modes. The structure of these modes, in the wave number range where they have westward group velocities and are nondispersive, resembles that of coastal Kelvin waves except that the meridional velocity component is not zero. Figure 3.13 shows the structure for a modified coastal Kelvin wave with frequency $\sigma = 0.5$ and zonal wave number -0.6. [The unit of time is $(\beta c)^{-1/2}$ and the unit of distance is the radius of deformation.] The existence of this very rapid westward-propagating mode in the Gulf of Guinea could enable the northern part of the gulf to adjust very rapidly to changes in the winds.

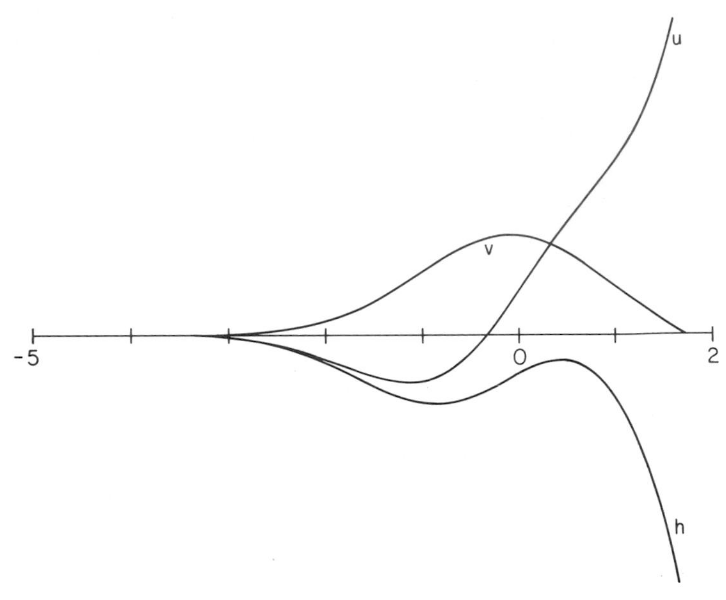

Figure 3.13. The structure of a Rossby-Kelvin mode for the point $\sigma = 0.5$, $k = -0.501$ in Fig. 3.12. The zonal velocity component (u) and thermocline depth (h) have maxima at the coast, as for coastal Kelvin waves, but the meridional velocity component (v) is nonzero. [From Cane and Sarachik (1979).]

3.4.5 *Reflections at Eastern Coasts*

The dispersion diagram (3.8) is strikingly asymmetrical about the $k = 0$ axis at subinertial frequencies. At periods from a week to a month only Kelvin and Rossby-gravity waves are possible and their group velocities are strictly eastward. Therefore energy accumulates at the eastern sides of equatorial ocean basins at these periods because waves with westward group velocities are unavailable for the westward reflection of energy. The wave numbers k of the waves that ought to be available for reflection can be calculated from the dispersion relation (3.32):

$$k = -\frac{\beta}{2\sigma} \pm \frac{1}{2}\left[\frac{\beta^2}{\sigma^2} - 4\left(\beta\frac{2n+1}{c} - \frac{\sigma^2}{c^2}\right)\right]^{1/2} \tag{3.56}$$

At frequencies close to $(\beta c)^{1/2}$—periods between a week and a month—the wave numbers k have complex values for all values of n. Since oscillations are assumed to have x dependence of the form $\exp(ikx)$, this result implies that a disturbance incident on an eastern boundary at $x = L$ excites coastally trapped waves. Far from the equator, for y large and positive, the sum of the coastally trapped disturbances asymptote to the expression

$$v = Ay^{1/2}\exp\left[i\left(\sigma t - \frac{\sigma y}{c} + \frac{\beta x}{2\sigma}\right) - \beta y\frac{L-x}{c}\right] \tag{3.57}$$

where A is a constant. This expression resembles a coastal Kelvin wave: it propagates poleward at speed c and is confined to a coastal zone with a width equal to that of the local radius of deformation $c/\beta y$. Although this width decreases with increasing latitude, the amplitude of the wave ($\sim y^{1/2}$) increases with increasing latitude so that energy is conserved. The wave differs from the coastal Kelvin wave of Eq. (3.50) because the velocity component normal to the coast is not zero and because the lines of constant phase are not normal to the coast. Similar waves are possible along the western sides of ocean basins where they propagate equatorward (Moore, 1968).

Analyses of sea level measurements along the western coasts of North and South America confirm the presence of coherent poleward-propagating disturbances, some of which are correlated with wind fluctuations over the equatorial Pacific Ocean (Enfield and Allen, 1980).

At frequencies near $(\beta c)^{1/2}$, Kelvin or Rossby-gravity waves incident on an eastern boundary excite only coastally trapped waves as shown in Fig. 3.8. As the frequency of the incident wave decreases, an increasing but always finite number of long Rossby waves become available for reflection. In Eq. (3.56) these waves are associated with the low values of n for which k is real. For large values of n, k is complex so that reflection involves a finite number of Rossby waves and an infinite number of coastally trapped waves. This means that there is always a poleward loss of energy at the eastern coast. This loss decreases with decreasing frequency. At very low frequencies the loss is negligible and an incident Kelvin wave of the form

$$u = \exp\left(-\frac{y^2}{2\lambda^2}\right)\cos\sigma\left(t - \frac{L-x}{c}\right)$$

reflects as long nondispersive Rossby waves (Eq. (3.44)). The sum of the incident and reflected waves can be written as (Cane and Moore, 1981)

$$u = -i\eta\tan(s)$$

$$v = i\sigma\eta y\sec 2(s)$$

$$\eta = \cos^{1/2}(s)\exp\left[i\sigma t + \frac{y^2}{2\lambda^2}\tan(s)\right] \tag{3.58}$$

where

$$s = 2\sigma(x-L)/c$$

These expressions are singular at $s = \pi/2, 3\pi/2, \ldots$, which are foci similar to those of Eq. (3.31) and Fig. 3.7.

The results of Eq. (3.58) can be used to determine the reflection of a Kelvin wave front or bore at an eastern coast. Let the incident front be described by

$$u = S(x - ct)\exp(-y^2/2\lambda^2)$$

where $S(x - ct)$ is a step function so that $S = 0$ if $x < ct$ and $S = 1$ if $x > ct$. As the front propagates eastward into a motionless region it suppresses the thermocline and introduces a steady, geostrophic eastward jet. At the eastern boundary the front excites poleward-traveling coastal waves that initially are trapped within a radius of deformation of the coast. With time, westward Rossby dispersion becomes possible (Anderson and Rowlands, 1976b). According to Eq. (3.57) this happens after a time $(2y/c)^{1/2}$ at a latitude y. The dispersion steadily increases the width of the coastal zone, more rapidly in low than in high latitudes because Rossby waves travel faster near the equator. After a long time the effect of all the reflected Rossby waves is to cancel the zonal current associated with the incident equatorial Kelvin front and to lower the thermocline uniformly everywhere (Cane and Sarachik, 1977). This asymptotic state is described by Eq. (3.58) in the limit $\sigma \to 0$:

$$u = v = 0; \qquad \eta = \sqrt{2} \tag{3.59}$$

3.4.6 Reflections at Western Coasts

The reflection of waves incident on the western boundary of an ocean basin involves a Kelvin or Rossby-gravity wave, depending on the symmetry of the incident wave. Because of this, reflection at a western boundary, unlike that at an eastern, is not associated with a poleward loss of energy. Consider a Rossby wave with meridional wave number N and frequency σ that is incident on a western coast at $x = 0$. In Eq. (3.56) the wave number of the incident wave corresponds to the plus sign for which group velocities are westward. The minus sign is appropriate for the reflected waves that have eastward group velocities. A recursive relation for the amplitude of the reflected waves can readily be written down (Moore and Philander, 1977); the solution has a number of important properties. The reflected waves are finite in number and have meridional wave numbers n that are all less than that of the incident wave. In other words, the incident wave reflects as a finite number of short Rossby waves plus a Kelvin or Rossby-gravity wave. The zonal wave numbers k of the reflected waves are all real so that the reflection does not involve coastally trapped waves. (Coastally trapped waves come into play when coastal Kelvin waves along the northern or southern coast are incident on the western coast or when they are excited by forcing along the western coast.) The Hermite functions D_n that describe the reflected waves all have $n < N + 1$, where N is the meridional wave number of the incident wave. This means that the reflected wave is at least as equatorially trapped as the incident wave.

At very low frequencies, reflections at western boundaries involve the short Rossby waves described by Eqs. (3.45). These nondivergent waves redistributed mass meridionally but they are not associated with a net zonal mass flux. Zonal mass flux into the western boundary is therefore returned eastward by the only other wave with an eastward group velocity, the equatorial Kelvin wave. This wave returns all the mass incident on the western coast, but it does not return all the energy incident on that coast. Suppose that the mass and energy flux incident on the western coast are associated with a long Rossby wave of meridional mode number N. At frequencies sufficiently low for Eqs. (3.45) to be valid, the fraction R of the energy flux [see Eq. (3.38)] returned eastward by the Kelvin wave is (Clarke, 1983)

$$R = 0.5, \qquad N = 1$$

$$R = \frac{(N-2)(N-4)\dots 1}{(N+1)(N-1)(N-3)\dots 2}, \qquad N = 3, 5, 7, \dots \tag{3.60}$$

At most half the energy of the gravest equatorially trapped Rossby mode returns eastward as a Kelvin wave. In the case of an incident wave that is antisymmetrical about the equator, there is no meridionally integrated mass flux into the western coast $x = 0$, no Kelvin wave is excited, and short Rossby waves transport mass across the equator in a western boundary current. (The energy flux ratio R is zero in this case.)

Reflections of Rossby waves off the western boundary of the Pacific Ocean are critically important in some coupled ocean–atmosphere models of the Southern Oscillation (Chapter 6). The Pacific of course does not have a continuous western boundary but if it is assumed that New Guinea, Irian Jaya, and Maluka form a barrier to westward-traveling equatorial Rossby modes then the slope of this barrier to meridians will affect reflections. In the Indian and Atlantic Oceans the coasts also slope relative to meridians. The condition that the incoming mass flux normal to the coast due to long Rossby waves must be returned by a Kelvin wave determines the reflections (Cane and Gent, 1984). Equatorial Kelvin waves can now be excited by Rossby waves that are symmetrical about the equator and also by those that are antisymmetrical. For the western equatorial Pacific it is estimated that the amplitude of reflected Kelvin waves is reduced on the order of 30%

from that which would be achieved if the boundary coincided with a meridian.

The higher the meridional mode number of equatorially trapped Rossby waves, the larger the number of zeroes of the zonal velocity component and the smaller the meridionally integrated zonal mass flux. This is why the amplitude of the reflected Kelvin wave is small. A disturbance with a large zonal mass transport that is incident on a western coast at a relatively high latitude, near 15°N say, will give rise to an equatorial Kelvin wave with the same mass flux even though each of the Rossby waves, whose sum describes the disturbance, will excite a Kelvin wave with a relatively small amplitude. It is the mass transport normal to the coast that matters.

3.4.7 Basin Modes

Figure 3.14 depicts the structure of one class of resonant modes of a closed ocean basin. The period is approximately the time it takes equatorial and coastal Kelvin waves to travel around the basin, clockwise in the Southern Hemisphere and anticlockwise in the Northern Hemisphere. (Allowance must be made for the time it takes Kelvin waves to turn corners.) The period of this class of modes must be close to $P = 2\pi(\beta c)^{-1/2}$ for the mode to involve only equatorial and coastal Kelvin waves. As the difference between P and the resonant period increases, Rossby (or inertia-gravity) waves come into play and the width of the equatorial zone in Fig. 3.14

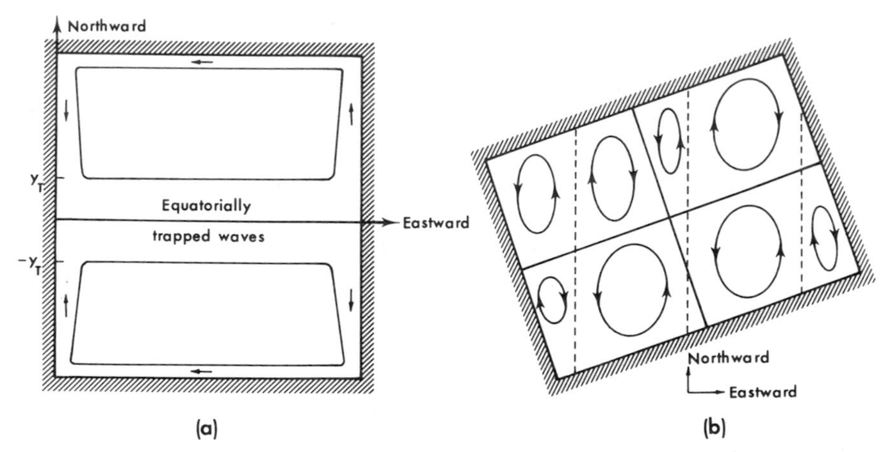

Figure 3.14. The structure of the modes of an ocean basin at frequencies (a) close to $\sigma_0 = (2\beta c)^{1/2}$ and (b) much lower than σ_0. In case (a) there is eastward phase propagation along the equator as Kelvin waves are the only equatorially trapped waves that are excited. In case (b) there are fixed nodal lines (the straight solid lines) and westward-propagating phase lines (the dashed lines).

widens. At very low (or high) frequencies it is possible to have modes without distinctive equatorial or coastal zones. Figure 3.14b shows the structure of such a low-frequency mode, which is a superposition of Rossby waves with eastward and westward group velocities (Moore, 1968). Low-frequency resonant modes of an ocean basin generally involve short Rossby waves with eastward group velocities. As noted earlier, the slow speeds and short scales of these waves make them prone to dissipation. A mode composed solely of equatorial Kelvin and long Rossby waves is therefore of special interest. Equation (3.58), which describes the sum of a Kelvin wave incident on an eastern coast ($x = L$) plus the long Rossby waves reflected there, also satisfies the condition $u = 0$ at a western ($x = 0$) wall provided (Cane and Moore, 1981)

$$P = 2\pi/\sigma = 4L/cm, \qquad m = 1, 2, 3, \ldots \tag{3.61}$$

For $m = 1$ this period is the time L/c it takes an equatorial Kelvin wave to propagate eastward across the basin, plus the time $3L/c$ it takes the gravest equatorially trapped Rossby wave to travel westward across the basin. In a shallow-water model this mode is excited by an abrupt intensification of the wind (Section 3.6).

3.4.8 Islands

The Gilbert and Galápagos Islands in the Pacific Ocean, and the Maldives in the Indian Ocean, fail to reflect or impede equatorial waves primarily because these islands have a latitudinal scale that is small relative to the equatorial radius of deformation (Yoon, 1981; Cane and du Penhoat, 1981; Rowlands, 1982). Small islands can and have been used as instrument platforms that do not affect the waves. Sea level measurements on the western side of the Galápagos Islands confirm the latitudinal structure of equatorial Kelvin waves (Ripa and Hayes, 1981). Even an island with a large latitudinal extent will fail to impede Kelvin waves with a frequency close to $(\beta c)^{1/2}$ because the coastally trapped waves excited at the island will propagate as shown in Fig. 3.15 and will regenerate eastward-traveling equatorial Kelvin waves.

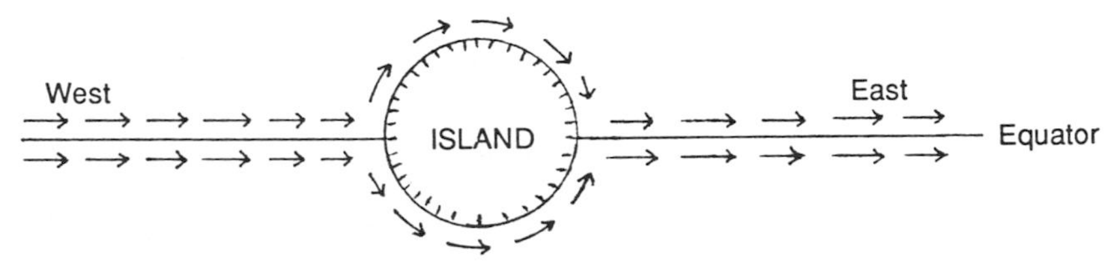

Figure 3.15. A schematic of Kelvin waves propagating around an island.

3.5 Generation of Sverdrup Flow

The mean surface currents in the tropical Pacific Ocean, except for those within a radius of deformation of the equator, are to a reasonable approximation in accord with the Sverdrup balance (3.9). These currents, shown in Fig. 2.1, have considerable seasonal and interannual variability. Assume that this variability is described by Eq. (3.8). A scale analysis of this equation yields interesting information about the oceanic response to variable winds in different frequency ranges. The ratio of the two terms on the left-hand side of the equation defines a time scale T where

$$T^2 = f^2 L/\beta c^2 = 2aL\Omega \sin^2\theta / c^2 \cos\theta \tag{3.62}$$

Latitude is denoted by θ. If the time scale of the wind fluctuations is T^* then the oceanic response depends critically on the ratio T to T^*.

$T^* \ll T$ and $T \gg 1/f$: At high frequencies, wind fluctuations do not excite Rossby waves because the term $(\beta\psi_x)$ in Eq. (3.8) is negligible. The response is local and the divergence of the Ekman drift determines vertical movements of the thermocline.

$T^* \sim T$: On time scales comparable to T the response is nonlocal because Rossby waves are important.

$T^* \gg T$: At low frequencies the first term in Eq. (3.8) is negligible and the oceanic response is a Sverdrup flow that is in phase with the slowly varying winds. The ocean is always in equilibrium with the winds and in effect passes through a succession of steady states.

If the forcing is at a fixed frequency, at a period of one year, say, so that $T^* = 1$ year, then the inequalities stated above define bands of latitude in which the oceanic response changes. In low latitudes, where T has a small value, $T^* \gg T$ and the response is equilibrium Sverdrup flow; at higher latitudes, where T has a longer value, Rossby wave propagation is evident; and farther north, local Ekman suction determines the vertical movements of the thermocline.

The time T, which is the time L/s it takes long Rossby waves with speed s to propagate a distance L, is the adjustment time of the ocean. To demonstrate this explicitly consider how the ocean adjusts to winds that suddenly start to blow and then remain steady. To simplify matters assume that the forcing has the latitudinal structure $\sin(ny)$ so that Eq. (3.8) after an appropriate redefinition of ψ becomes

$$(\psi_{xx} - r^2\psi)_t + \beta\psi_x = A \tag{3.63}$$

where

$$r^2 = n^2 + f_0^2/c^2$$

If the winds, which have a constant curl A, are assumed to be zonal then they can be viewed as an idealization of the winds that drive the subtropical gyre: westerly winds north of 30°N, say (where $y = 0$ and $f = f_0$), and easterly winds to the south of this latitude. These winds start to blow suddenly at time $t = 0$ and then remain steady. The oceanic response can be written

$$\psi = \psi^{\mathrm{I}} + \psi^{\mathrm{LR}} + \psi^{\mathrm{SR}}$$

where ψ^{I} is the initial response in the oceanic interior, far from coasts. This response is independent of longitude:

$$\psi^{\mathrm{I}} = -At/r^2 \tag{3.64}$$

This steady vertical movement of the thermocline is caused by the divergence of the Ekman drift. Associated with this displacement of the thermocline is an accelerating geostrophic zonal current u $(= (g'/f_0)n\psi \cos ny)$ that satisfies the boundary conditions on neither the western ($x = 0$) nor eastern ($x = L$) coasts. To meet the boundary conditions it is necessary to superimpose on the particular integral (3.64) the free modes, namely Rossby waves. [All other waves are filtered from Eq. (3.8).] At the eastern coast, Rossby waves ψ^{LR} with westward group velocities are excited. Assume that these waves are long so that they satisfy the hyperbolic equation

$$-r^2\psi_t^{\mathrm{LR}} + \beta\psi_x^{\mathrm{LR}} = 0$$

so that

$$\psi^{\mathrm{LR}} = F(x + \beta t/r^2)$$

where F is a function which satisfies the equation

$$\begin{aligned} F(L + \beta t/r^2) &= 0 && \text{for } t < 0 \\ &= At/r && \text{for } t > 0 \end{aligned} \tag{3.65}$$

It follows that

$$\begin{aligned} \psi^{\mathrm{I}} + \psi^{\mathrm{LR}} &= -At/r^2 && \text{for } t < r^2(L - x)/\beta \\ &= -A(L - x)/\beta && \text{for } t > r^2(L - x)/\beta \end{aligned} \tag{3.66}$$

This equation describes a dramatic change in the motion, from an accelerating zonal current to steady Sverdrup flow:

$$v = \psi_x = A/\beta = \mathrm{curl}_z\,\boldsymbol{\tau} \tag{3.67}$$

This happens after a time $t = (L - x)r^2/\beta$, which is how long it takes a long nondispersive Rossby wave to travel from the wall at $x = L$ to the point x. West of the front the accelerating flow is strictly zonal but to the east of the front the Sverdrup flow has a meridional component given by Eq. (3.67). Since the flux across a circle of latitude must be zero in an ocean basin, the southward flux east of the front returns northward in the discontinuity at the front.

To satisfy the boundary conditions at the western coast $x = 0$ it is necessary to invoke short Rossby waves ψ^{SR} with eastward group velocities. Under the assumptions already made these waves satisfy Eqs. (3.45). The solution that satisfies the boundary condition at $x = 0$ and that merges with the interior solution (3.64) is

$$\psi^{I} + \psi^{RS} = Ar^{-2}\left[-t + \left(\frac{t}{\beta x}\right)^{1/2} J_1(2\sqrt{\beta x t})\right] \tag{3.68}$$

At this stage three regimes characterize the oceanic response: a western boundary current described by Eq. (3.68); an interior region where the zonal current accelerates according to Eq. (3.64); and an eastern region where the westward-expanding Sverdrup balance given by Eq. (3.66) obtains. In due course—in the time it takes a long Rossby wave to propagate westward across the basin—the solution in Eq. (3.68) is inappropriate because the western boundary layer must merge, not with the accelerating zonal current, but with the Sverdrup balance, which is now established across the entire basin. Motion is now described by the expression

$$\begin{aligned}\psi &= \psi^{I} + \psi^{RS} + \psi^{LR} \\ &= -\frac{AL}{\beta}\left[1 - x/L - J_0(2\sqrt{\beta x t})\right]\end{aligned} \tag{3.69}$$

Far from the western coast there is steady Sverdrup flow. If the width of the western boundary current is taken to be the first zero of the Bessel function J then this width decreases with increasing time. However, the thinning current must return northward the Sverdrup transport $\int_0^L A\,dx$ that flows southward across a circle of latitude. It follows that the speed of the western boundary current must increase as its width decreases—this is evident in the solution shown in Fig. 3.16—so that the vorticity v_x in the western boundary increases steadily. The wind imparts vorticity uniformly over the basin but it accumulates near the western coast, where a singularity develops with increasing time. Friction can be invoked to dissipate the vorticity near the coast (Stommel, 1948), but measurements do not show high levels of dissipation underneath the Gulf Stream, for example. A realistic alternative is to permit the western boundary current to become unstable, because of its shear. In that case the equilibrium response to steady winds is steady Sverdrup flow except for an unstable time-dependent

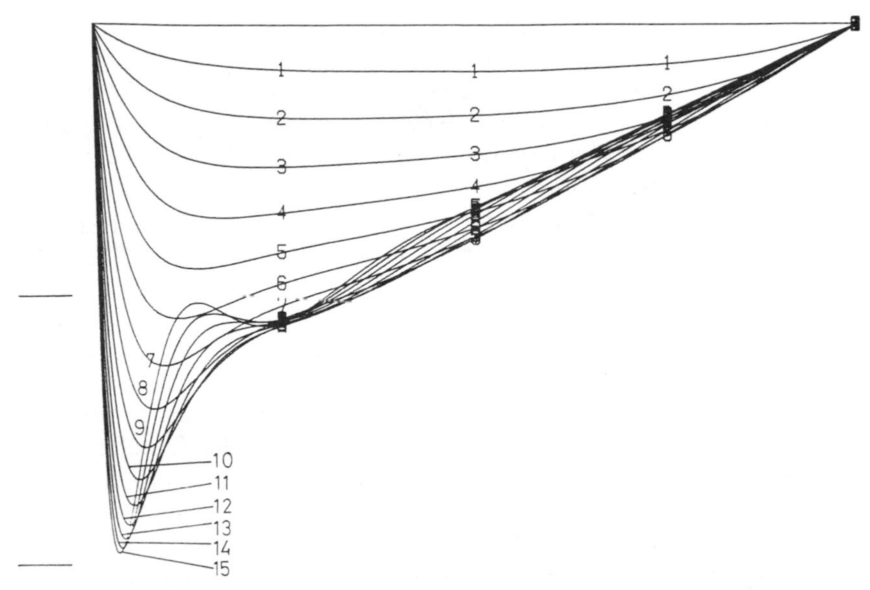

Figure 3.16. Displacements of the thermocline at different times (marked 1 to 15) after the wind turns on at time zero over an ocean initially at rest. In the center of the basin, Ekman pumping at first deepens the thermocline until the arrival of long Rossby waves from the east arrests the deepening and establishes a Sverdrup balance so that the thermocline has a zonal slope. The western boundary current becomes thinner and more intense with increasing time. [From Anderson and Gill (1975).]

western boundary current. Waves radiated by this current can be dissipated in regions remote from the current.

The time it takes the ocean to return to a state of equilibrium after a sudden change in the winds is essentially the time it takes a long Rossby wave to propagate from east to west across the basin of width L and is given by Eq. (3.62). This adjustment time decreases rapidly with decreasing latitude, as shown in Fig. 3.5, so that it takes far longer to generate, from rest, midlatitude currents such as the Gulf Stream than low-latitude currents such as the Somali Current or those in Fig. 2.1. Measurements indicate that the time it takes the ocean to adjust to a change in the winds is indeed shorter in low than in high latitudes. There have also been attempts to compare measurements with the solutions of Eqs. (3.8) and (3.63) in a more quantitative manner. In studies of the North Equatorial Countercurrent in the Pacific (Meyers, 1980) and Atlantic (Garzoli and Katz, 1983; Katz, 1987b), seasonal vertical movements of the thermocline have been explained in terms of Rossby waves near the southern boundary of the current and local Ekman suction near 10°N, consistent with the analysis of

Eq. (3.63). This agreement between measurements in low latitudes and the solution to Eq. (3.63) is a puzzle because some of the conditions that have to be satisfied for (3.63) to be valid are violated. For example, the assumption that the waves that effect the oceanic adjustment have a constant north–south wave number requires that the value of the Coriolis parameter f be relatively constant. This is a poor assumption in a latitude as low as 10°N. If variations in f are taken into account then the waves propagate not with a fixed but with a variable north–south wave number. This results in Rossby wave dispersion such as that shown in Fig. 3.7. It follows that the waves excited near one meridian will fail to reach certain parts of the ocean because of refraction. In these shadow zones a Sverdrup balance is impossible and local Ekman pumping determines thermocline movements. Along 10°N there ought to be shadow zones between certain meridians while Rossby waves should be evident in other regions. If this is not observed then other factors, such as the presence of mean currents, must come into play. The point is that Eq. (3.63) is not valid in the region of the North Equatorial Countercurrent. The apparent agreement that has thus far been found between the measurements and the theory is perplexing, not reassuring. The matter is pursued in Section 3.9.

The expression for the adjustment time of the ocean in Eq. (3.62) is singular at the equator because the approximations that were made in the derivation of this equation become invalid. Neglected factors that are important close to the equator include the change in the dispersion relation for Rossby waves, the existence of Kelvin waves, and the equatorial jet that can be generated by winds parallel to the equator.

3.6 Equatorial Adjustment

Near the equator the sudden onset of spatially uniform zonal winds at first generates the accelerating equatorial jet described in Section 3.3. Ultimately the zonal winds maintain a pressure gradient while the ocean approaches a state of rest (Eq. (3.10)). The adjustment from the initial state to the final equilibrium state is effected by the waves discussed in Section 3.4. In the case of a nonrotating tank of water, only gravity waves are available for the adjustment, but in the case of a rotating spherical shell of fluid, the gravity waves can be completely unimportant. This is because the inertia-gravity waves have periods strictly less than a week $[2\pi(2\beta c)^{-1/2})$ near the equator]. It follows that if the onset of the winds is gradual so that they attain full strength after a week or two, then the winds excite practically no inertia-gravity waves. Assume that the winds behave in this manner. Kelvin and Rossby waves are then responsible for the oceanic adjustment so that the motion at any moment is a superposition of these waves and the wind-driven jet described by Eq. (3.16):

$$u = u^{\mathrm{JET}} + u^{\mathrm{K}} + u^{\mathrm{R}} \tag{3.70}$$

The superscripts K and R indicate Kelvin and Rossby waves. The amplitudes of the waves are determined by the boundary conditions at the meridional coasts:

$$u = 0 \qquad \text{at } x = 0 \text{ and } L \tag{3.71}$$

Only waves with eastward group velocities are excited at the western boundary $x = 0$. The short, slow Rossby waves, as pointed out in Section 3.4, are important only near this coast where they redistribute mass meridionally (alongshore). They do not transport mass zonally at low frequencies so that the Kelvin wave is solely responsible for returning eastward any westward mass flux into the coast (Cane and Sarachik, 1977):

$$\int_{-\infty}^{\infty} u^{\mathrm{JET}}\, dy = -\int_{-\infty}^{\infty} u^{\mathrm{K}}\, dy \qquad \text{at } x = 0 \tag{3.72}$$

This condition determines the function F in the expression for the Kelvin wave in Eq. (3.48):

$$\begin{aligned} F(-ct) &= 0 \qquad \text{for } t < 0 \\ &= -\alpha t \tau^{x}/H \qquad \text{for } t > 0 \end{aligned}$$

It follows that

$$\begin{aligned} u^{\mathrm{K}}(x, y, t) &= 0 && \text{for } t < x/c \\ u(x, y, t) &= \frac{\alpha \tau^{x}}{cH}(x - ct)\exp(-y^{2}/2\lambda^{2}) && \text{for } t > x/c \end{aligned} \tag{3.73}$$

where α is a constant with a value of 0.84. This expression describes a front or bore that is excited at the western coast at time $t = 0$ and that propagates eastward at speed c. The front dramatically changes the initial response of the ocean to the zonal winds: in the wake of the front there is a sharp reduction in the acceleration of the zonal jet and in the intensity of the equatorial upwelling:

$$u = \tau^{x} x/Hc \qquad \text{on } y = 0 \text{ for } t > x/c \tag{3.74}$$

This happens because the front introduces a steady zonal pressure gradient that balances the windstress:

$$\eta_{x} = \frac{\alpha \tau^{x}}{H}\exp(-y^{2}/2\lambda^{2}) \tag{3.75}$$

In the wake of the front the zonal momentum balance near the equator changes from $u_t = \tau^x/H$ to $g'\eta_x \sim \tau^x/H$. The front also causes the jet to become horizontally divergent so that the poleward Ekman drift is no longer maintained by equatorial upwelling: the equation for the conservation of mass changes from $\eta_t + hv_y = 0$ to $H(u_x + v_y) \sim 0$.

Consider next the waves with westward group velocities that are excited at the eastern coast. Of most importance are the long nondispersive Rossby waves

$$u^{\mathrm{R}} = \frac{\tau^x}{cH} \sum_n \alpha_n S[x - L + ct/(2n+1)] R_n(y) \tag{3.76}$$

where R_n describes the latitudinal structure of the nth Rossby mode, which is given by Eq. (3.36) with $\sigma = -ck/(2n+1)$. The function S has the property $S(x) = 0$ if $x < 0$ and $S(x) = x$ if $x > 0$. The constants α_n are chosen such that the sum in Eq. (3.76) is equal to the expression for the equatorial jet [Eq. (3.16)]. The westward-traveling Rossby wave fronts, like the eastward-traveling Kelvin wave front, modify the equatorial jet by introducing zonal gradients. The Rossby waves, however, can extend to high latitudes, whereas the Kelvin wave affects only a region within a radius of deformation of the equator. The most rapid Rossby wave travels at speed $c/3$ and influences a narrow equatorial zone. The other Rossby modes travel more slowly, at speeds $c/7$, $c/11, \ldots,$ but they extend to higher latitudes. Far from the equator the initial motion is predominantly meridional Ekman drift $v = \tau^x/fH$. The long Rossby waves that emanate from the eastern coast ultimately eliminate both components of the horizontal flow so that the zonal wind maintains the pressure gradient of Eq. (3.10). The farther from the equator, the longer it takes to attain this equilibrium state. Figure 3.17 shows schematically how the oceanic adjustment proceeds: the initial response, an accelerating equatorial jet, persists longest in region I; a Kelvin front introduces steady motion in region II; westward-traveling Rossby fronts affect region III similarly; and very short Rossby waves are important in the western boundary layer IV.

After a time $3L/4c$ the Kelvin and Rossby fronts meet at the meridian $x = 3L/4$. There is now a zonal pressure gradient all along the equator, and the equatorial jet is practically steady at all meridians. The jet now starts to decelerate in the wake of the Kelvin front that propagates into the region already affected by the Rossby front, and similarly in the wake of the Rossby front as it propagates into the region $x < 3L/c$. Figure 3.18 clearly shows how the motion changes after the passage of the fronts. The Kelvin front incident on the eastern boundary at time L/c excites a new set of westward-traveling Rossby fronts. Subsequently there are repeated reflections at both coasts as fronts propagate back and forth. Each reflection at

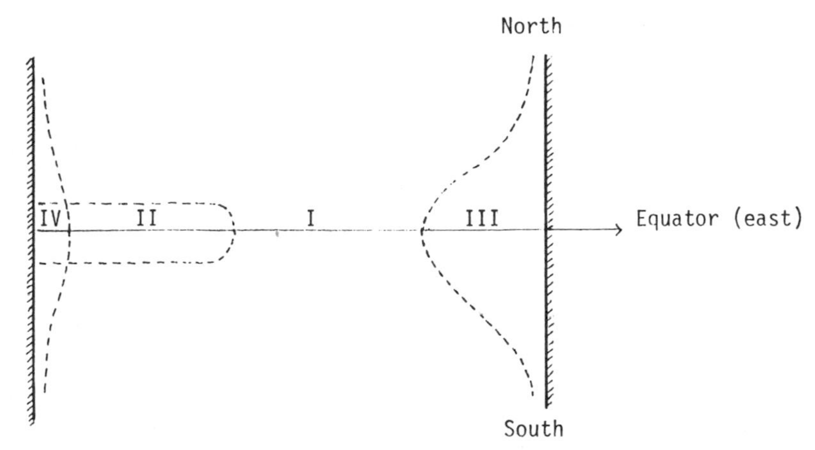

Figure 3.17. Schematic diagram that shows the distinct regions that characterize the oceanic adjustment to a sudden change in the winds.

the eastern boundary results in a loss of energy to high latitudes. Because of this the kinetic and potential energies of the equatorial region approach their equilibrium values rapidly.

Figure 3.19 shows that the adjustment time for the equatorial zone is, for practical purposes, $4L/c$, the time it takes a Kelvin wave to propagate eastward across the basin plus the time for the reflected Rossby wave to travel back westward across the basin. [This time $4L/c$ is also the period of the oscillations in the potential energy in Fig. 3.19 and is the period of the resonant equatorial mode of Eq. (3.61).]

The adjustment time of the ocean—the time it takes to establish equilibrium conditions after the sudden onset of steady winds—depends on the width of the basin and on latitude as shown in Fig. 3.5. Outside the equatorial zone the adjustment proceeds from the eastern coast and is effected by long Rossby waves as described in Section 3.5. Equation (3.62) gives the approximate adjustment time for that region. For the equatorial zone it is $4L/c$. For the Pacific Ocean the adjustment time is of the order of a decade in midlatitudes and decreases to approximately 450 days near the equator. The width of the equatorial Atlantic is only 5000 km, one-third that of the equatorial Pacific, so that its adjustment time is approximately 150 days. (The proximity of the northern coast of Brazil and the northern coast of the Gulf of Guinea to the equator probably reduces this estimate as mentioned earlier.)

Once the inviscid ocean is in equilibrium with steady, spatially uniform winds, it is motionless and has a sloping thermocline with which is associated a zonal pressure gradient that balances the wind. Suppose that the

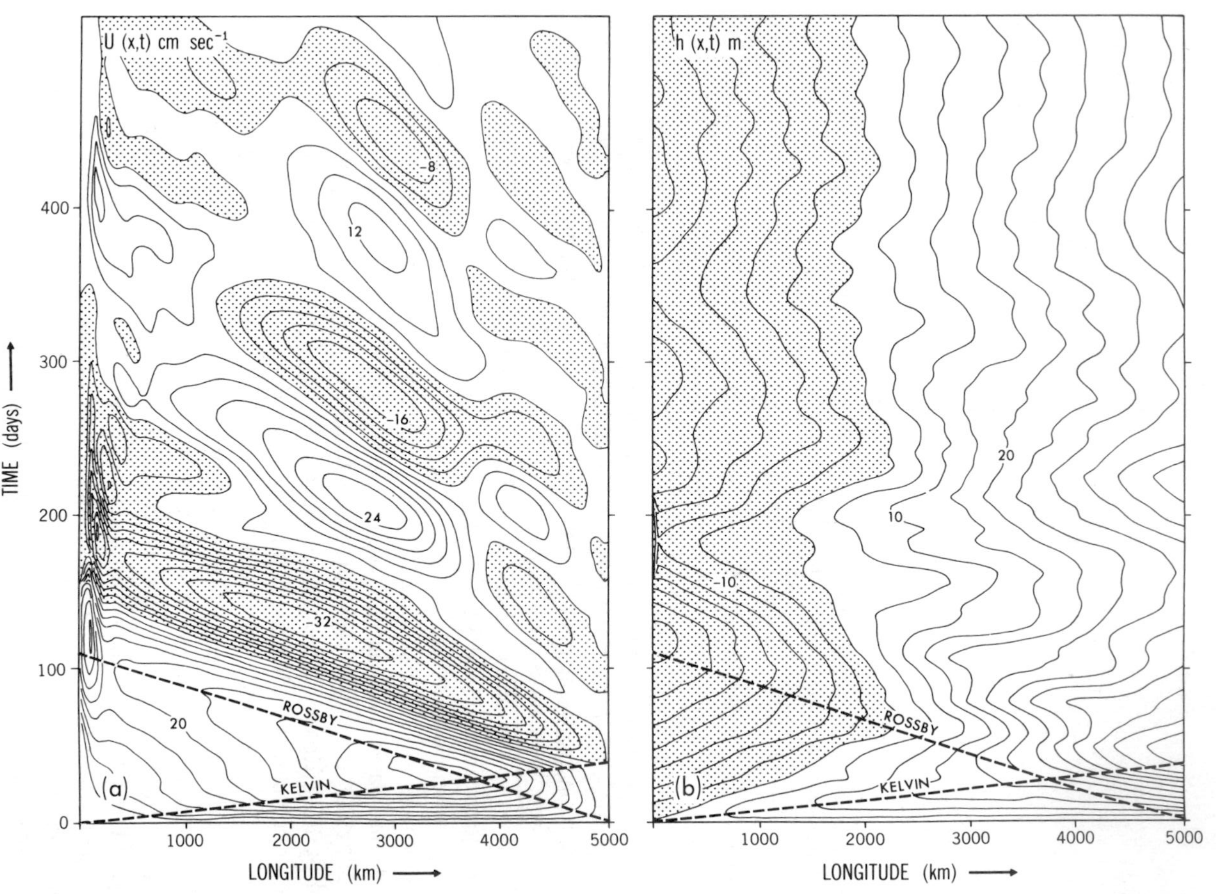

Figure 3.18. Changes in the zonal velocity component (centimeters per second) and in departures from the mean depth of the thermocline along the equator after the sudden onset of spatially uniform eastward winds. The dashed lines indicate the speeds at which Kelvin and the gravest Rossby mode propagate. The thermocline is elevated and motion is westward in shaded areas.

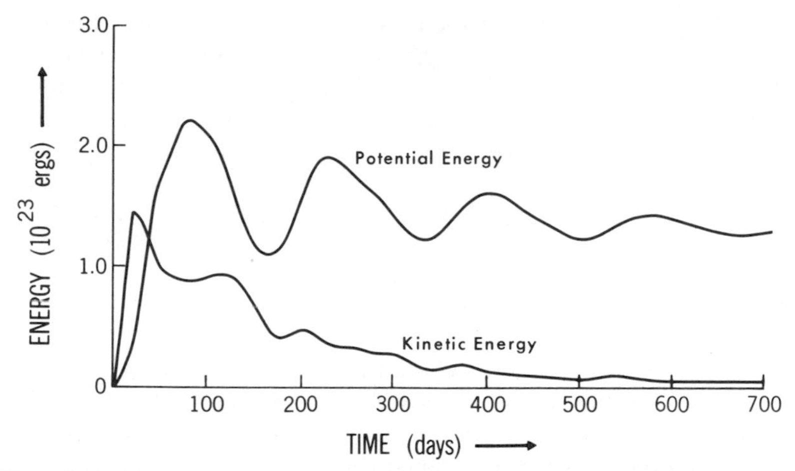

Figure 3.19. The kinetic and potential energy for the latitude band 5°N to 5°S for the motion in Fig. 3.18.

winds at this stage suddenly stop blowing. The pressure force will be left unbalanced and will initially accelerate the fluid in a direction opposite to that in which the wind had been blowing. The oceanic adjustment, back to a motionless state in which the thermocline is horizontal, will proceed exactly as before: waves excited at the coast will eliminate the zonal pressure gradient. This response to winds that suddenly stop blowing should be contrasted with the response (discussed in Section 3.3) of a zonally unbounded ocean to a similar change in the winds.[5]

3.7 Response to Remote Forcing

The winds over the ocean vary spatially, a factor not taken into account in the analysis thus far. To determine some of the effects associated with this complication, assume that the ocean is zonally unbounded and that spatially uniform eastward winds suddenly start to blow at time $t = 0$, but only between meridians A at $x = 0$ and B at $x = L$. There is no wind over the regions $x < 0$ and $x > L$.

The response is composed of a wind-driven equatorial jet between A and B—it is similar to the jet described in Section 3.3—plus Kelvin and Rossby waves excited at A and B to ensure continuity of the zonal velocity component and of the thermocline displacements at A and B. In the forced region, between A and B, the Kelvin front from A and the Rossby fronts from B introduce zonal pressure gradients that balance the windstress so that the initial acceleration of the jet stops. The early evolution of the flow is therefore similar to that described in Section 3.6. In the absence of north–south walls from which waves reflect, there is no further adjustment so that the final equilibrium state includes a steady zonal jet (McCreary, 1976; Cane and Sarachik, 1977):

$$u = \frac{\tau^x L}{\sqrt{2}\, Hc} \exp(-y^2/2\lambda^2) \tag{3.77}$$

$$\eta = \frac{\tau^x}{c^2}\left(x - L + \frac{L}{\sqrt{2}} \exp(-y^2/2\lambda^2)\right), \qquad 0 < x < L \tag{3.78}$$

The eastern unforced region ($x > L$) is motionless and has a horizontal thermocline until the arrival of the Kelvin front excited initially at B. This front introduces an accelerating equatorial jet and a steadily deepening thermocline that has a constant zonal slope:

$$u = -\frac{\tau^x}{Hc}(x - L - ct)\exp(-y^2/2\lambda^2)$$

$$\eta = -cu/g', \qquad x > L;\ t > (x - L)/c \tag{3.79}$$

At a fixed point these conditions persist until the arrival of the second Kelvin front excited at A. This front arrests the acceleration of the jet, stops the deepening of the thermocline, and eliminates the slope of the thermocline. The winds between A and B therefore maintain a steady jet and a deepened thermocline to the east of B in a region that steadily expands eastward:

$$u = \frac{\tau^x L}{\sqrt{2}\, Hc} \exp(-y^2/2\lambda^2), \qquad x > L;\ t > x/c \tag{3.80}$$

In the region to the west of B, where Rossby waves excited at A and B introduce an eastward jet and alter the topography of the thermocline, the latitudinal structure of the flow is more complex than in the east but the temporal evolution is similar.

The effect of a meridional boundary to the east of $x = L$ can be calculated by using the results of Section 3.4. From Eq. (3.58), which describes the long-term effect of a Kelvin front incident on an eastern boundary, it can be inferred that reflections at an eastern boundary will ultimately eliminate the equatorial jet established by the Kelvin waves in the region east of B but will leave the thermocline deeper than it originally was:

$$u = v = 0$$

$$\eta = \sqrt{2}\,\tau^x L/Hc, \qquad x > L,\ t \gg x/c \tag{3.81}$$

Figure 3.20 shows changes in the depth of the thermocline in response to winds with a limited zonal extent. After a long time the thermocline is seen to have a slope in the forced region only. The uniform deepening of the thermocline to the east of the forced region gradually spreads poleward along the coast.

If the eastward winds, over a region with a zonal extent L, last for a time T, and if these winds start and stop abruptly, then the region to the east is affected by four Kelvin waves: two are excited at the extremes of the forced region when the winds start to blow and two more are excited when the winds stop blowing. From the analysis that leads to Eqs. (3.80) and (3.81) it follows that there is an eastward current, and a deepening of the thermocline, for a time $T + L/c$ at a point east of the forced region. From measurements in the eastern side of the basin it is impossible to distinguish between winds that persist for a long time over a small region and winds that persist for a short time over a large region. The oceanic response depends on both the zonal extent of the forced region and the length of time for which the winds blow.

The remote response to winds with a complex structure can be calculated as follows. Expand the forcing function and the dependent variables as a series of Parabolic Cylinder Functions $D_n(y)$ after the introduction of new variables

$$q = u + g'\eta/c = \sum q_n(x,t) D_n(y)$$

$$r = -u + g'\eta/c = \sum r_n(x,t) D_n(y)$$

Equations for q_n, r_n, and v_n can readily be derived by exploiting the orthogonality of the Cylinder Functions. The equation for q_0 is

$$\frac{\partial q_0}{\partial t} + c\frac{\partial q_0}{\partial x} = X_0(x,t) \tag{3.82}$$

where X_0 is the projection of the zonal windstress onto a Gaussian (the Cylinder Function D_0). Equation (3.82) describes the Kelvin waves excited by the wind and has the solution

$$q_0(x,t) = \int_{-\infty}^{t} X_0[x + c(t'-t), t']\,dt' \tag{3.83}$$

For winds that are described by a Gaussian in longitude and time,

$$X = \exp\left[-(2x/L)^2 - (2t/T)^2\right] \tag{3.84}$$

the solution to (3.82) is

$$q = \frac{1}{4}\left[1 + \operatorname{erf}\left(\frac{2\varepsilon x + 2ct/\varepsilon}{(L^2 + c^2T^2)^{1/2}}\right)\right]\exp\left(\frac{-4(x-ct)^2}{L^2 + c^2T^2}\right)$$

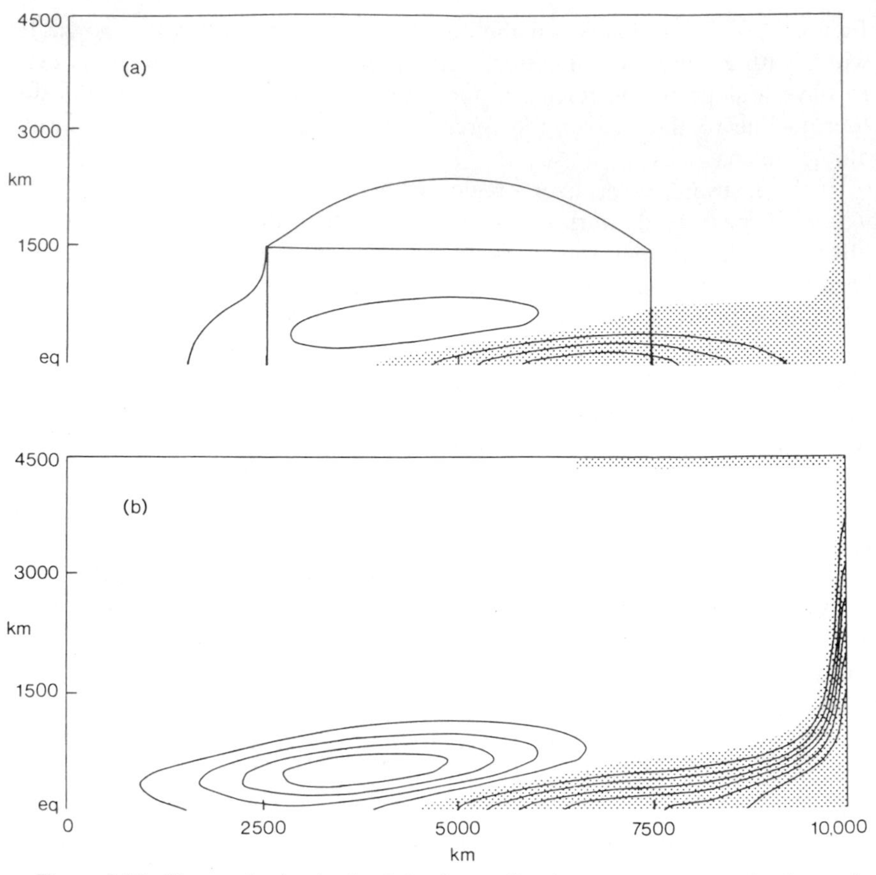

Figure 3.20. Changes in the depth of the thermocline in response to a patch of easterly winds with the zonal and latitudinal structure shown in (a). The maximum value of the windstress is 0.5 dyne/cm^2. The instantaneous pictures show conditions after (a) 1 month, (b) 3 months, (c) 13 months, and (d) 60 months. Shaded regions indicate an elevation of the thermocline. The contour interval for curves that show departures from the mean depth of the thermocline is 10 m. [From McCreary and Anderson (1984).] (*Figure continues.*)

where $\varepsilon = cT/L$. Far to the east of the forced region the response is an eastward-traveling Gaussian pulse. At a fixed point it lasts a time $(T^2 + L^2/c^2)^{1/2}$. Fig. 3.21 shows a pulse, generated in the western Pacific, that traveled eastward nondispersively for several thousand kilometers along the equator.

The oceanic response is complicated if the areal extent of the forcing function gradually expands. This happened in 1982 when eastward winds

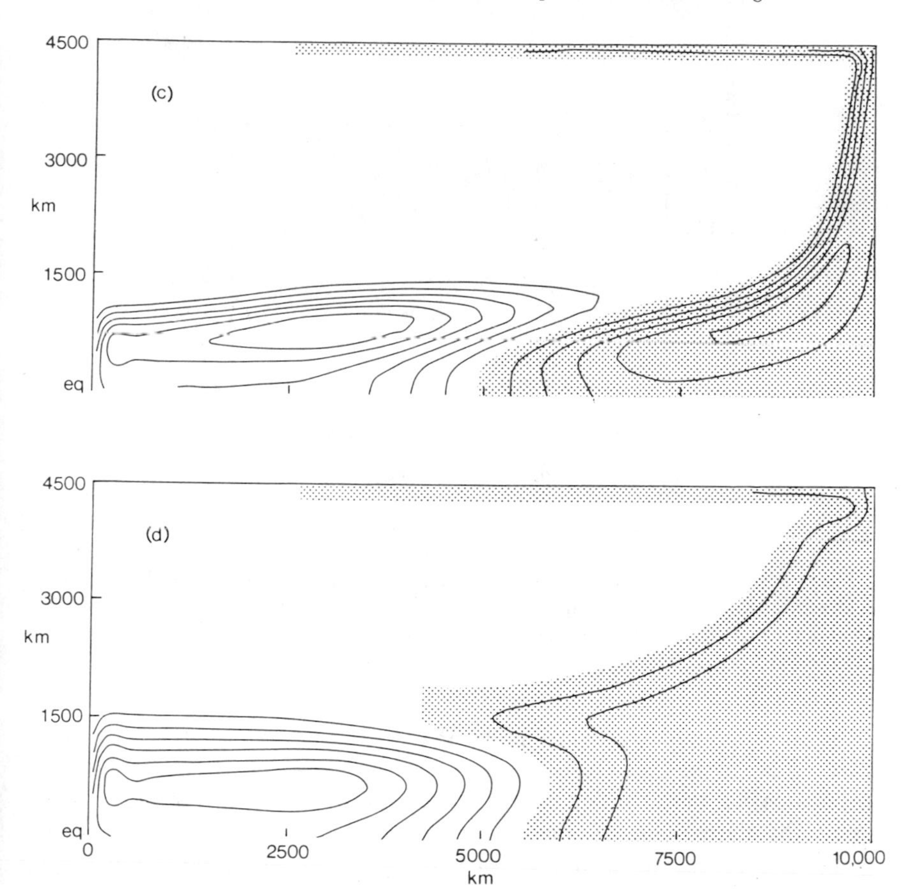

Figure 3.20 (*Continued*)

slowly penetrated farther and farther eastward in the tropical Pacific. Tang and Weisberg (1984) show that the response to a patch of westerly winds increases in amplitude if the patch moves eastward instead of remaining stationary. If the patch moves at the speed of a Kelvin wave then the response grows linearly with time. The eastward expansion of westerly winds during El Niño of 1982–1983 contributed to its large amplitude. In the tropical Atlantic the seasonal intensification of the southeast trades starts in the east and progresses westward. Rossby waves can then be forced resonantly (Weisberg and Tang, 1983, 1985; McCreary and Lukas, 1986).

Figure 3.20 depicts the oceanic response to eastward winds over a band of meridians in the middle of the basin. At first the winds deepen the thermocline to the east of the forced region but elevate it to the west. The

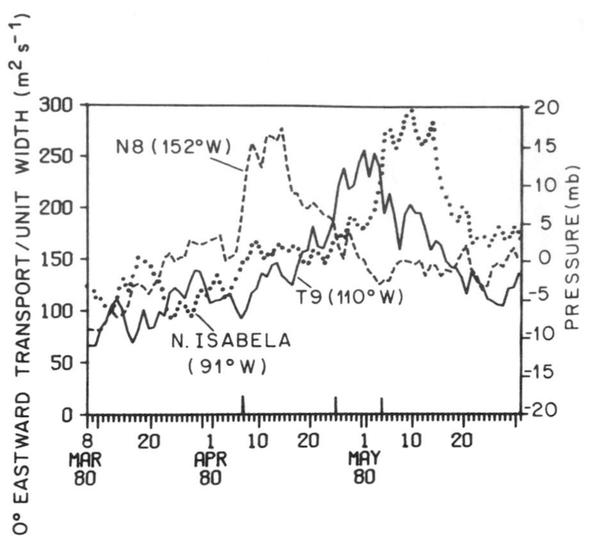

Figure 3.21. Eastward transport per unit width (essentially the average zonal current over the depth range 0 to 250 m) on the equator at 152°W and 110°W and sea level at Isabela Island in the Galápagos. A pulse progresses eastward. Its leading edge reaches 152°W on April 7, 110°W on April 25, and Isabela on May 3. [From Knox and Halpern (1982).]

elevation propagates westward as Rossby waves and reflects off the western coast an an equatorial Kelvin wave that elevates the thermocline as it propagates eastward. In due course this elevation cancels the initial deepening of the thermocline in the forced region. This feature of the oceanic response is of enormous importance in some coupled ocean–atmosphere models of the Southern Oscillation (Chapter 6).

3.8 The Effects of Dissipation

In the absence of any mixing processes, steady, spatially uniform, zonal winds maintain a zonal pressure gradient but do not drive any currents (Eq. (3.10)). This property of linear inviscid models can disappear if nonlinear or dissipative processes are taken into account. In the shallow-water equations (3.1), include Rayleigh damping represented by a coefficient a and Newtonian cooling represented by a coefficient b:

$$\begin{aligned} u_t - \beta y v + g'\eta_x &= -au + \tau^x/H \\ v_t + \beta y u + g'\eta_y &= -av + \tau^y/H \\ g'\eta_t + c^2(u_x + v_y) &= -b\eta \end{aligned} \tag{3.85}$$

If $b = 0$ and if a steady state is assumed so that a stream function ψ can be introduced, then

$$a\nabla^2\psi - \beta\psi_x = \mathrm{curl}_z(\tau/H) \tag{3.86}$$

This equation implies that a wind with a curl that is zero does not maintain steady currents. The state of no motion that characterizes linear, inviscid models can persist even in the presence of Rayleigh damping. It is not the mixing of momentum but the mixing of heat that permits spatially uniform winds to drive steady currents. This is true for a bounded ocean. In a zonally unbounded ocean, spatially uniform zonal winds drive a steady equatorial jet (Yamagata and Philander, 1985):

$$\begin{aligned} u &= \frac{\tau^x}{aH}(1 - \xi Q) \\ v &= -\left(\frac{a}{b}\right)^{1/4}\frac{\tau^x}{H}(\beta c)^{-1/2}Q \\ \eta &= -(ab)^{1/2}\tau^x\frac{1}{c}Q_\xi \end{aligned} \tag{3.87}$$

where

$$\xi = y/\lambda' \qquad \text{and} \qquad \lambda' = (agh/b\beta^2)^{1/4}$$

This jet is a steady version of the one in Section 3.3 so that its structure—the function Q—is that shown in Fig. 3.4. Note the change in the latitudinal scale, the radius of deformation. It decreases as a approaches zero so that the jet becomes very narrow.

If north–south walls are present then the waves that are excited at the coast and that effect the oceanic adjustment attenuate as they propagate away from the coast. In the case of the Kelvin wave (McCreary, 1981b),

$$u = D_0(\xi)\exp\left[-(ab)^{1/2}x/c\right] \tag{3.88}$$

where D_0 is a Gaussian. If $a = b = 0$ then the steady solution in the presence of north–south walls is a state of no motion in which a zonal pressure gradient balances the spatially uniform wind. If $a \neq 0$, $b = 0$ then the steady state is again one of no motion. However, if $a = 0$, $b \neq 0$ then

$$\begin{aligned} u &= -\frac{2c^2\tau^x}{by^4\beta^2H} + \left(\frac{c^2}{b\beta y^2} + L - x\right)\exp(\beta b y^2(x - L)/c^2) \\ v &= -\frac{\tau^x}{\beta H y}[1 - \exp(by^2\beta(x - L)/c^2)] \\ \eta &= -\frac{\tau^x}{b\beta y^2}[1 - \exp(by^2\beta(x - L)/c^2)] \end{aligned} \tag{3.89}$$

The waves attenuate as they propagate across the basin and the pressure gradient they establish is too weak to balance the wind. At the equator the flow is singular unless the mixing of momentum is taken into account. In other words, nonzero but finite-amplitude steady currents depend on the mixing of both heat and momentum. The steady equatorial currents described by this model are generally unrealistic because nonlinearities, discussed in the next chapter, are important.

3.9 The Effects of Mean Currents

The ratio of the speed of waves to that of mean currents is a measure of the influence currents have on waves. The equatorial Kelvin wave, which travels at 140 cm/sec, is faster than the observed equatorial currents and is least affected by the currents. Rossby waves, on the other hand, have a maximum speed of 50 cm/sec, which is less than the speed of the eastward equatorial jet in Fig. 3.1, and which is barely comparable to the speed of the North Equatorial Countercurrent. A simple model with which to study how these currents affect the waves assumes that the mean flow $U(y)$ depends on latitude only and is in geostrophic balance so that the thermocline depth $H(y)$ is given by

$$\beta y U + g' H_y = 0$$

Linear waves in the presence of the mean flow satisfy the equations

$$\begin{aligned} u_t + U u_x - (f - U_y) v + g' \eta_x &= 0 \\ v_t + U v_x + f u + g' \eta_y &= 0 \\ \eta_t + U \eta_x + (Hu)_x + (Hv)_y &= 0 \end{aligned} \tag{3.90}$$

Let $1/\sigma$ denote the time scale of the motion, L a zonal scale, and $\overline{U}$ a zonal velocity scale, then the assumptions $\sigma \ll f$ and $\overline{U}/fL \ll 1$, neither of which is valid near the equator, permit derivation of the vorticity equation

$$\frac{D}{Dt}\left[\eta_{xx} + R^{-2}\left(R^2 \eta_y\right)_y - R^{-2}\eta\right] + B\eta_x = 0 \tag{3.91}$$

where

$$\frac{D}{Dt} = \frac{\partial}{\partial t} + U\frac{\partial}{\partial x}$$

The effective β is

$$B = fQ_y/Q$$

and the effective radius of deformation is R:

$$R^2 = g'/Qf$$

The potential vorticity is

$$Q = (f - U_y)/H(y)$$

If the quasi-geostrophic approximations are adopted then the radius of deformation is the constant $\lambda = (c/f)^{1/2}$ and Eq. (3.91) simplifies to

$$\frac{D}{Dt}\left[\nabla^2\psi - \frac{1}{\lambda^2}\psi\right] + \left[\beta + \frac{U}{\lambda^2} - U_{yy}\right]\psi_x = 0 \tag{3.92}$$

From this equation there follows a dispersion relation for plane waves of the form $\exp[i(kx + ny - \sigma t)]$, where the latitudinal scale $1/n$ is much smaller than that of the mean current:

$$\begin{aligned}\sigma &= kU - k\frac{\beta - U_{yy} + (U/\lambda^2)}{k^2 + n^2 + (1/\lambda^2)} \\ &= \frac{Uk(k^2 + n^2) - \beta k + kU_{yy}}{k^2 + n^2 + (1/\lambda^2)}\end{aligned} \tag{3.93}$$

Long waves ($k^2, n^2 \ll 1/\lambda^2$ and $k < (U/\beta)^{1/2}$) have the dispersion relation

$$\sigma = -k\lambda^2(\beta - U_{yy})$$

and are unaffected by a mean current without shear. Contrary to expectations, the frequency σ is not equal to that of a Rossby wave in the absence of mean flow, Doppler shifted by kU. The Doppler shift is canceled by the effect of the mean thermocline slope on the vorticity gradient. This shows that the ratio of the speed of the wave to that of the current is not a reliable indicator of the effect of currents on waves.

The quasi-geostrophic equations assume a constant value for the Coriolis parameter and its derivative. To explore how the latitudinal variations of the Coriolis parameter, as well as the presence of mean currents, affect the propagation of Rossby wave packets, it is necessary to revert to Eq. (3.91). If the nonconstant coefficients of this equation are assumed to be slowly varying functions of latitude in comparison with the meridional scale $1/n$ of the waves, then the application of the WKB method leads to the dispersion relation (Chang, 1988; Chang and Philander, 1988)

$$\sigma = kU - Bk/(n^2 + k^2 + R^{-2})$$

where

$$B = \left(\beta - U_{yy} + U/R^2\right)\frac{f}{f - U_y}$$

This dispersion relation can be used to calculate ray paths for packets of waves with a fixed frequency σ and zonal wave number k provided the initial meridional wave number n is specified. As a simple example consider the case in which the shear of the mean flow is so small that $U_y \ll f$ and $U_{yy} \ll \beta$ so that

$$R^2 \sim g'H(y)/f^2(y)$$

and

$$B \sim \beta + U/R^2$$

If the long (zonal) wave approximation is made then

$$\sigma \sim -\beta k g'H(y)/f^2$$

Although the curvature of the mean flow may not modify the β effect, the slope of the thermocline associated with the mean geostrophic current strongly influences Rossby wave propagation. The shallower the thermocline, the slower the waves. A westward current, and the associated equatorward shoaling of the thermocline, can cause the zonal speed of Rossby waves to decrease with decreasing latitude. (In the absence of mean currents this speed increases as the equator is approached.) Eastward mean currents, on the other hand, magnify the zonal speed of Rossby waves.

Realistic mean currents strongly affect ray paths as is evident from Fig. 3.22. Waves with a period of one year, and with the indicated wavelengths, are seen to have critical layers near 10°N and 3°N, where their phase speed equals that of the mean flow and where the mean current absorbs the waves. This absorption can prevent the waves from reaching certain regions. Figure 3.22 clearly shows that the waves from the Southern Hemisphere have difficulty penetrating farther north than approximately 3°N while those from the north cannot penetrate much farther south than 10°N. Westward currents such as the North Equatorial Current enhance the westward speed of Rossby waves but inhibit their meridional propagation. In Fig. 3.22, waves to the north of 10°N are therefore capable of crossing a basin as wide as the Pacific before they reach the critical layer near 10°N. (In the absence of mean currents—see Fig. 3.7—the waves have a far larger meridional group velocity.) This could explain why, in the measurements, waves to the north of 10°N appear to propagate across the Pacific without significant equatorward refraction (Pazan *et al*., 1986).

The oceanic adjustment to a change in the winds is effected by waves described by Eq. (3.91). In the absence of mean currents there is an infinite set of discrete latitudinal modes. The fastest ones are equatorially trapped; slower ones propagate to higher latitudes. In the presence of mean currents the slower waves have critical layers and the set of waves that effects the

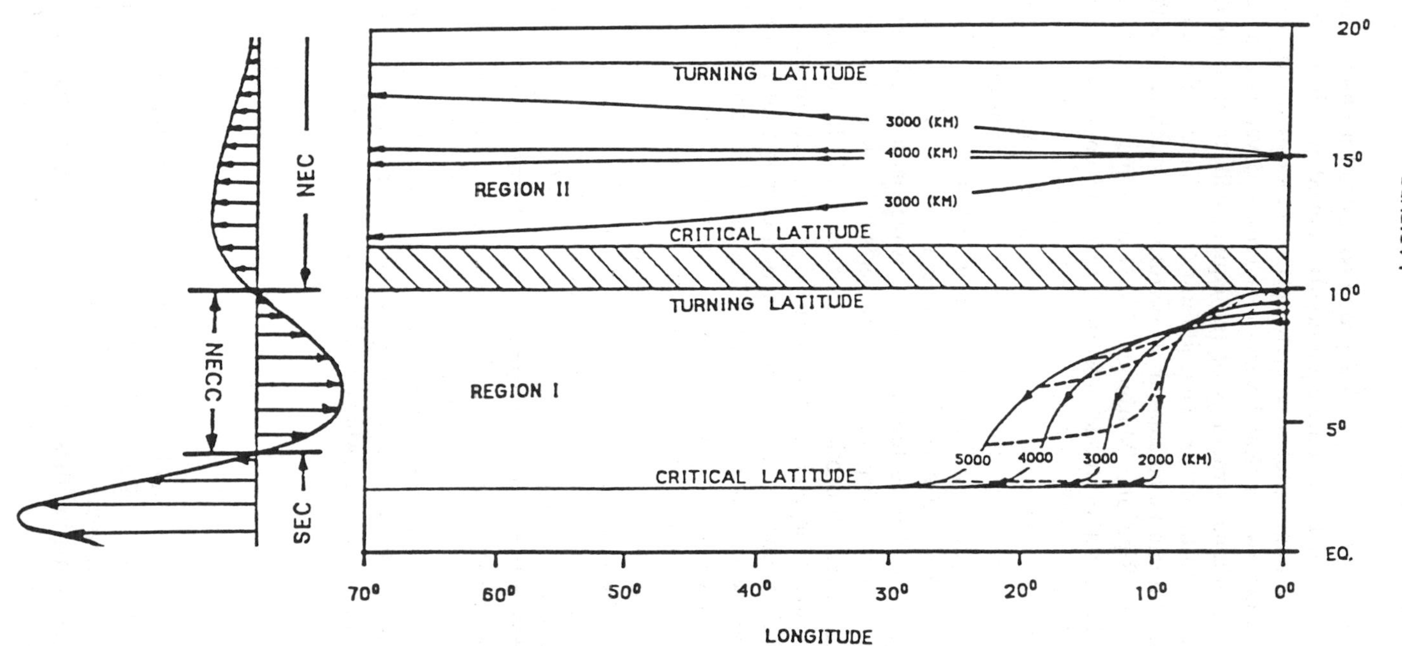

Figure 3.22. Ray paths of wave packets with a period of one year and with the indicated zonal wavelength in the presence of the realistic mean currents shown in the left-hand panel. The dashed lines are constant phase contours at intervals of 30 days. [From Chang and Philander (1988).]

adjustment falls into two groups. One group consists of a finite number of discrete latitudinal modes that are equatorially trapped and the other group consists of a continuum (not a discrete set) of modes each of which is bounded by either two critical layers or a critical layer and a turning latitude. The meridional structures of the equatorially trapped modes no longer correspond to the gravest Hermite functions but are affected by the mean currents and can be calculated by perturbation methods (McPhaden and Knox, 1979; Ripa and Marinone, 1983) or numerically (Philander, 1979a). This finite set of modes together with the continuum forms a complete set. Chang (1988) has developed a formalism that describes the oceanic response to wind variations in terms of these two groups of waves. The adjustment near the equator involves the modified equatorially trapped modes and is very similar to that described in Section 3.6. The adjustment far from the equator involves the continuum whose members all have critical layers. If the forcing function excites waves with small meridional group velocities—waves with long zonal wavelengths satisfy this condition—then critical layer absorption is unimportant and the waves, whose structure and speed depend on the mean flow, succeed in effecting an adjustment. This happens if the forcing function corresponds to an abrupt change in winds that are otherwise steady, or if the winds are periodic and have a low frequency. The waves have a significant meridional group velocity in the case of the annual and semiannual cycles. This means that critical layer absorption is important so that the waves fail to reach certain regions, which therefore respond strictly to local winds. In such regions, vertical movements of the thermocline are dictated by local Ekman suction and do not correspond to a Sverdrup balance. Figure 3.22 suggests that the neighborhood of 10°N is such a region, a result consistent with measurements (Meyers, 1980; Katz, 1987b). The same figure shows that eastward currents enhance meridional propagation and hence the speed with which waves reach critical layers. For this reason it is expected that the western part of the North Equatorial Countercurrent will not have a Sverdrup balance. Measurements with which to check this result are unavailable at this time.

3.10 Instabilities

Currents can become unstable under certain conditions that permit small perturbations to amplify. The conditions necessary for stability can readily be determined for currents $U(y)$ that are zonal and that are a function of latitude only. Assume that these currents are in geostrophic balance so that perturbations to the flow are described by Eqs. (3.90). The current $U(y)$ is stable to infinitesimal perturbations provided there exists a constant α such that (Ripa, 1983)

$$(\alpha - U)Q_y > 0 \tag{3.94a}$$

$$(\alpha - U)^2 < g'H(y) \tag{3.94b}$$

for all y, where $Q = (\beta y - U_y)/H(y)$ is the potential vorticity of the mean flow. A well-known special case of this general stability condition requires that the flow be nondivergent $[g' \to \infty,\ Q_y \to (\beta - U_{yy})/H]$ so that condition (3.94b) is trivially satisfied. By choosing α to be outside the range of U, stability is then assured provided the potential vorticity gradient Q_y does not change sign.

Stability analyses indicate that eastward equatorial jets with widths similar to those of the Equatorial Undercurrent are unstable provided their speeds exceed 1.5 m/sec. (The jets are assumed to be steady.) This sinuous mode of instability causes the jet to have eastward-propagating meanders about the equator. Its period to close to a month and its wavelength is of the order of 1000 km (Philander, 1976). Such instabilities have been excited in numerical models of the oceanic circulation but have not yet been observed in the ocean because eastward equatorial jets seldom attain a sufficiently high speed for a prolonged period. (There is also a varicose mode of instability in which perturbations remain symmetrical about the equator but the growth rate of this mode is slower than that of the sinuous mode.)

Instabilities associated with the shear of the westward South Equatorial Current and the eastward North Equatorial Countercurrent are common in the Pacific and Atlantic Oceans and are described in Section 2.7. Calculations for the observed currents indicate that they are stable in March and April, when both the South Equatorial Current and the North Equatorial Countercurrent are weak. They are unstable during the rest of the year, primarily because of the westward jet just north of the equator. Figure 3.23 shows a dispersion diagram for waves caused by instabilities of the surface currents in Fig. 2.1. The period, structure, and wavelength of the most unstable wave are in reasonable agreement with the measurements. This analysis, based on Eqs. (3.90), simplifies the vertical structure of the flow considerably. Calculations with a model that has realistic vertical structure indicate that the instabilities draw on both the kinetic and the potential energy[6] of the mean flow (Cox, 1980). Realistic General Circulation Models include an additional complication—zonal and temporal inhomogeneities of the mean flow. The simulations, shown in Fig. 2.24, are strikingly similar to sea surface temperature patterns in satellite photographs. Further analyses of the results from the models are necessary.

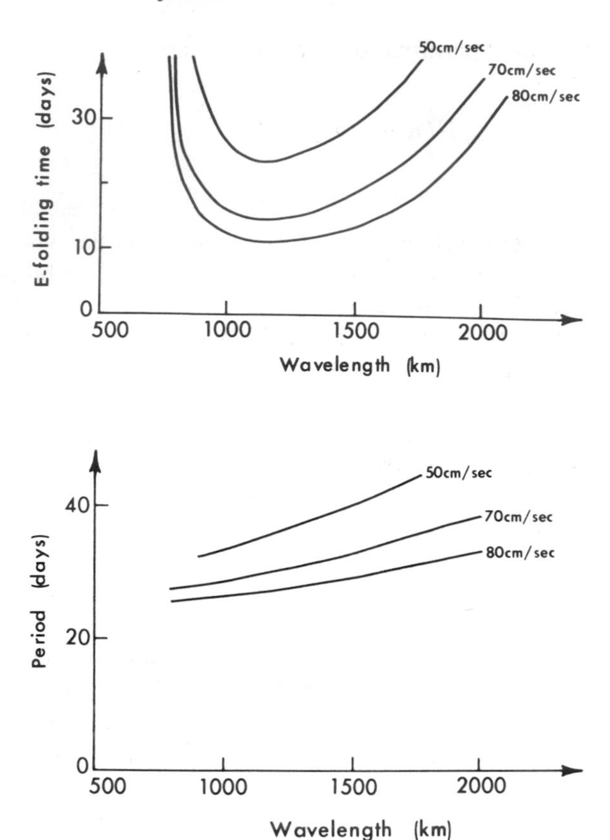

Figure 3.23. The e-folding time and period as a function of wavelength of waves associated with instabilities of the surface currents shown in Fig. 2.1. The curves correspond to different amplitudes of the zonal current. The shape of the profile $U(y)$ is the same for all the calculations, but the amplitude, which is taken to be the maximum speed of the South Equatorial Current, is varied. [From Philander (1978b).]

3.11 Discussion

The shallow-water model is one of the most powerful tools available to oceanographers. It provides answers to the questions raised by the measurements depicted in Figs. 3.1 and 3.2, but the measurements, which motivated the analyses presented here, prove too meager for a rigorous test of the theoretical results. Because the data are sparse, different interpretations of the measurements are possible. If the onset of the westerly winds over the Indian Ocean is taken to be instantaneous, then Kelvin and Rossby wave fronts are excited at the western and eastern extremes of the forced region, respectively. The Kelvin wave establishes a zonal pressure gradient and can be invoked to explain why, in Fig. 3.1, the acceleration of the jet stops a few weeks after the sudden onset of the winds. The Rossby wave could have caused the subsequent deceleration of the jet. Figure 3.2 seems to confirm the presence of the Rossby wave but there are no measurements to confirm that a Kelvin wave actually arrested the initial acceleration of the jet. This opens the door to a different interpretation of the data provided it is taken into account that the wind changes are not instantaneous but occur over a period of weeks. This is far longer than the few days it takes a Kelvin wave to propagate from the western extreme of the forced region to Gan, where the measurements in Fig. 3.1 were made. (The westerly winds do not extend far west of Gan.) A zonal pressure gradient to balance the wind is therefore established within a matter of days. Hence the high correlation between the intensifying winds and the accelerating jet in Fig. 3.1 can be interpreted as an equilibrium response. The wind at each moment is balanced by a pressure force; the acceleration of the jet stops when the wind becomes steady; and the jet decelerates several weeks later when the eastward winds relax. However, if the relaxation is too sudden then the westward pressure force that the winds had maintained is left unbalanced and it decelerates the jet. This cannot be the entire story because Fig. 3.2 suggests that a Rossby wave played a role. A unique interpretation of the data in Fig. 3.1 clearly requires more measurements.

The shallow-water model is useful not only for studying idealized situations but also for simulating certain aspects of the oceanic response to the observed winds. The model is reasonably good at reproducing seasonal and interannual variations in the depth of the thermocline (Busalacchi and O'Brien, 1980, 1981; Busalacchi *et al.*, 1983). In certain parts of the ocean, the eastern tropical Pacific for example, there is a high correlation between sea surface temperature and thermocline depth variations so that the shallow-water model provides some information about sea surface temperatures. It is possible to go a step further by introducing thermodynamics in a simplified manner, so that sea surface temperature is an explicit variable, while retaining the one-layer formalism for dynamical purposes. An example of an equation for the temperature T of the upper ocean, from Anderson and McCreary (1985), is

$$(\eta T)_t + (u\eta T)_x + (v\eta T)_y = Q/\rho c_p + wT' - \overline{w}T \qquad (3.95)$$

where T' is the specified constant temperature of the deep ocean, $\overline{w}$ is a specified vertical velocity component that brings cold, deep water into the upper ocean, and w is an entrainment velocity given by

$$\eta(T - T')w = 2r - \eta Q/\rho c_p$$

where Q is the prescribed heat flux across the ocean surface and r is a specified rate of potential energy increase due to mechanical stirring by the wind. Schopf and Cane (1983) have a somewhat more elaborate procedure for calculating the temperature of the upper ocean. These simple models have proved valuable in studies of the interactions between the ocean and atmosphere (Chapter 6). They provide reasonable descriptions of sea surface temperature variations in response to certain windstress patterns but, because of the various approximations that are made, are sometimes unrealistic. To improve on these models it is necessary to take into account the continuous stratification of the ocean. The next chapter examines this subject.

Notes

1. Equations (3.1) in a spherical coordinate system, and with the appropriate forcing function, are Laplace's Tidal Equations. The Cartesian coordinate system of Section 3.2, known as the equatorial β-plane coordinates system, is an approximation to spherical coordinates and is accurate provided motion in a thin shell on a large sphere is confined to the tropics. Gill (1982) and Pedlosky (1987) derive these equations formally and state the approximations explicitly.

2. Matsuno (1966) provided the first consistent description of the properties of equatorially trapped waves in a shallow-water ocean. Lindzen (1967) first discussed their vertical propagation (Section 4.4). Some oceanographers refer to Rossby-gravity modes as Yanai waves, presumably after Professor Yanai, whose analysis of meteorological data revealed the presence of this mode in the atmosphere (Yanai and Marayama, 1966). If these oceanographers were consistent they would refer to equatorial Kelvin waves as Wallace–Kousky (1968) modes.

3. The first few Hermite polynomials are

$$H_0 = 1, \qquad H_1 = 2\xi$$

$$H_2 = 4\xi^2 - 2, \qquad H_3 = 8\xi^3 - 12\xi$$

$$H_4 = 16\xi^4 - 48\xi^2 + 12, \qquad H_5 = 32\xi^5 - 160\xi^3 + 120\xi$$

Note that

$$\frac{dH_n}{d\xi} = 2nH_{n-1}$$

$$\xi H_n = nH_{n-1} + 0.5H_{n+1}$$

4. The nonlinear interactions between various equatorial waves have been studied by Ripa (1982), Boyd (1980a and b), and others. One goal of these efforts is to explain the continuous spectrum of relatively high-frequency fluctuations in the ocean. In regions far from the equator where the fluctuations correspond to inertia-gravity waves, nonlinear interactions redistribute energy so that the spectrum, between the Brunt–Väisälä frequency and the inertia frequency, has a universal shape known as the Garrett and Munk (1979) spectrum. Close to the equator the fluctuations below the thermocline correspond to equatorially trapped waves but the spectrum is apparently not a universal one (Eriksen, 1981, 1985).

5. The seminal paper on oceanic adjustment is Lighthill's (1969) study of the generation of the Somali Current from a state of rest. The role of the equatorial Kelvin wave, which Lighthill overlooked, was pointed out by D. W. Moore and by Gill (1975), who investigated the generation of the Equatorial Undercurrent, and by McCreary (1976) and Hurlburt *et al.* (1976), who investigated the oceanic response to the relaxation of the trades during El Niño. The elliptically written series of papers by Cane and Sarachik (1976, 1977, 1979, 1981, 1983a) is a valuable and exhaustive study of the adjustment of a shallow-water ocean to various windstress patterns.

6. Baroclinic instability is not possible in a shallow-water model. In a two-layer system, currents with a given vertical shear become baroclinically more stable as their mean latitude decreases until, near the equator, there is no baroclinic instability. This is an artifact of the two-layer model. In a continuously stratified model the vertical scale of the baroclinically unstable waves decreases with decreasing latitude—it is zero at the equator—so that a two-layer model fails to resolve the waves near the equator (Held, 1978). Baroclinic instability near the equator has received little attention.

Chapter 4 Oceanic Adjustment: II

4.1 Introduction

The sea surface temperature patterns in the tropical Pacific and Atlantic Oceans, shown in Fig. 1.10, are distinctly asymmetrical relative to the equator. The thermal equator, the line of maximum sea surface temperature, does not coincide with the geographical equator but lies within the basinwide band of warm surface waters between approximately 3°N and 10°N. Sea surface temperatures in the eastern tropical Pacific are low outside this band, especially off the coast of the Americas and along the equator. A sharp thermal front near 3°N in the eastern Pacific separates the cold water near the equator from the warmer water farther north. To explain this sea surface temperature pattern it is necessary to take into account two factors not discussed in Chapter 3: the continuous vertical stratification of the ocean, and the cross-equatorial component of the wind. (The southeast trades penetrate into the Northern Hemisphere.)

The southeast and northeast trades have components that are parallel to the western coast of the Americas. The oceanic response to these alongshore winds includes a distinctive, narrow coastal zone that has much in common with the equatorial zone. For example, it is possible for alongshore winds to generate intense, narrow, surface coastal jets in the direction of the wind, to drive offshore Ekman drift that induces coastal upwelling, and to excite coastally trapped waves that establish alongshore pressure gradients and coastal undercurrents. These phenomena, all of which have their equatorial counterparts, can occur along both the eastern and western boundaries of ocean basins on relatively short time scales of days and weeks. On longer time scales, fascinating differences emerge between the eastern and western coasts: currents along the eastern boundaries of ocean basins tend to be slow and broad, and those along the western boundaries swift and narrow. Thus the equatorward currents along the western coasts of Africa and the Americas—the Benguela, Canary, California, and Peru Currents—have speeds on the order of 10 cm/sec and have widths that can be measured in hundreds of kilometers. The Somali and North Brazil Currents, on the other hand, attain speeds in excess of 1 m/sec and have widths of approximately 100 km. These western boundary currents veer offshore sharply near 5°N, where their effect on the thermal structure of the ocean extends only a few hundred kilometers from the coast. The much slower eastern boundary currents separate from the coast gradually as they merge with the westward South Equatorial Current, and they influence the sea surface temperatures for thousands of kilometers offshore.

The continuous vertical stratification of the ocean strongly influences sea surface temperatures in regions where upwelling is important, for example, the eastern halves of the tropical Atlantic and Pacific Oceans. Continuous stratification also introduces a number of phenomena that are impossible in models that regard the thermocline as the interface between two fluids each of constant density. For example, it becomes possible for waves to propagate not only horizontally but also vertically. Since waves can establish horizontal pressure gradients that drive currents, it is then possible to have different vertical structures for currents that are driven directly by the wind and currents that are driven by pressure gradients. Consider westward winds that maintain an eastward pressure force that extends through the thermocline. Away from the equator the wind drives poleward Ekman drift near the surface, whereas at deeper levels the pressure gradient maintains equatorward geostrophic flow. Near the equator the Coriolis force fails to balance the pressure gradient, which then drives the convergent fluid eastward in the Equatorial Undercurrent. (Coastal undercurrents have similar explanations.) The critical feature of this model, and of almost all models of the Equatorial Undercurrent, is the difference between the vertical structures of the directly wind-driven surface current and that of the pressure force maintained by the wind.[1] An equatorial undercurrent is absent from the shallow-water model of Chapter 3 because it does not permit such a difference.[2] An undercurrent is possible in a continuously stratified ocean that permits the downward propagation of waves, which establish pressure gradients below the surface layers.

Vertical wave propagation can have a direct effect on sea surface temperature variations. Consider the deepening of the thermocline in the wake of a Kelvin wave front excited by a sudden burst of westerly winds over the far western equatorial Pacific. (This problem is discussed in Section

3.7.) The current introduced by the wave advects warm water eastward and can be expected to increase sea surface temperatures in the east (Harrison and Schopf, 1984). This need not happen in a continuously stratified ocean, however, because the wave in such an ocean propagates both eastward and downward. It can disappear from the surface layers of the ocean by the time it reaches the eastern Pacific, in which case its direct effect on the sea surface temperatures in the east is negligible. Figure 2.14 illustrates an occasion when this happened. The relaxation of westerly winds over the western equatorial Pacific towards the end of 1982 excited an eastward-traveling pulse that elevated the thermocline without affecting sea surface temperatures on the equator near 95°W.

This chapter explores the properties of continuously stratified models of the tropical oceans and describes two different methods for solving the linear equations of motion. One method, presented in Section 4.3, exploits the mathematical completeness of the set of vertically standing modes that are possible in an ocean with a flat floor. This method, though powerful, is inconvenient for the discussion of vertically propagating waves. A different method that exploits the completeness of latitudinal modes is useful for the discussion of such waves that are excited by forcing at a fixed frequency and zonal wave number. Both these methods of solution apply only to the linear equations. To solve the fully nonlinear equations it is necessary to resort to General Circulation Models, which are discussed in Section 4.9.

4.2 A Continuously Stratified Model

The nonlinear equations that describe motion on an equatorial plane, simplified by making the hydrostatic and Boussinesq approximations, are referred to as the primitive equations:

$$\frac{Du}{Dt} - fv + \frac{1}{\rho_0} P_x = F^x \tag{4.1a}$$

$$\frac{Dv}{Dt} + fu + \frac{1}{\rho_0} P_y = F^y \tag{4.1b}$$

$$-\rho g + P_z = 0 \tag{4.1c}$$

$$u_x + v_y + w_z = 0 \tag{4.1d}$$

$$\frac{D\rho}{Dt} = G \tag{4.1e}$$

The symbols, with the exception of the following ones, are defined in Section 3.2. The coordinate z measures distance vertically upward from the ocean floor at $z = -H$. The ocean surface is at $z = 0$. The vertical velocity component in that direction is w. The total derivative D/Dt includes advective terms. The terms F and G represent body forces, heat sources, and dissipative processes. For example, for vertical diffusion,

$$F^x = (K_M u_z)_z \tag{4.2a}$$

$$F^y = (K_M v_z)_z \tag{4.2b}$$

$$G = (K_H \rho_z)_z \tag{4.2c}$$

where K_M and K_H are the coefficients for vertical momentum and density diffusion, respectively. The density of the ocean is given by the sum $\rho_0 + \rho(x, y, z, t)$ where ρ_0 has the constant value 1 g/cm^3 and $|\rho|$ is of the order of 0.025 g/cm^3. The equation of state describes how the density depends on the temperature T, salinity S, and pressure p. In simple models it is approximated by the expression

$$\rho = \rho_0(1 - \alpha T)$$

where $\alpha = 0.0002$ (C°)$^{-1}$ The boundary conditions involve thermal and freshwater fluxes at the ocean surfaces so that equations for S and T are necessary. Except for a change of the variable, they are identical to Eq. (4.1e).

To linearize Eqs. (4.1) it is necessary to assume that the ocean, in its basic state, is motionless and has density $\rho = \bar{\rho}(z)$, which is a function of depth only. Motion and its associated density variation $\rho'(x, y, z, t)$ are considered as small perturbations to this basic state. The density equation then becomes

$$\rho_t - N^2 w/g = (K_H \rho_z)_z \tag{4.3}$$

where

$$N = (-g\bar{\rho}_z/\rho_0)^{1/2} \tag{4.4}$$

is the Brunt–Väisälä frequency.

If, for the sake of mathematical expediency, it is assumed that

$$K_M = K_H = A/N^2$$

where A is a constant (McCreary, 1981a and b) then a separation of variables

$$u = \sum U_m(x, y, t) R_m(z) \tag{4.5a}$$

$$v = \sum V_m(x, y, t) R_m(z) \tag{4.5b}$$

$$\frac{p}{g} = \sum \eta_m(x, y, t) R_m(z) \tag{4.5c}$$

$$w = \sum W_m(x, y, t) S_m(z) \tag{4.5d}$$

simplifies the linear unforced equations of motion to

$$U_t - fV + g\eta_x = -AU/gh \tag{4.6a}$$

$$V_t + fU + g\eta_y = -AV/gh \tag{4.6b}$$

$$\eta_t + h(U_x + V_y) = -A\eta/gh \tag{4.6c}$$

and

$$S_{zz} + \frac{N^2}{gh}S = 0 \tag{4.7}$$

where h is the constant of separation and

$$R = -ghS_z \tag{4.8}$$

Equation (4.7) describes the vertical structure of the flow and Eq. (4.6) the horizontal structure. The latter set is identical to the shallow-water equations (3.1), which suggests that many of the results obtained in chapter 3 are also valid in continuously stratified oceans. There are also important differences between a shallow-water system and a continuously stratified ocean. In Chapter 3 the depth of the thermocline (interface) determines the equivalent depth h, which in turn determines the gravity wave speed $c = (gh)^{1/2}$ and the length scale $\lambda = (c/\beta)^{1/2}$, the radius of deformation. A continuously stratified ocean has not one but an infinity of values for h so that there is a wide range of gravity wave speeds and radii of deformation. It is evident from Eq. (4.7) that h is related to the vertical scale of the motion, which is therefore related to the horizontal scales such as the radius of deformation. For example, the larger the vertical scale, the wider the equatorial waveguide.

For a continuously stratified fluid it is possible to determine the equivalent depths h as either eigenvalues of (4.7), which must be solved subject to boundary conditions on the ocean floor and ocean surface—this is the approach taken in Section 4.3—or as eigenvalues of the horizontal structure equations (4.6)—this is the approach taken in Section 4.4.

4.3 Vertically Standing Modes

The vertical structure of motion in a linear, continuously stratified model satisfies Eq. (4.7) and the following boundary condition at the rigid, horizontal ocean floor:

$$w = S = 0 \qquad \text{at } z = -H \tag{4.9}$$

At the free upper surface the boundary condition is

$$w - \frac{1}{\rho_0 g} p_t = S - hS_z = 0 \qquad \text{at } z = 0 \tag{4.10}$$

Equations (4.7) and (4.9) determine an infinite set of eigenfunctions and eigenvalues $[S_m, h_m; m = 0, 1, 2, 3, \ldots]$. The value of h_m decreases monotonically as m increases and h is positive for all m. The set of eigenfunctions S_m is complete so that expansions such as that in Eq. (4.5) are indeed possible. The forcing function too, if it acts as a body force, and if its horizontal and vertical structures are separable, can be expressed as a sum of vertical modes. To determine the oceanic response to given forcing it is then necessary to solve, for each mode S_m, the shallow-water equations (4.6) with the appropriate value for the equivalent depth h and with the appropriate projection of the forcing function onto that vertical mode. This method of solution is extremely powerful. To calculate the functions S_m it is necessary to know only the vertical stratification of the ocean. It does not matter whether or not the forcing excites vertically propagating waves. If it does, then each S_m describes a possible vertically standing mode, which is a superposition of upward-and downward-propagating waves. If no waves are excited, if the response attenuates with increasing distance from the forcing, for example, then the complete set S_m is a useful mathematical artifact with which to describe that response.

The stratification shown in Fig. 4.1a is used in several modeling studies described later in this chapter. It is representative of the observed stratification of the tropical oceans, which varies considerably with time and position, but which is always such that the Brunt–Väisälä frequency has a maximum in the thermocline at a depth of approximately 100 m. Figure 4.1b shows the structure of the first few vertical modes and Table 4.1 lists the first few eigenvalues h, the corresponding gravity wave speeds

Table 4.1

Properties of the Barotropic ($m = 0$) and First Few Baroclinic Modes

m	h (cm)	c (cm/sec)	λ (km)	T (days)
0	400000	20000	3000	0.2
1	60	240	325	1.5
2	20	140	247	2.0
3	8	88	197	2.6
4	4	63	165	3.1
5	2	44	131	3.6

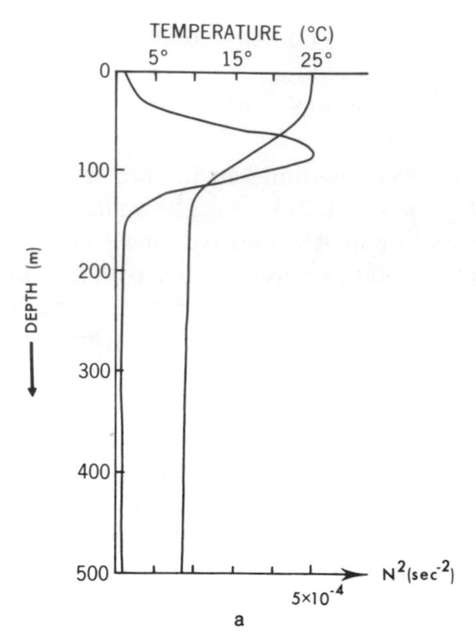

Figure 4.1. (a) A representative profile of the temperature and Brunt–Väisälä frequency in the tropics. This profile, with the temperature decreasing linearly to zero below 500 m, is used as initial condition in the model that provides the results in Figs. 4.5 to 4.9 and 4.16. (b) The structure of the three gravest vertical modes associated with the stratification in (a). (*Figure continues.*)

$c = (gh)^{1/2}$, the equatorial radii of deformation $\lambda = (c/\beta)^{1/2}$, and the equatorial inertial time $T = (\beta c)^{-1/2}$.

As the vertical mode number m increases, the number of nodes of the mode increases so that m can be regarded as a vertical wave number. The mode $m = 0$, known as the barotropic mode, has an equivalent depth h_0 equal to the actual depth H of the ocean. Since $h_0 \gg N^2H^2/g$, Eq. (4.7) simplifies to $S_{zz} = 0$. It follows that the barotropic mode is unaffected by the stratification of the ocean, that its vertical velocity component is a linear function of depth, and that its horizontal velocity components are independent of depth. The radius of deformation of this mode is so large that the equatorial β-plane approximations are of questionable validity.

The vertical modes for which $m > 1$ are known as the baroclinic modes and have equivalent depths much smaller than the total depth of the ocean: $h \ll H$ for $m > 1$. This simplifies the boundary condition (4.10) to

$$w = S = 0 \qquad \text{at } z = 0 \tag{4.11}$$

an approximation that implies that thermal changes in the ocean have a very small effect on sea level. This small effect is negligible for modeling

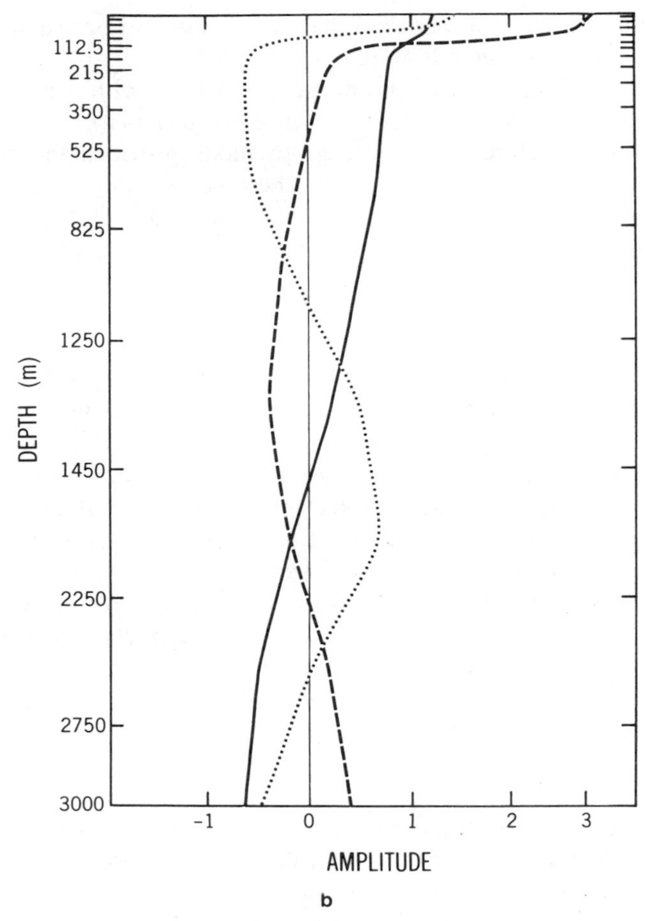

Figure 4.1 (*Continued*)

purposes but it is of enormous practical importance because it permits inferences about thermal changes in the ocean from sea level measurements. Water expands when its temperature increases so that sea level is a measure of the vertically averaged temperature of the water column. This expansion does not affect the pressure on the ocean floor, so that bottom pressure measurements are necessary to determine to what extent sea level variations are caused by thermal changes as opposed to barotropic (depth independent) changes. The larger the vertical scale of an internal mode, the larger its effect on sea level. Thus tide-gauge records give a good indication of motion associated with the first few baroclinic modes but not of motion associated with high vertical modes; this instrument effectively acts as a filter. The eastward movement of warm surface waters in the equatorial Pacific during El Niño of 1982–1983 in Fig. 1.20 involves high baroclinic modes and is at a speed slower than that of the eastward-propagating signal detected in tide-gauge records (Lukas *et al.*, 1984).

The horizontal structure of the motion associated with a given vertical mode is described by the shallow-water equations (4.6). The results of Chapter 3 are therefore valid for a given mode provided the equivalent depth is assigned its appropriate value. There is, for example, a dispersion diagram identical to that in Fig. 3.8 for each mode. The temporal and spatial scales to make the units of that figure dimensional are different for each mode and are given in Table 4.1. The vertical modes satisfy the unforced equations of motion so that the results of Fig. 3.8 can immediately be used to describe the response of a linear, continuously stratified ocean to remote forcing. Suppose, for example, that an unbounded ocean is forced to the west of the longitude X by the sudden onset of winds, which then remain steady. In the shallow-water model of Section 3.7, a single equatorial Kelvin wave introduces motion east of X. In a continuously stratified ocean, not one but an infinity of Kelvin waves are excited at X. Each corresponds to a different vertical mode m. As m increases, the wave speed c and the latitudinal scale of the wave decrease. Thus, at a point east of X, the barotropic mode arrives first, followed by the first baroclinic mode, then the second mode, and so on. The amplitude of a given mode depends on the projection of the forcing function onto that mode. If the winds, instead of starting to blow suddenly, are periodic, then many modes are superimposed and the approach discussed in the next section may provide a more convenient description of the motion.

The assumption that the ocean floor is flat, in the boundary condition (4.9a), is unrealistic, especially in the Atlantic, where the mid-Atlantic ridge causes the depth of the ocean to vary by thousands of meters. If the vertical modes are dependent on reflections at the ocean floor then modes are unlikely in large parts of the Atlantic. It is possible, however, for the topography of the ocean floor to be unimportant because variations in the stratification of the ocean could cause waves to reflect internally before they reach the ocean floor. For example, if the Brunt–Väisälä frequency decays exponentially with depth then vertically standing modes are possible even if the ocean is infinitely deep. Internal reflections may be particularly important in determining the structure of the second baroclinic mode, which, in Fig. 4.1, has a large amplitude primarily above the thermocline.[3] Its vertical structure is similar to that of the intense currents in the upper ocean so that this mode is likely to be preferentially excited when those currents change in response to a change in the winds. The value of the equivalent depth for this mode, $h_2 = 20$ cm, is also the value of h for the shallow water of Chapter 3, where the sharp tropical thermocline, as a first approximation, is regarded as the interface between two layers each of constant density. This reinforces the expectation that the second baroclinic mode is the most important one in the adjustment of the upper ocean to changes in the wind.

4.4 Vertically Propagating Waves

Atmospheric waves with a period of approximately 5 days and a wavelength of 2000 km are observed to propagate westward along the ITCZ. These fluctuations of the surface winds excite oceanic waves with the same period and wavelength. The instabilities of the surface currents above the thermocline, discussed in Section 2.7, similarly determine the period and wavelength of the free waves excited below the thermocline. These waves are observed to propagate downward (Weisberg *et al.*, 1979). Their vertical structure is described by Eq. (4.7) but does not correspond to that of a vertically standing mode. In other words, the equivalent depth h associated with these waves is not an eigenvalue of Eq. (4.7) and the boundary conditions (4.9). This suggests an alternative to the procedure followed in Section 4.3. Rather than determine the constant of separation h by solving the vertical structure equation (4.7) as an eigenvalue problem, determine h from the horizontal structure equation (4.6). The eigenvalues of this equation, subject to the condition that the eigenfunctions be bounded as $|y| \to \infty$, are given by the dispersion relation (3.32). Previously this was regarded as a relation between the frequency σ and wave number k for a given value of h. Now h is the unknown and σ and k are specified by the periodic forcing. From (3.32) it follows that

$$(gh)^{1/2} = \frac{-\beta(2n+1) \pm \left[\beta^2(2n+1)^2 + 4(\beta k/\sigma + k^2)\sigma^2\right]^{1/2}}{2(k^2 + \beta k/\sigma)} \tag{4.12}$$

Once h is known, Eq. (4.7) can be solved to determine how the properties of a vertically propagating wave change as it travels through variable stratification. If the Brunt–Väisälä frequency N varies on a scale that is large in comparison with the vertical scale of the wave then the WKB methods can be used. If the changes in N are abrupt—this is the case in the sharp tropical thermocline—then internal reflections are possible (see Note 2).

It follows from Eq. (4.12) that two sets of waves are excited in response to forcing at a specified wave number and frequency. One set, denoted by $[h_n^+]$, corresponds to inertia-gravity waves and the other set $[h_n^-]$ to Rossby waves. The eigenfunction D_n associated with an eigenvalue h_n is a Hermite function as before but the set of functions $[D_n, h_n]$ differs from the set discussed in Chapter 3. This is so because the eigenvalue of Eq. (3.19) is now h_n, not σ or k as before. The previous orthogonality relation (3.35) for the set $[D_n, \sigma_n]$ is now replaced by the relation

$$\int_{-\infty}^{\infty} (\sigma^2 - \beta^2 y^2) D_n D_m dy = 2^n n! \left(\frac{\pi c}{\beta}\right)^{1/2} \left[\sigma^2 - \frac{2n+1}{2} (gh\beta^2)^{1/2}\right] \delta_{mn} \tag{4.13}$$

which is used to determine the projection of the forcing function onto each vertically propagating latitudinal mode. The set of eigenfunctions $[D_n]$ associated with eigenvalues h_n that are all positive do not form a complete set. If $h_n > 0$ then motion corresponds to vertically propagating waves. Suppose that the oceanic response decays exponentially with increasing distance from the forced region. For Eq. (4.7) to describe such a response it is necessary that $h < 0$. It is one of the flaws of the unbounded equatorial β-plane that it excludes eigenfunctions for which $h < 0$. [Equation (4.12) does not have solutions with the property $h < 0$. If the original equations are analyzed on a sphere then these eigenvalues are recovered.[4]] It is clear that the method of solution discussed in this section has serious drawbacks and does not provide a convenient description of the oceanic response to certain types of forcing. This method is most useful when dealing with vertically propagating waves. The different types of waves that can be excited are discussed next.

4.4.1 Inertia-Gravity Waves

Equation (4.12) gives the vertical scales $h_n^{\pm}$ of the two sets of waves that are excited when the ocean is forced at a certain wavelength and frequency. The set h_n^{+}, $n > 1$, corresponds to inertia-gravity waves for which the equivalent depth, and hence the vertical scale, decreases as the meridional wave number n decreases. The gravest equatorially trapped modes, those with the smallest values of n, therefore have the largest vertical scales. The waves observed far from the equator have large meridional wave numbers n and short vertical scales. For large values of n,

$$\sqrt{gh_n^{+}} = \frac{\sigma^2}{\beta(2n+1)} \tag{4.14}$$

It then follows from Eq. (3.21) that the turning latitude Y for these waves is the inertial latitude

$$Y^2 \to \frac{\sigma^2}{\beta^2} \qquad \text{for } n \gg 1 \tag{4.15}$$

Poleward of this latitude, which can be close to the equator if the frequency is low, amplitudes decay exponentially.

Figure 4.2 depicts the latitudinal structure of the gravest inertia-gravity waves excited by the atmospheric disturbances, with a period of 5 days and

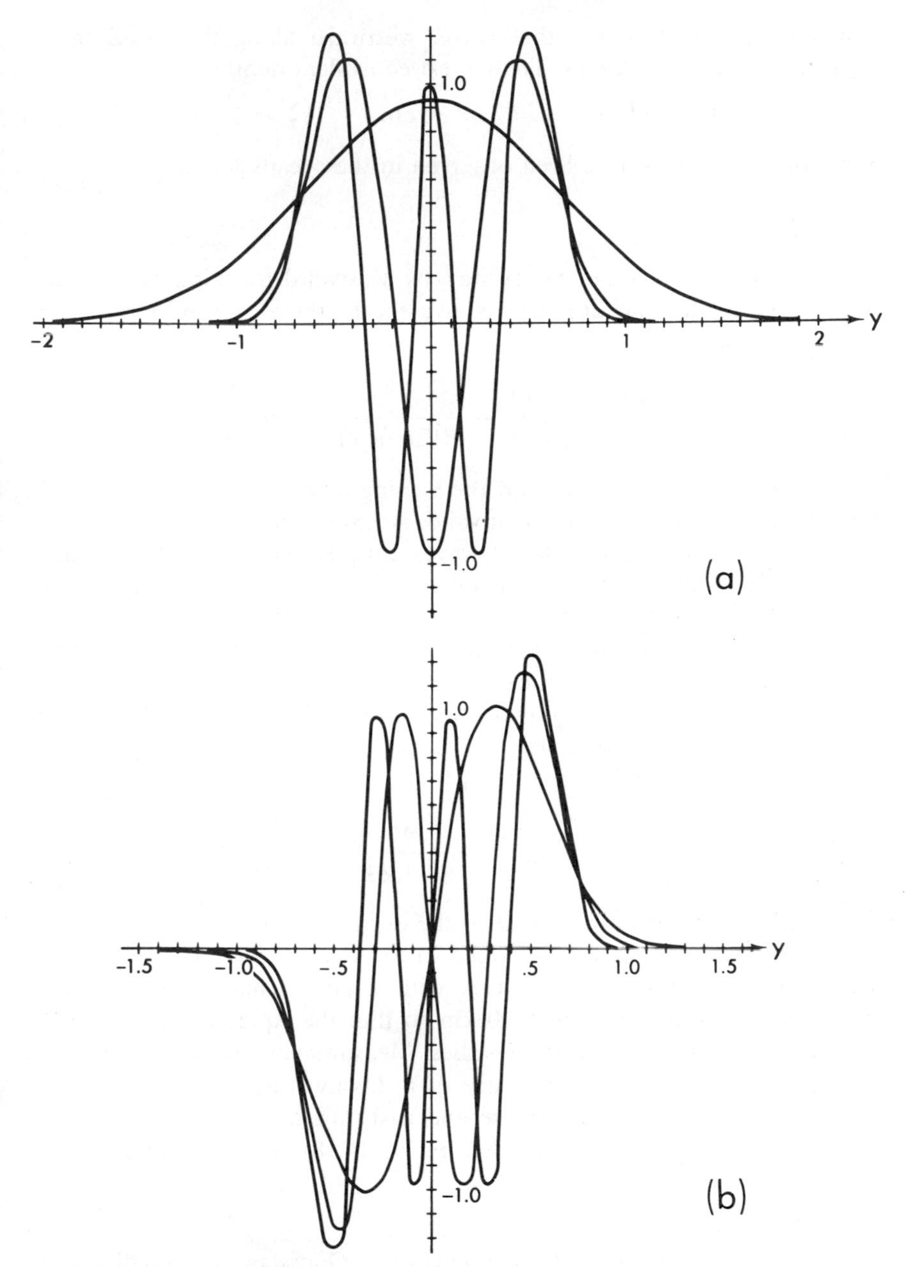

Figure 4.2. Latitudinal structure of the meridional velocity component of the gravest (a) symmetric and (b) antisymmetric westward-propagating inertial-gravity waves with a period of 5 days and a wavelength of 2000 km. Distance from the equator is in units of 1000 km.

a wavelength of 2000 km that travel westward along the ITCZ in the Atlantic and Pacific Oceans. The largest equivalent depths are

$$h_1^+ = 65 \text{ cm}, \qquad h_2^+ = 30 \text{ cm}, \qquad h_3^+ = 15 \text{ cm} \tag{4.16}$$

Some of these waves have been observed in the oceans.[5]

4.4.2 *Rossby Waves*

Rossby waves, which can be excited by westward-traveling disturbances that satisfy condition (3.26), have equivalent depths h_n^- given by Eq. (4.12). For large values of n,

$$\sqrt{gh_n^-} \sim \frac{\beta(2n+1)}{k^2 + \beta k/\sigma} - \frac{\sigma^2}{\beta(2n+1)} \qquad \text{for } n \gg 1 \tag{4.17}$$

As n increases, h_n^- increases and the turning latitude increases so that the β-plane approximations become invalid. Accurate calculations on a sphere show that only a finite number of waves are possible. As a rule of thumb, waves with turning latitudes that exceed the radius of the earth are not permissible (see Note 4).

The increase in equivalent depth as the meridional wave number n increases implies that the vertical scale of Rossby waves that propagate to high latitudes is large. The gravest modes, those trapped near the equator, have the shortest vertical scales.

For the $n = 0$, the Rossby-gravity mode

$$\sqrt{gh_0^-} = \frac{\sigma^2}{\beta + \sigma k} \tag{4.18}$$

This is the principal wave excited in the deep ocean by the instabilities of the intense surface currents described in Section 2.6. The instabilities give rise to westward-propagating waves with a period near 3 weeks and a wavelength of approximately 1000 km so that the equatorial ocean below the thermocline is forced at these scales. Measurements in the Atlantic and Pacific Oceans confirm the presence of a downward-propagating Rossby-gravity wave, with westward phase and eastward group velocity and with the equivalent depth of 30 cm given by Eq. (4.18) (Weisberg *et al.*, 1979).

4.4.3 *Kelvin Waves*

The latitudinal structure of Kelvin waves is a Gaussian and the dispersion relation is

$$(gh)^{1/2} = \sigma/k \tag{4.19}$$

If the zonal wavelength is fixed, then the lower the frequency, the slower the waves and the smaller the latitudinal and vertical scales of the waves.

4.4.4 Beams

Measurements show that the zonal currents below the thermocline, below the Equatorial Undercurrent, reverse direction repeatedly as is evident in Fig. 4.3. At a given depth these zonal jets are coherent over large zonal distances and persist for a considerable time. This suggests that the jets may be associated with waves that have low frequencies and low zonal wave numbers. If the forcing that excites the waves at a fixed frequency is confined to one part of the ocean then beams of equatorially trapped waves will radiate zonally and downward from the forced region. The inclination α of the ray paths to the vertical is the ratio of the vertical to the zonal group velocity. Long Rossby waves propagate at a small angle α to the horizontal,

$$\alpha \sim (2n + 1)\sigma/N, \qquad n = 1, 2, 3, \ldots \tag{4.20}$$

In the case of Kelvin waves,

$$\alpha = \sigma/N \tag{4.21}$$

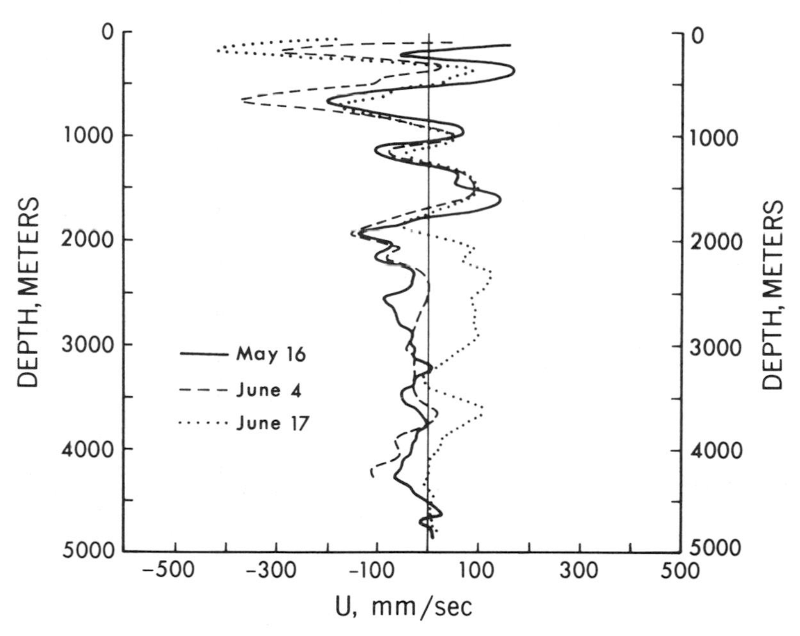

Figure 4.3. Profiles of the zonal velocity component on the equator at 53°E in the Indian Ocean on the indicated dates in 1976. [From Luyten and Swallow (1976).]

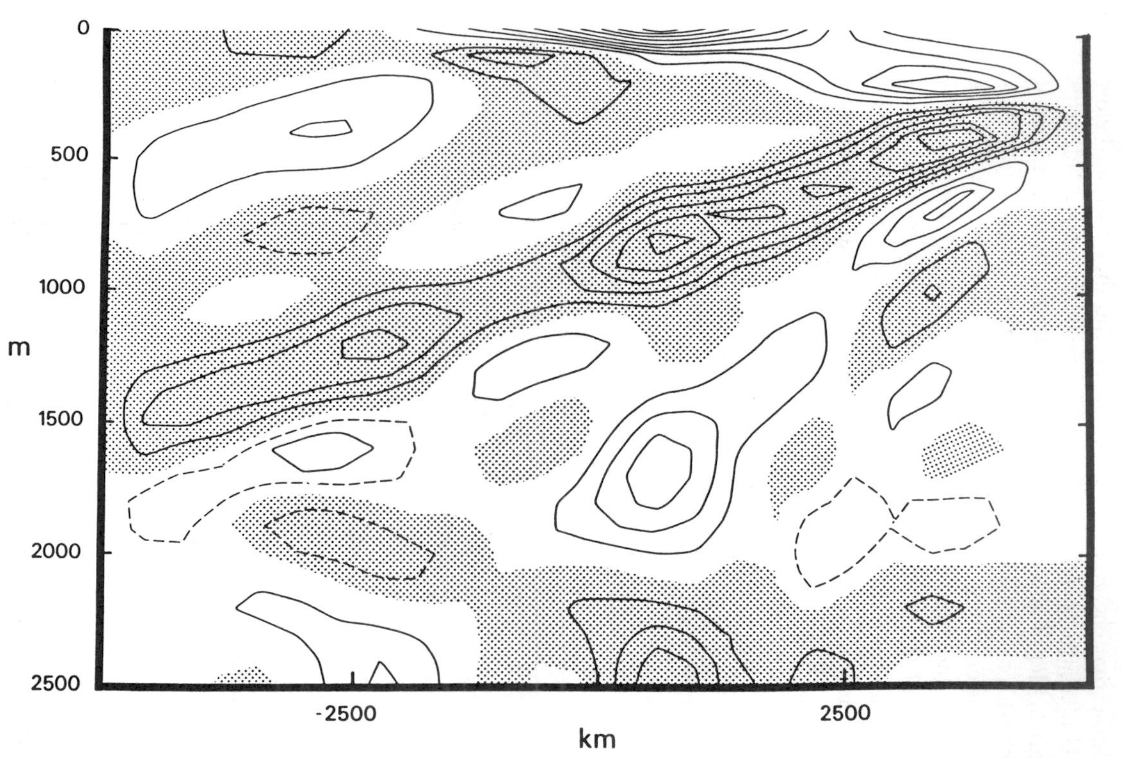

Figure 4.4. An equatorial section of the zonal velocity component—the contour interval is 5 cm/sec and motion is westward in shaded areas—in a linear model forced by oscillating zonal winds with a period of 1 year. The winds are confined to the band of longitudes ± 2500 km. The section coincides with the time of maximum eastward winds. The Brunt–Väisälä frequency N is a constant. Eastward Kelvin and westward Rossby waves are clearly evident. [From McCreary (1984).]

Low-frequency waves have very shallow beams. A Kelvin wave with a period of one year, forced near the surface in the far west can cross the Pacific without reaching the ocean floor. Even the grave Rossby waves excited by reflection of the Kelvin beam at the coast of South American can travel back westward across the Pacific without reaching the ocean floor as can be seen in Fig. 4.4.

An explanation for the deep jets in terms of beams is inconsistent with the measurements, over a 15-month period, that show no vertical phase propagation (Firing, 1987). Presently there is no explanation for the deep jets, but it is known why fluctuating surface winds are unlikely to generate downward-propagating beams as outlined above. One reason is the absorption of the waves at critical layers where the speed of the mean currents, the Equatorial Undercurrent, for example, equals that of the waves (McPhaden *et al.*, 1986). Another reason is related to the nonlinear response of the upper equatorial ocean to the winds. Not only are the near-surface currents nonlinear but the changes in the thermal field involve large vertical displacements of the thermocline. Its large-scale zonal slope changes seasonally in the Atlantic and Indian Oceans, and interannually in the Pacific. This means that disturbances reach the deep ocean not be propagating through a specified time-independent thermocline as assumed in linear theories, but by being forced directly by vertical movements of the thermocline. Fluctuations with a period of 1 year in the Pacific (Eriksen, 1981) and 6 months in the Indian Ocean (Luyten, 1982) are present at depth near the equator but it is unlikely that these fluctuations can be explained as the linear response to surface forcing at the same periods.

4.5 Equatorial Surface Jets

Consider the oceanic response to spatially uniform winds that suddenly start to blow at time $t = 0$ and then remain steady. If the windstress acts as a body force in a surface layer of depth D then the function F in Eq. (4.1) is

$$\mathbf{F} = (F^x, F^y) = (\tau^x/D, 0), \qquad -D \leqslant z \leqslant 0 \tag{4.22a}$$

$$= (0,0), \qquad -H \leqslant z < -D \tag{4.22b}$$

If zonal variations are neglected ($\partial/\partial x = 0$) then the linear, inviscid equations of motion can be written

$$\frac{\partial}{\partial z}\left[\frac{1}{N^2}\frac{\partial}{\partial z}(v_{tt} + f^2 v)\right] + v_{yy} = -\frac{\partial}{\partial z}\left[\frac{1}{N^2}\frac{\partial}{\partial z}(fF^x)\right] \tag{4.23}$$

This equation is similar to Eq. (3.11) and many of the results inferred from that equation in Chapter 3 remain valid. For example, at large distances L from the equator, westward winds drive poleward Ekman drift $\tau^x/f_0 D$ after a local inertial period $1/f_0$ has elapsed. (It is necessary that $t \gg 1/f_0$ and $L \gg N_0 D/f_0$, the local radius of deformation. Here N_0 denotes a typical value of N and D is the vertical scale of the motion.) The Ekman drift away from the equator implies a distinctive equatorial zone where upwelling is intense. The width of this zone, in which the third term on the left-hand side of Eq. (4.23) is important, is the equatorial radius of deformation

$$\lambda = \left(\frac{N_0 D}{\beta}\right)^{1/2} \tag{4.24}$$

If $D = 100$ m is the depth of the sharp tropical thermocline and $N_0 = 0.02$ sec^{-1} is the value of N in the thermocline, then $\lambda \sim 300$ km.

In addition to strong upwelling, the equatorial zone has an accelerating jet because the zonal momentum balance, in the absence of zonal gradients, becomes

$$u_t = F^x \qquad \text{at } y = 0 \tag{4.25}$$

From the expression for F it follows that, at the equator, the jet is confined to the surface layer of depth D. Off the equator there are horizontal currents below a depth D but they are relatively weak and are driven by the divergence of the surface flow. For a mathematical description of the motion it is convenient to use the method of Section 4.3 that involves vertical modes so that the wind, a body force, is projected onto the modes, Equations (3.16) describe the solution for each of the modes and these have to be summed as in Eq. (4.5).

If the wind acts as a stress on the ocean surface then mixing processes must be taken into account as in Eq. (4.1), where the body force F is now zero. The appropriate boundary conditions are

$$K_M u_z = \tau^x; \qquad K_M v_z = 0 \qquad \text{at } z = 0 \tag{4.26}$$

The equations that now describe the steady Ekman flow far from the equator are

$$-fv = \frac{\partial}{\partial z}\left(K_M \frac{\partial u}{\partial z}\right) \qquad \text{and} \qquad fu = \frac{\partial}{\partial z}\left(K_M \frac{\partial v}{\partial z}\right) \tag{4.27}$$

The vertically integrated Ekman drift is again $-\tau^x/f$ in a surface layer of depth $\sqrt{\nu/f}$. That the deeper layers have meridional flow opposite to the

Ekman drift follows from the equation for the conservation of mass

$$v_y + w_z = 0 \tag{4.28}$$

which can be integrated from the ocean floor to the surface to yield

$$\int_{-H}^{0} v\,dz = 0 \tag{4.29}$$

provided the flow is symmetrical about the equator ($v = 0$ at $y = 0$). The equation for the flow at depth, where mixing processes are unimportant, is

$$\left(\frac{f^2}{N^2} v_z\right)_z + v_{yy} = 0 \tag{4.30}$$

on time scales much longer than an inertial period. A scale analysis indicates that the deep flow is confined to a layer of depth $f(L/N)$, where L is the meridional scale imposed by the divergence of the surface flow. The depth of this layer decreases, but that of the Ekman layer increases as the equator is approached. The two layers meet near the equator, where intense upwelling links the convergent flow at depth with the divergent surface flow. The width of the distinctive equatorial zone now depends on the eddy viscosity and diffusivity. Many aspects of the motion near the equator are unexplored. An example of results obtained numerically is shown in Fig. 4.5, which depicts the structure of the equatorial jet. At the equator it diffuses downward steadily in accord with the equation

$$u_t = (K_M u_z)_z \qquad \text{at } y = 0 \tag{4.31}$$

The meridional circulation that is superimposed on the zonal equatorial jet has a strong nonlinear effect on the jet. In the case of eastward winds the convergent surface flow and the equatorial downwelling cause the jet to be narrower and deeper than its linear counterpart. In the zonal momentum equation on the equator,

$$u_t = -wu_z + (K_M u_z)_z \tag{4.32}$$

both terms on the right are positive so that nonlinearities intensify the jet. In the case of westward winds, equatorial upwelling brings water with little zonal momentum into the surface layers, thus causing the jet to be weaker and shallower. The divergent Ekman drift broadens it. The results in Fig. 4.5 confirm these differences between nonlinear eastward and westward jets. Hysteresis is possible if the winds first blow in one direction and then the other because the deep, intense eastward jet generated during a period of eastward winds will persist below the shallow, weak westward surface current driven by subsequent westward winds.

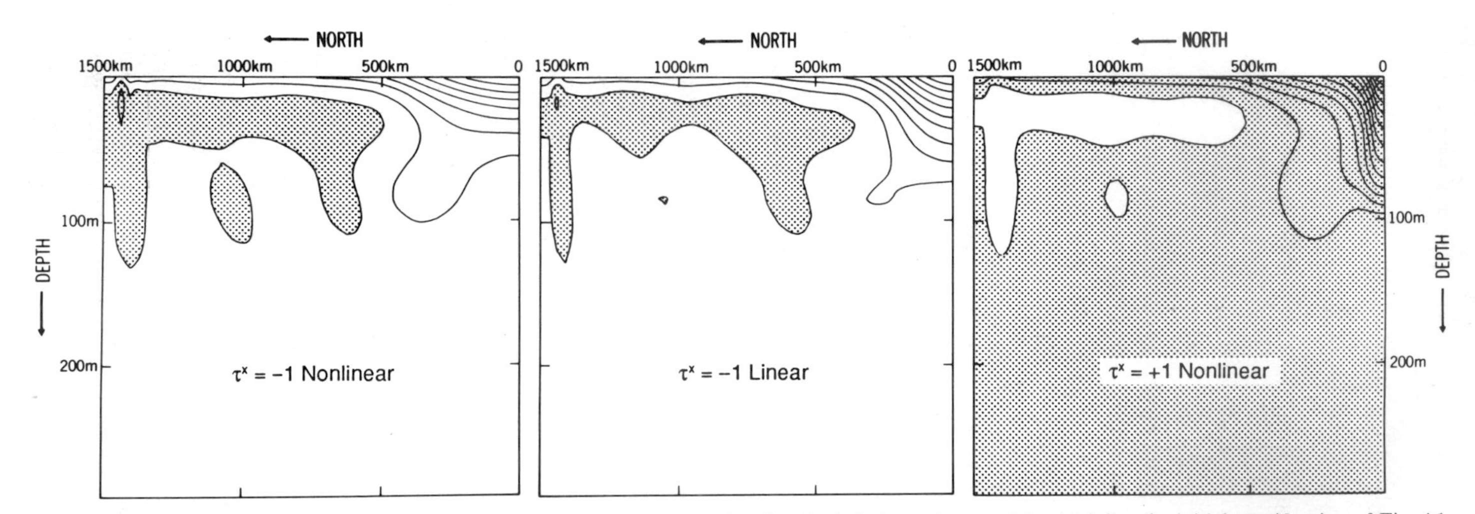

Figure 4.5 . The effects of nonlinearities on the zonal velocity component in a longitude-independent model, which has the initial stratification of Fig. 4.1. The picture shows conditions 7.5 days after the sudden onset of zonal winds of intensity 1 dyne/cm^2. The contour interval is 20 cm/sec and motion is eastward in shaded areas. [From Philander (1979b).]

4.6 The Equatorial Undercurrent

Immediately after the onset of uniform zonal winds, the oceanic motion is independent of longitude and includes the poleward Ekman drift, equatorial upwelling, and the accelerating equatorial jet described in Section 4.5 and depicted in Fig. 4.5. This motion does not satisfy the boundary conditions at the north–south coasts,

$$u = 0 \qquad \text{at } x = 0 \text{ and } L$$

so that waves, which satisfy the unforced equations of motion, are excited at the coasts. The situation is analogous to the adjustment described in Chapter 3: Kelvin waves from the western and Rossby waves from the eastern boundary introduce zonal pressure gradients, arrest the acceleration of the jet, and ultimately establish equilibrium conditions. In a continuously stratified ocean the waves propagate not only horizontally but also vertically. Vertical modes are rapidly established so that a succession of wave fronts, the barotropic, first baroclinic, second baroclinic modes, and so on, emanate from the coasts. (In a shallow-water model there is only one vertical mode.) At a longitude in the interior of the basin the passage of the first baroclinic mode Kelvin wave, for example, reduces the acceleration of the surface jet by establishing an opposing pressure force, and introduces deep currents and deep pressure gradients with the vertical structure of that mode, which is shown in Fig. 4.1. The passage of the subsequent modes modifies the vertical structure of the motion. If the windstress acts as a body force in the upper ocean then, in the absence of mixing processes, the equilibrium state is a motionless one in which the zonal pressure gradient balances the windstress. Dissipation affects motion with small horizontal and vertical scales most severely so that the high vertical modes fail to make their contribution to the adjustment. The equilibrium state in that case includes surface currents and undercurrents. The structure and the intensity of the final currents depend on the mixing processes and on the relative amplitudes of the vertical modes that are excited. To determine these amplitudes it is necessary to project the accelerating equatorial jet of Fig. 4.5 onto the various vertical modes. In the middle panels of Fig. 4.6, which shows the evolution of the motion along the equator in a linear numerical model[6] with the realistic initial stratification of Fig. 4.1, it is striking that the adjustment is accomplished by essentially one mode, the second baroclinic mode. That mode has by far the largest amplitude of all the modes because its vertical structure is closest to that of the accelerating equatorial jet. (This result suggests that in shallow-water models the equivalent depth should be that of the second mode. That is indeed the value obtained for h in Chapter 3 on the basis of the observed depth of the thermocline.[7])

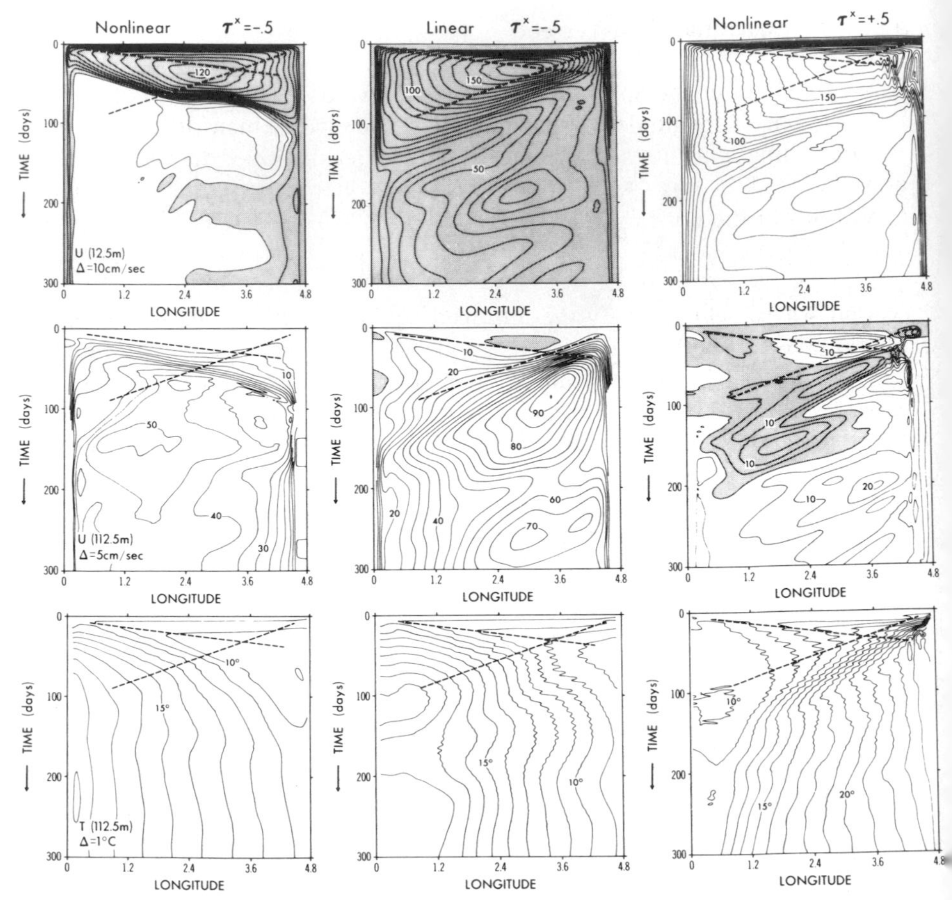

Figure 4.6. The response of a model bounded in the west and east and with the initial stratification of Fig. 4.1 to the sudden onset of spatially uniform winds of intensity 0.5 dyne/cm^2. The left-hand panels show the nonlinear response to westward winds, the central panels the linear response to the same winds, and the right-hand panels the nonlinear response to eastward winds. The figures show the evolution of the zonal velocity component (centimeters per second, westward motion in shaded areas) at depths of 12.5 and 112.5 m along the equator and the temperature at a depth of 112.5 m. Longitude is measured in units of 1000 km. [From Philander and Pacanowski (1980).]

In Fig. 4.6 the acceleration of the surface jet practically stops after the passage of the second baroclinic mode Kelvin and Rossby fronts so that the surface adjustment is very similar to that of the shallow-water model shown in Fig. 3.18. The vertical structures of the second mode and that of the surface jet do not coincide exactly and it is this discrepancy that permits the existence of the Equatorial Undercurrent as shown schematically in Fig. 4.7. In Fig. 4.6, zonal currents appear at the depth of the thermocline only

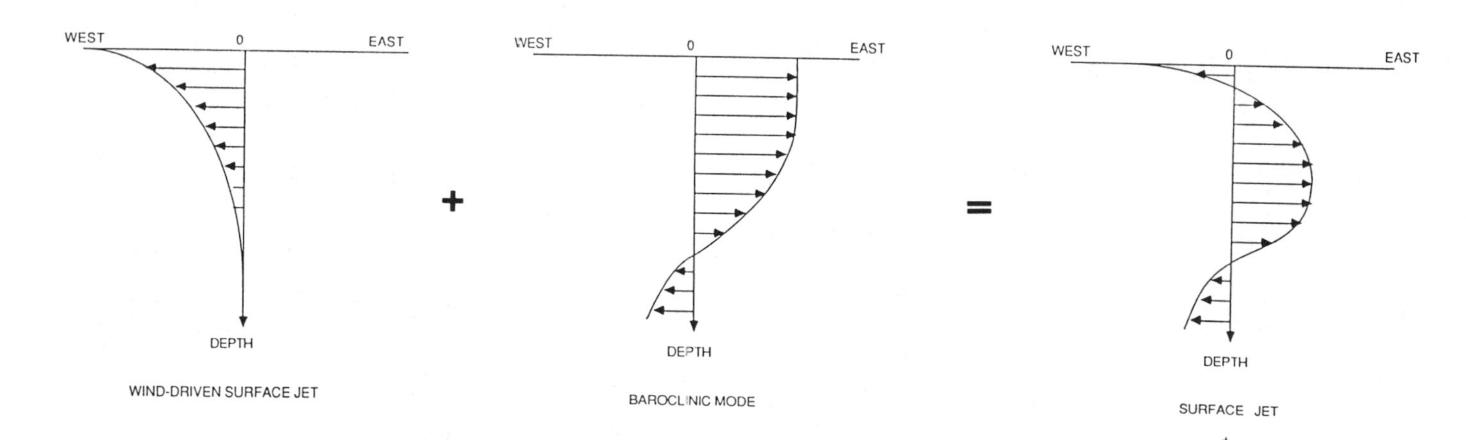

Figure 4.7. A schematic diagram that shows how the motion, initially in the direction of the wind and trapped near the surface, is modified by the arrival of a baroclinic mode that establishes the Equatorial Undercurrent. Figure 4.11(c) shows the similar development of a coastal undercurrent.

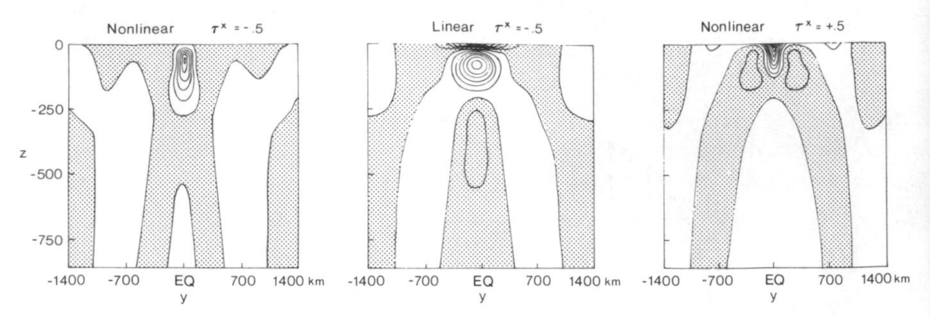

Figure 4.8. The zonal velocity component (the contour interval is 10 cm/sec and motion is westward in shaded areas) along the central meridian of the ocean basin after equilibrium conditions are attained in the calculations of Fig. 4.6. The nonlinear eastward equatorial surface jet penetrates downward and eliminates the subsurface equatorial undercurrent. In the case of westward winds, upwelling of eastward momentum from the Equatorial Undercurrent eliminates the westward surface flow at the equator. [From Philander and Pacanowski (1980).]

after the passage of the wave fronts that establish zonal pressure gradients and that inhibit upwelling. (Initially westward winds induce equatorial upwelling so that the temperature at a depth of 112 m, for example, falls, but in the wake of the Kelvin front the temperature starts to increase, especially in the west.) Ultimately the waves establish equilibrium conditions. For a basin that is 5000 km wide this happens after approximately 150 days, the time it takes a second baroclinic Kelvin plus Rossby wave to cross the basin. Figure 4.8 shows the zonal currents in the equilibrium state. Analytical methods to determine the linear steady state for certain parameterizations of the mixing processes are available (McCreary, 1981a; McPhaden, 1981).

The second baroclinic mode plays the dominant role in the adjustment of the upper ocean. The adjustment below the thermocline is more complicated and is slower because many more vertical modes are involved.

Westward winds maintain an eastward Equatorial Undercurrent. Such a current is one of the most prominent features of the circulation of the tropical Atlantic and Pacific Oceans, where westward winds prevail. Linear models predict that the currents driven by eastward winds are identical to those driven by westward winds except that their directions change. Hence a westward undercurrent should be dominant where eastward winds prevail. Yet the salient feature of the circulation of the Indian Ocean during periods of eastward winds is an eastward equatorial surface jet. The flow below the jet is weak and variable. It appears that the most prominent equatorial current, irrespective of the direction of the wind, is always eastward. This indicates that nonlinearities are of prime importance, a conclusion that is readily confirmed by a scale analysis based on the intensity and width of the observed equatorial currents. Section 4.6 describes how the meridional circulation determines the effect nonlinearities have on the motion when longitudinal variations are negligible. The meridional circulation changes in the presence of zonal pressure gradients that support equatorward geostrophic flow at depth. The geostrophic balance fails at the equator, where the Coriolis force vanishes, but the pressure gradient does not. A scale analysis indicates that the new term that comes into play in the zonal momentum balance is the meridional advection of zonal momentum (Fofonoff and Montgomery, 1955; Cane, 1980),

$$vu_y - \beta y v + \frac{1}{\rho_0} p_x = 0 \tag{4.33}$$

From this equation, which assumes that zonal and vertical gradients are small in comparison with meridional gradients, it follows that

$$v \frac{\partial}{\partial y}(\beta y - u_y) = 0 \tag{4.34}$$

This is a statement that a fluid parcel conserves the vertical component of its vorticity. If the parcel starts at latitude Y with negligible speed and shear, then Eq. (4.34) can be integrated twice to conclude that

$$\beta y - u_y = \beta Y \tag{4.35}$$

and

$$u(y = 0) = \tfrac{1}{2}\beta Y^2 \tag{4.36}$$

A parcel that moves equatorward gains relative vorticity while it loses planetary vorticity and in the process is accelerated eastward. If it starts at rest at 2.7°N it arrives at the equator with a speed of 1 m/sec. The source of momentum is the eastward pressure force. This follows from Eq. (4.33), which, upon integration, gives

$$u(y = 0) = -\frac{1}{2}\beta Y^2 + \int_0^Y \frac{1}{v\rho_0} p_x \, dy \tag{4.37}$$

In the absence of a zonal pressure gradient the equatorward-moving parcel conserves its angular momentum and arrives at the equator moving westward. A pressure force increases the angular momentum of the fluid parcel as it travels equatorward. Vorticity conservation implies that

$$\beta Y v = \frac{1}{\rho_0} p_x \tag{4.38}$$

If this is substituted in Eq. (4.37) then the result in (4.36) is recovered. The conclusion is that a nonlinear eastward Equatorial Undercurrent should be more intense than a linear one. This is not so in Fig. 4.8 because equatorial upwelling transfers eastward momentum to the surface layers where the westward jet practically disappears. Note that the deceleration of the westward surface flow in the top left-hand panel of Fig. 4.6 starts after the passage of the Kelvin wave, once the Equatorial Undercurrent has come into existence.

In the case of eastward winds the equatorward flow near the surface does not conserve vorticity because frictional effects are dominant. Downwelling at the equator now carries the eastward momentum to the thermocline so that the westward undercurrent disappears as shown in Fig. 4.8.

The Equatorial Undercurrent is driven by an eastward pressure force maintained by westward winds. Yet it is present in the Gulf of Guinea in the Atlantic, and sometimes east of the Galápagos Islands in the Pacific, regions where the winds have an eastward component that maintains a westward pressure force in the thermocline. Wacongne (1988) analyzed data from a General Circulation Model of the tropical Atlantic Ocean in a study of the dynamical balance in the Equatorial Undercurrent, not only near its terminus but all along the equator. The results, obtained by averaging the data from the model over a seasonal cycle, are summarized in Fig. 4.9. Different dynamical balances obtain in the different regions:

$$\text{(I)} \quad vu_y + wu_z = -\frac{1}{\rho_0} p_x + (\nu u_z)_z$$

$$\text{(II)} \quad uu_x + vu_y + wu_z = -\frac{1}{\rho_0} p_x$$

$$\text{(III)} \quad uu_x = A\nabla^2 u$$

$$\text{(IV)} \quad uu_x + vu_y + wu_z = -\frac{1}{\rho_0} p_x + A\nabla^2 u \sim 0$$

In the surface layers, in region I, the westward momentum imparted by the wind diffuses downward. This effect is countered by the eastward pressure force maintained by the wind. The upwelling of eastward momentum causes a further weakening of the westward surface flow. This complex dynamical balance in the upper ocean above the core of the Equatorial Undercurrent is captured by the model of Charney (1960). (Charney assumed that the zonal velocity component vanishes at the depth of the thermocline, which he regarded as a rigid lid. It is better to assume that the vertical shear vanishes at that depth.) Below the core of the Equatorial Undercurrent the dynamical balance changes strikingly from west (region II) to east (region III). In the west, where the thermocline is deep, the pressure force is large at depth and accelerates an inertial undercurrent eastward as in Pedlosky's (1987) model. In the central and eastern part of

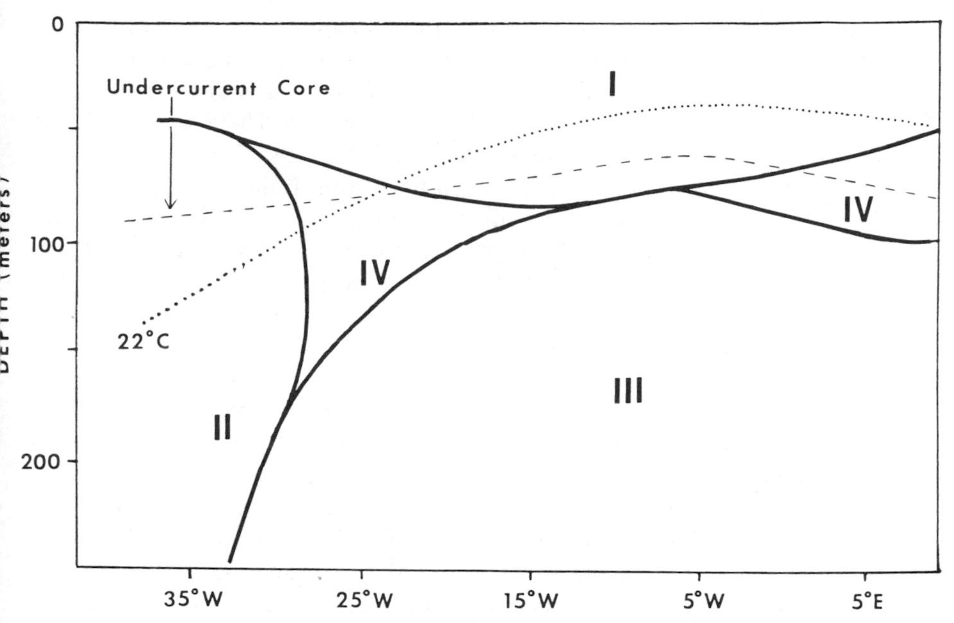

Figure 4.9. Regions of different dynamical balances for the Equatorial Undercurrent. [From Wacongne (1988).]

the basin the thermocline is above the core of the Equatorial Undercurrent so that the pressure force is relatively unimportant in and below the core. Horizontal diffusion now causes the inertial jet to decelerate gradually.

These results from a General Circulation Model provide a persuasive explanation for the different mean speeds of the Equatorial Undercurrent in the Atlantic and Pacific Oceans. Region II, in which the Equatorial Undercurrent accelerates in a downstream direction, has a far larger longitudinal extent in the Pacific than in the Atlantic so that greater speeds are attained in the Pacific. The results also explain why the Equatorial Undercurrent can penetrate close to the coasts of Africa and South America, where the pressure force in the thermocline is westward: the core of the Equatorial Undercurrent in those regions is below the thermocline, at depths where pressure gradients are small.

4.7 Response to Time-Dependent Forcing

The eastward Equatorial Undercurrent is driven by an eastward pressure force and hence flows downhill. An eastward equatorial surface jet, on the other hand, flows uphill because the eastward winds that drive the jet also maintain a westward pressure force that opposes the jet. The oceanic response to a sudden relaxation of the prevailing winds is therefore interestingly different for eastward and westward winds. If eastward winds should abruptly stop blowing then the unbalanced pressure force will decelerate the eastward jet and will generate westward currents. This is part of the reason for the reversal of the currents in the central Indian Ocean (Fig. 3.1). If westward winds that maintain the Equatorial Undercurrent should abruptly stop blowing, then the unbalanced pressure force will accelerate the undercurrent eastward and it will become a surface current. In the Pacific this happens seasonally as shown in Fig. 2.25. The seasonal relaxation of the winds in the Pacific is for a few weeks only. How does the duration of the relaxation affect the response? How does the response to an abrupt relaxation differ from the response to a gradual relaxation?

Two sets of time scales determine the oceanic response to time-dependent forcing: the time scale of the forcing and time scales intrinsic to the ocean. To generate an equatorial jet in the ocean, winds in one direction have to prevail for several days at least. Hence oscillatory winds with a period of 1 day excite inertia-gravity waves in the ocean but are unlikely to generate significant currents. Winds with a period of 100 days, however, excite Rossby and Kelvin waves and generate intense equatorial surface jets and undercurrents.

Let $2\pi T$ be the period of the wind fluctuations. The response of the upper ocean then changes as follows when the value of T changes. (To obtain numerical values the second baroclinic mode is assumed to effect the adjustment.)

$0 < T < (\beta c)^{-1/2}$: If T is shorter than the equatorial inertial time, approximately 1.5 days, then the oceanic response does not include a distinctive equatorial zone and the winds excite primarily inertia-gravity waves that propagate from high southern to high northern latitudes.

$(\beta c)^{-1/2} < T < L/c$: On time scales longer than 2 days but shorter than approximately 1 month, shorter than the time L/c it takes a Kelvin wave to cross the basin of width L, the winds generate equatorial jets in the surface layers of the ocean and much weaker currents at greater depths. The Kelvin waves that are superimposed on the currents have relatively short wavelengths and establish zonal pressure gradients with a scale much smaller than the width of the basin. The superposition of an oscillatory surface jet and Kelvin waves distorts the phase propagation of the waves. Suppose that the zonal windstress is described by $\sin(t/T)$. In a shallow-water model the zonal velocity component can then be written

$$u = U_1(y)\sin(t/T) + U_2(y)\sin(kx - t/T)$$

where U_1 is the amplitude of the jet and U_2 is that of the Kelvin wave that propagates with speed $c = 1/Tk$. Near the equator $y = 0$,

$$u \sim 2U_1 \sin(kx/2)\cos(t/T - kx/2) \tag{4.39}$$

provided U_1 is equal to U_2. Phase propagation is therefore at twice the speed of Kelvin waves. This is one of the reasons why observations of eastward phase propagation along the equator are generally not at the expected Kelvin wave speed (Eriksen *et al.*, 1983). The situation is even more complicated if the period of the winds is sufficiently long, longer than approximately 1 month, to excite Rossby waves. The superposition of a large number of waves can lead to odd phenomena such as the foci in Fig. 3.7 (Cane and Sarachik, 1981). For this reason phase propagation associated with the annual cycle may not correspond to the presence of a single wave.

$T > L/c$: As the period of the forcing increases beyond a few months the zonal scale of the zonal pressure gradient increases and intense undercurrents appear in the thermocline. If the period is much longer than $4L/c$, which is the time it takes the upper equatorial ocean to come into a state of equilibrium after the sudden onset of steady winds, then the ocean at each moment is in equilibrium with the waves at that moment. Waves play a critical role in maintaining equilibrium conditions but there is no explicit evidence of their presence, no phase propagation, for example.

The tides provide a marvelous test for these results concerning changes in the oceanic response as the period of the forcing changes. The tidal force penetrates throughout the water column so that the response is essentially barotropic. At the periods of the semidiurnal and diurnal tides, only the gravest equatorially trapped waves, excluding Rossby waves, are excited. [In the dispersion diagram, Fig. 3.8, the forcing is near the frequency $(\beta c)^{1/2}$.] The response should therefore have a structure similar to that in Fig. 3.14a. This is indeed the case. The tidal amplitudes in the Pacific are at a maximum along the coasts and the equator and decay towards the interior of the ocean basin. Phase propagation is (anti-) clockwise around the basin in the (Northern) Southern Hemisphere. The phase speed and attenuation scale are consistent with the barotropic values for the gravity-wave speed and radius of deformation in Table 4.1. The forcing of the long-period tides, near 14 and 30 days, is at sufficiently long periods for Kelvin waves to be relatively unimportant in the oceanic response, which should resemble that shown in Fig. 3.14b. This appears to be the case and westward phase propagation, for example, is a feature of the long-period tides (Wunsch, 1967; Luther, 1980). At much longer periods the oceanic response is in the form of equilibrium tides.

The most important example of periodic forcing is the seasonal cycle. Unlike the tidal forcing, which induces a depth-independent response, the winds act in the surface layers of the ocean and induce a response with a complex vertical structure that involves intense currents, undercurrents, and waves. In a large basin such as the equatorial Pacific, the time scale T of the seasonal cycle is such that $T < 4L/c$ so that the response to seasonal winds is far from an equilibrium one as explained in Section 2.3. The relatively small equatorial Atlantic Ocean, however, has an adjustment time $4L/c$ on the order of 150 days so that the response to the seasonal forcing should be close to an equilibrium one. Katz and collaborators (1977) indeed found that seasonal changes in the slope of the thermocline in the western equatorial Atlantic are practically in phase with the changes in the wind.

The annual harmonic may be dominant in the seasonal cycle but there are also nonsinusoidal variations. The intensification of the southeast trades when the ITCZ moves poleward in May can be very abrupt so that some aspects of the response correspond to the adjustment after the sudden onset of winds (Katz and Garzoli, 1982). In the western equatorial Atlantic there is a slight phase lag between the intensification of the winds and the strengthening of the zonal pressure gradient (Weisberg and Tang, 1985). The transients excited by the intensification of the winds in May, which is most prominent in the western equatorial Atlantic, contribute to the secondary maximum of the eastward surface currents in the Gulf of Guinea in October. (Figure 2.21 shows the semiannual cycle in the eastern equatorial Atlantic.) The winds in the west therefore affect oceanic conditions in the Gulf of Guinea. This influence is symmetrical about the equator and affects both hemispheres in the east (McCreary *et al.*, 1984). Yet the suppression of upwelling along the African coast during the warm event of 1984 was confined to the Southern Hemisphere (Hisard *et al.*, 1986). Thus remote winds may influence oceanic conditions in the eastern tropical Atlantic but the winds over the Gulf of Guinea, especially the meridional component to be discussed in the next section, are also important.

The response of the deep ocean to periodic forcing over a confined region is in the form of beams similar to those in Fig. 4.3 but, as pointed out in Section 4.4, the relation between the forcing and the deep motion is complicated because of the nonlinear response of the upper ocean.

The response to transient wind changes, a relaxation that lasts a time T say, depends on the value of T as outlined above. For a temporary weakening of the winds to affect the Equatorial Undercurrent, T must exceed L/c, where L is the zonal extent of the region over which the winds relax. Figure 4.10 shows how a nonlinear model ocean in equilibrium with steady, spatially uniform westward winds responds to a temporary relaxation of the winds over the central part of the basin. (The initial state corresponds to conditions on day 300 in Fig. 4.6.) In this case $T > L/c$ but T is not sufficiently long for the response to be an equilibrium one. The

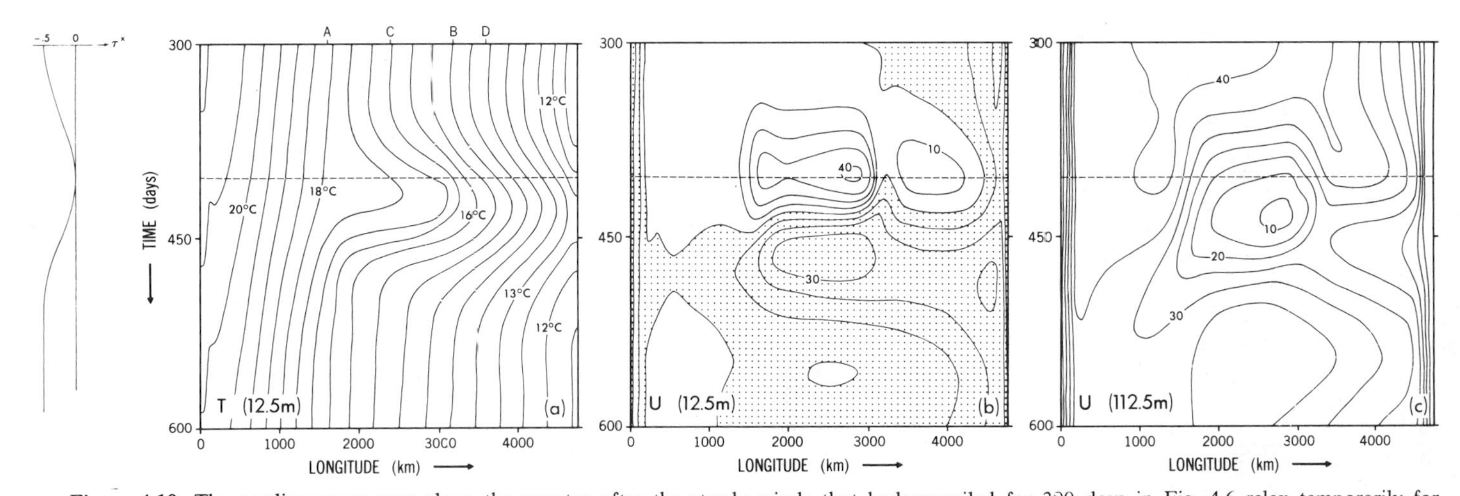

Figure 4.10. The nonlinear response along the equator after the steady winds that had prevailed for 300 days in Fig. 4.6 relax temporarily for approximately 100 days as shown in the left-hand panel. The winds, which are zero at the time of the horizontal dashed line, relax between meridians A and B only. The currents, at 12.5 and 112.5 m, are westward in shaded areas and are in centimeters per second.

adjustment of the zonal pressure gradient is more gradual than the change in the wind so that it is at first unbalanced and accelerates the surface current eastward, but later it is too weak to balance the intensifying wind, which then accelerates the surface current westward. The initial weakening and subsequent strengthening of the zonal pressure gradient causes the Equatorial Undercurrent to decelerate and then accelerate. The sea surface temperature is affected by both advection and upwelling so that its changes are not simply related to that of the surface currents but there clearly is a temporary eastward migration of isotherms. This surface warming can be absent from the eastern side of a large ocean basin (such as the Pacific) if the winds relax in the west because the Kelvin waves that introduce changes in the east propagate downward as they proceed eastward.

4.8 The Response to Cross-Equatorial Winds

The southerly component of the wind, as mentioned in Section 4.1, gives rise to a number of interesting phenomena: the thermal front near 3°N, intense narrow western boundary jets such as the Somali Current, and slow, broad, cold currents such as the Peru Current. To investigate the dynamics of these phenomena, consider spatially uniform meridional winds that suddenly start to glow at time $t = 0$ over a linear, continuously stratified ocean. The response can be described as the sum of vertical modes, each of which satisfies the shallow-water equations

$$u_t - fv + g'\eta_x = 0 \tag{4.40a}$$

$$v_t + fu + g'\eta_y = \tau^y/H \tag{4.40b}$$

$$g'\eta_t + c^2(u_x + v_y) = 0 \tag{4.40c}$$

4.8.1 The Extra-Equatorial Response

Let the northward winds τ^y be confined to a band of latitudes $0 < y < Y$ that is sufficiently small and sufficiently far from the equator for the Coriolis parameter to be regarded as a constant. The response is described by the following solution, which is valid on time scales much longer than the local inertial time and which satisfies the condition that the zonal current u has to vanish at the eastern coast $x = 0$:

$$u = -\frac{\tau^y}{fH}[1 - \exp(x/\lambda)] \tag{4.41a}$$

$$v = \frac{1}{c}\xi(y, t)\exp(x/\lambda) \tag{4.41b}$$

$$\eta = \xi(y, t)\exp(x/\lambda) \tag{4.41c}$$

where $\lambda = c/f$ and

$$\xi_t + c\xi_y = c\tau^y/H \tag{4.42}$$

This solution describes how the offshore Ekman drift

$$u = \tau^y/fH \tag{4.43}$$

is maintained by coastal upwelling in a zone whose width is the radius of deformation. One part of the solution to (4.42) describes an accelerating alongshore jet that is in geostrophic balance:

$$\xi = ct\tau^y/H, \qquad 0 < y < L \tag{4.44}$$

The jet is discontinuous at the boundaries $y = 0$ and Y of the forced region so that waves, solutions to the unforced equations of motion, must be introduced to ensure continuity. The only waves that are available at subinertial frequencies are the nondispersive Kelvin waves of Section 3.4. These are solutions to the homogeneous version of Eq. (4.42):

$$\xi = F(y - ct) \tag{4.45}$$

The function F is arbitrary. In the Northern Hemisphere the waves propagate with the coast on their right, poleward along the eastern boundary of an ocean basin. Winds confined to the region $0 < y < Y$ therefore have no effect equatorward of that region but do influence the region that is poleward ($y > Y$). The condition that the alongshore flow must be continuous at $y = 0$ determines the function F:

$$F(y - ct) = 0, \qquad t < y/c$$

$$F(y - ct) = \frac{\tau^y}{H}\left(t - \frac{y}{c}\right), \qquad t > \frac{y}{c} \tag{4.46}$$

Hence the motion in the forced region is given by Eqs. (4.41) and (4.44) for $t < y/c$. When the Kelvin wave arrives at time $t = y/c$ the solution changes to

$$v = y\frac{\tau^y}{c}e^{x/\lambda}$$

$$g'\eta = y\tau^y e^{x/\lambda} \tag{4.47}$$

while the zonal flow remains unchanged. The Kelvin wave establishes an alongshore pressure gradient that balances the wind so that the acceleration of the jet stops. Upwelling also stops because the alongshore divergence of the steady jet sustains the offshore Ekman drift (Allen, 1976).

In the unforced region to the north ($y > Y$), the motion first introduced by the Kelvin wave excited at $y = Y$ is an accelerating alongshore jet

$$g'\eta/c = v = \tau^y\left(t - \frac{y - Y}{c}\right)e^{x/\lambda}, \qquad \frac{y - Y}{c} < t < \frac{y}{c} \tag{4.48}$$

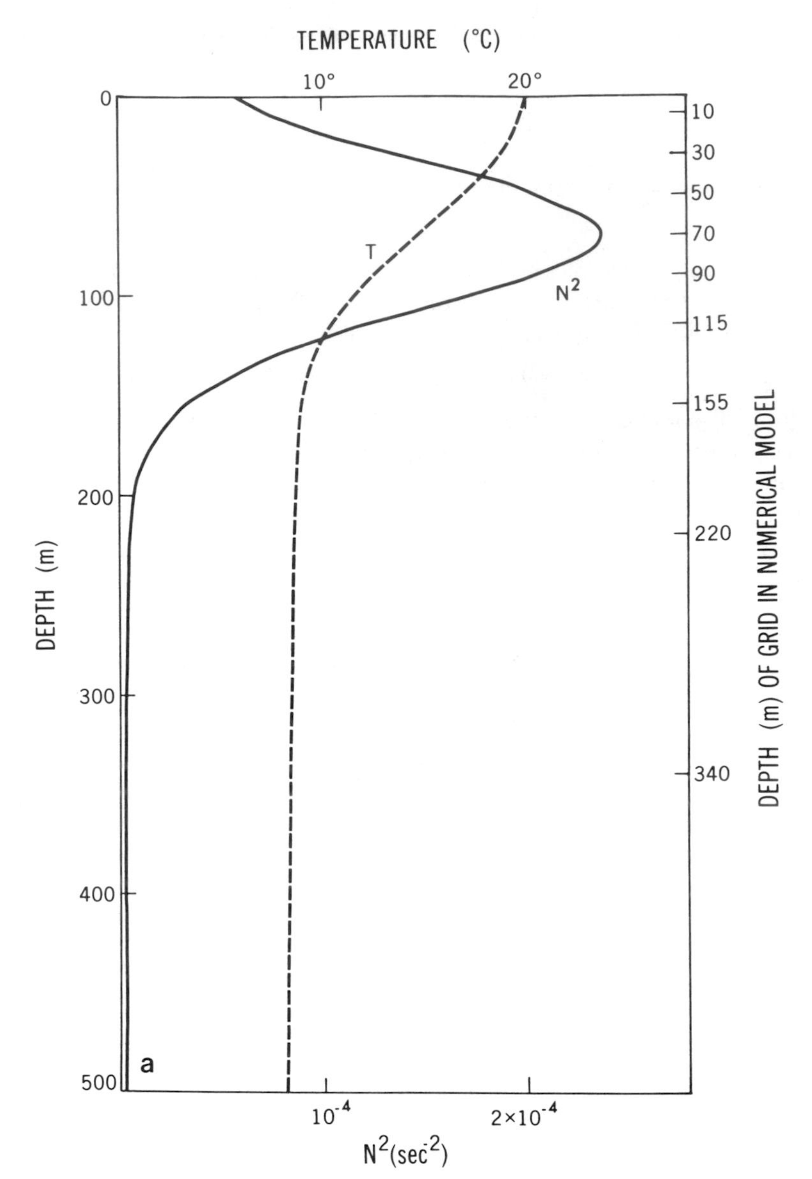

Figure 4.11. (a) The basic stratification (temperature and Brunt–Väisälä frequency) in a model to study the response of coastal zones to alongshore winds. Below 500 m the temperature decreases linearly to zero at the ocean floor. (b) The structure of the vertical modes associated with the stratification in (a). The first few modes have the equivalent depths 15.2, 6.5, and 2.8 cm. (c) The development of a coastal undercurrent after the sudden onset of spatially uniform southward winds, at a distance of 3.8 km from the eastern boundary of an ocean basin and 500 km from the southern boundary of the forced region. The Coriolis parameter has a constant value in the model. The vertical profiles are at times t_n that correspond to the arrival of the nth Kelvin mode. Up to time t_0 the flow is in the direction of the wind and there is no coastal undercurrent. The subsequent arrival of waves establishes a coastal undercurrent. [From Yoon and Philander (1982).] (*Figure continues.*)

which becomes steady after the passage of the Kelvin wave from $y = 0$,

$$g'\eta/c = v = Y\frac{\tau^y}{c}e^{x/\lambda}, \qquad t > \frac{y}{c} \tag{4.49}$$

Consider next an inviscid continuously stratified ocean forced by alongshore winds that act as a body force in a shallow surface layer of depth D. The accelerating alongshore jet is now confined to the surface layer because the momentum equation, in the absence of an alongshore pressure gradient, is

$$v_t = \frac{\tau^y}{\rho_0 H}, \qquad -D < z < 0$$

$$v = 0, \qquad z < -D \tag{4.50}$$

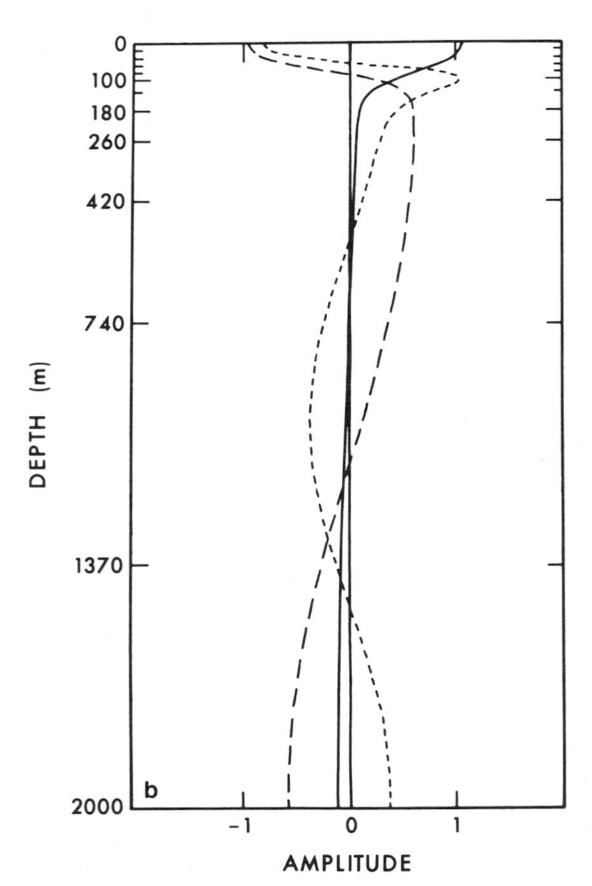

Figure 4.11 (*Continued*)

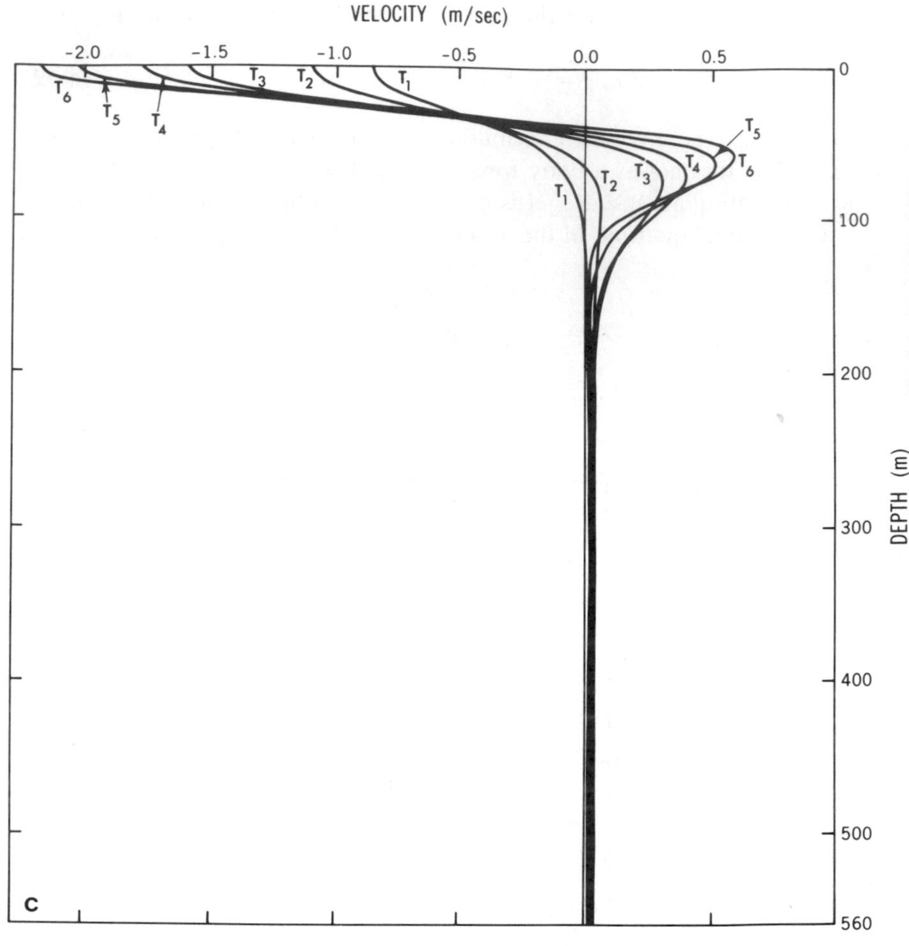

Figure 4.11 (*Continued*)

at the coast $x = 0$. There is offshore Ekman drift in the surface layers, and onshore flow below that. This follows from the vertical integral of the two-dimensional equation for the conservation of mass, $u_x + w_z = 0$.

The next phase of the oceanic adjustment is the arrival of a succession of Kelvin waves, corresponding to the first, second, and so on, baroclinic modes. The modes extend below the surface jet and hence introduce coastal undercurrents as is evident in Fig. 4.11 (Yoon and Philander, 1982). From the vertical structure of the motion in the unforced region, shown in Fig. 4.12, it appears that the first baroclinic mode is dominant for the chosen stratification. If the fluid has N layers each of constant density, then there

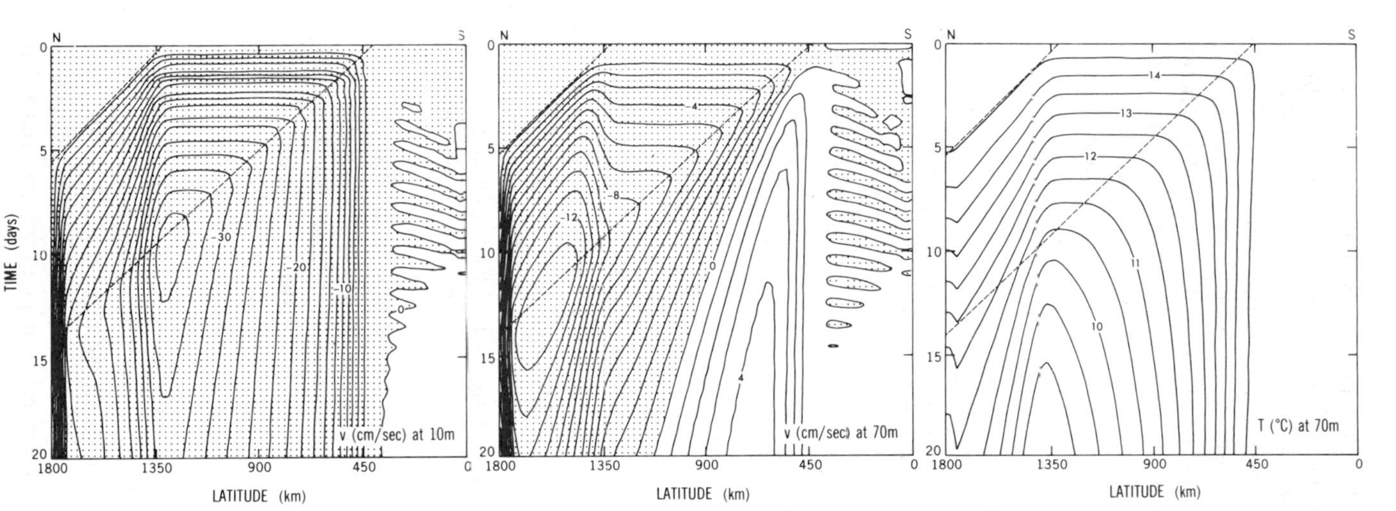

Figure 4.12. The development of alongshore currents in centimeters per second at depths of 10 and 70 m and of temperature in degrees Celsius at 70 m at a distance of 6 km from the coast. The steady winds are southward and blow only in the band of latitudes between 450 and 1350 km from the southern boundary. The Coriolis parameter is constant in this linear model, which has the basic stratification of Fig. 4.11(a). The dashed line corresponds to the first baroclinic Kelvin wave excited at the extremes of the forced region. [From Yoon and Philander (1982).]

are exactly N vertical modes and the motion is steady after the passage of the Nth mode.[8] If the ocean is continuously stratified so that the number of modes is infinite, then a singularity develops near the coast as $t \to \infty$ because the very high modes take infinitely long to propagate to the latitude under consideration. These modes have short radii of deformation and are therefore confined to the coast. A modest amount of friction to attenuate these high-order modes eliminates the problem and results in a steady surface jet and coastal undercurrent (McCreary, 1981b).

The time it takes a coastal zone to adjust to a change in the winds is $T_K = L/c$, where L is an alongshore scale over which the ocean is forced. If L is of the order of 500 km, then T_K is approximately a few days. The response of a coastal zone to fluctuating winds changes at this period. Let the winds fluctuate sinusoidally with a period P. The response of a shallow-water model is described by the solution to Eq. (4.42):

$$\xi = \frac{Pc\tau^y}{\pi H}\sin\left(\frac{\pi}{Pc}y\right)\sin\left[\frac{\pi}{Pc}(y - 2ct)\right], \qquad 0 < y < L$$

In the unforced region $y > L$,

$$\xi = \frac{Pc\tau^y}{\pi H}\sin\left(\frac{\pi L}{Pc}\right)\sin\left[\frac{2\pi}{Pc}\left(y - ct - \frac{L}{2}\right)\right]$$

The dependence of the response on the period P of the forcing has a number of interesting properties. (1) As P increases, the amplitude of the response at first increases until the factor $\sin(\pi y/Pc)$ comes into play. At that stage, when $P \gg T_K$ the amplitude becomes independent of the period. This means that a wind with a white spectrum—all the Fourier components have the same amplitude—forces motion with a red spectrum at high frequencies and a white spectrum at frequencies much lower than $1/T_K$. (2) The alongshore scale of the motion is determined by the function $\sin(\pi y/Pc)$ and increases as the period of the forcing increases. This is evident in Fig. 4.13 and the reason is the increase in the wavelength of the coastal Kelvin wave with increasing period. (3) The alongshore phase propagation in the forced region is at twice the speed of Kelvin waves.

A feature of Fig. 4.13 not explained by the solutions that assume that the Coriolis parameter is a constant is the increase in the offshore scale as the period of the forcing increases. This feature depends on the β-effect, which introduces Rossby waves. Their dispersion relation is given by Eq. (3.25) and is depicted in Fig. 3.6. No waves are possible at high frequencies $\sigma > \beta c/2f$. The corresponding time scale is $T_R = 4\pi f/\beta c$. If c has the value 2 m/sec, then $T_R \sim 100$ days near 45°N. At periods shorter than T_R the Rossby wave dispersion relation gives an imaginary value k_i for the

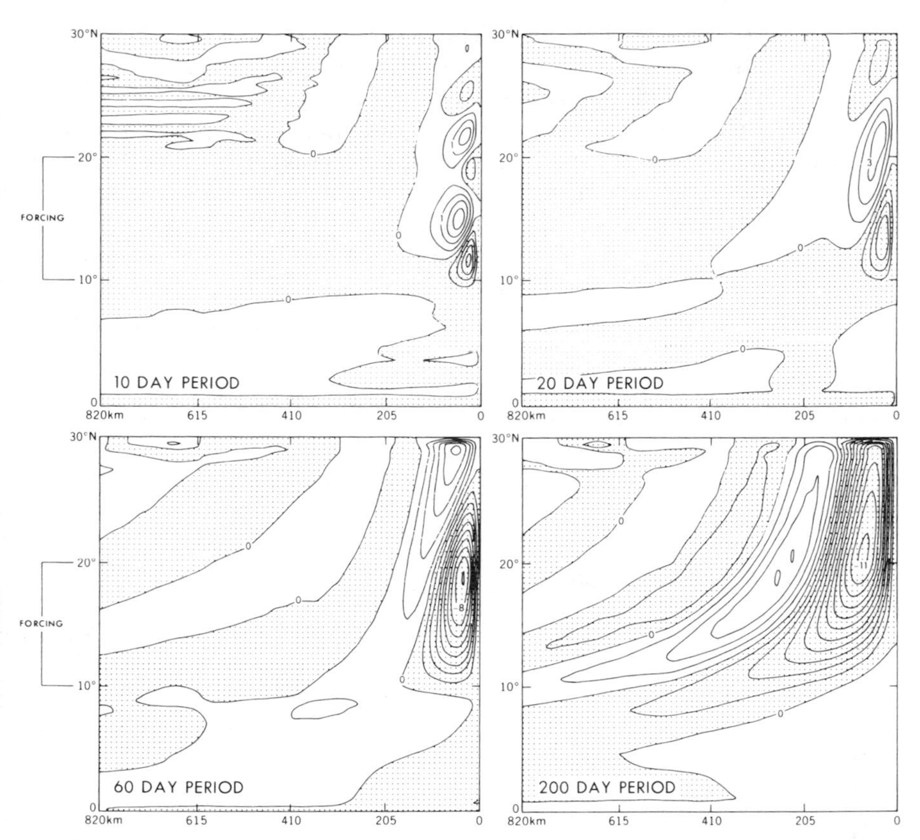

Figure 4.13. Changes in the structure of the oceanic response to fluctuating alongshore winds as the period of the winds changes. The winds are confined to the indicated band of latitudes but otherwise have no spatial structure. The figure shows the meridional velocity component in the eastern half of the model ocean at the time when the winds are southward and are at their maximum. (This happens after one-quarter cycle.) Flow is southward in shaded areas. In the shallow-water equations, which are solved numerically, the Coriolis parameter varies with latitude. [From Philander and Yoon (1982).]

zonal wave number

$$k_i^2 = \left(1 - \frac{\beta^2\lambda^2}{4\sigma^2}\right)\bigg/\lambda^2$$

This determines the offshore scale of the oceanic response if the period of the fluctuating winds is shorter than T_R. The offshore scale is the radius of deformation at very short periods but it increases as the period increases. When the period P exceeds T_R then the offshore scale is the distance

Rossby waves travel in the time P. This result is evident in Fig. 4.13, which also shows that fluctuating winds with a low frequency, even if they are spatially uniform in the offshore direction, generate currents with a complex structure. The alternating northward and southward currents in Fig. 4.13 extend farther offshore the lower the latitude because the Rossby wave speed increases with decreasing latitude.

The low sea surface temperatures along the western coasts of the Americas and Africa in Fig. 1.10 extend far offshore because Rossby waves disperse the upwelling zone westward. The currents near the coast, the California and Peru Currents for example, have a complex offshore structure with the direction of the alongshore flow reversing direction several times. The mechanism that causes the reversing currents in Fig. 4.13 also contributes to the structure of the currents off the western coast of the Americas but additional factors, the curl of the wind for example, are also important (Anderson and Gill, 1975; Philander and Yoon, 1982; McCreary *et al.*, 1987).

4.8.2 The Equatorial Response

The zonal Ekman drift τ^y/fH in response to northward winds intensifies with decreasing latitude until a maximum is reached near the equator. At the equator the zonal currents vanish. (This follows from symmetry arguments.) A scale analysis indicates the presence of a distinctive equatorial zone with a width equal to the radius of deformation $(c/\beta)^{1/2}$. Equatorward of this latitude the meridional pressure gradient rather than the Coriolis force balances the northward windstress in the momentum equation (4.40b). At the equator

$$g'\eta_y = \tau^y/H \qquad \text{at } y = 0 \tag{4.51}$$

Northward winds elevate the thermocline south of, and depress the thermocline north of, the equator while driving zonal currents that have maximum speeds at a radius of deformation from the equator. Figure 4.14 depicts this steady, inviscid response, from which meridional currents are absent and that appears an inertial time $(\beta c)^{1/2}$ after the sudden onset of the winds.

In a continuously stratified ocean, forced in a surface layer of depth D, the flow can have a complex vertical structure because motion at depth is driven by the divergence of the surface flow and by pressure gradients. For example, the southward pressure force maintained by northward winds can penetrate below the surface layers and can drive deep geostrophic zonal currents that have a direction opposite to that of the surface currents. Figure 4.15 shows such zonal currents—they are relatively weak—below

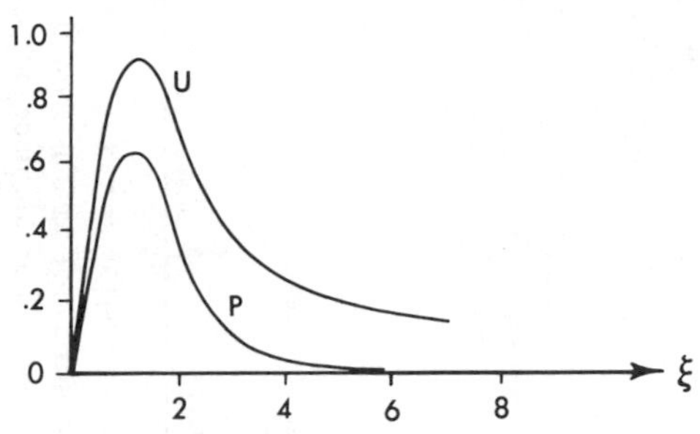

Figure 4.14. The steady zonal velocity component U and thermocline depth perturbation P of a zonally unbounded shallow-water model when forced with steady, spatially northward winds. The coordinate ξ measures latitude in radii of deformation. Another possible solution to the equations of motion is a state of no motion and a pressure force that balances the wind (see Note 9). [From Moore and Philander (1977).]

the strong surface Ekman drift in a linear continuously stratified model forced with spatially uniform, steady northward winds. The surface waters flow northward across the equator, sink near 3°N, return southward at depth, and well up near 3°S, where the thermocline is elevated. The steady subsurface currents and the meridional circulation depend on the presence of mixing processes because, in an inviscid ocean, the motion can be described in terms of vertical baroclinic modes each of which has horizontal motion similar to that in Fig. 4.14. In a steady-state inviscid model, motion is therefore confined to the forced layer of depth D and is strictly zonal.

The meridional circulation is of great importance in a nonlinear model. The northward flow in the surface layer advects the westward current into the Northern Hemisphere. Downwelling causes it to merge with the subsurface westward current in that hemisphere as is evident in Fig. 4.16. Upwelling south of the equator brings the eastward undercurrent near 3°S close to the surface. From Figs. 4.15 and 4.16 it follows that nonlinearities intensify the eastward surface jet north of the equator. Conservation of angular momentum is the important factor (Cane, 1979a). A zonal ring of fluid that is moved poleward, in a rotating coordinate frame, appears to gain eastward momentum. These results indicate that the North Equatorial Countercurrent is to some extent forced by the southerly component of the winds. The westward current just south of the eastward jet in Fig. 4.16 has a structure similar to that of the observed one (Fig. 2.1) so that the southerly winds must in part be responsible for that current too.

The discussion thus far has concerned the x-independent response that is observed far from the north–south coasts shortly after the sudden onset of the uniform northward winds. Consider next the response close to the

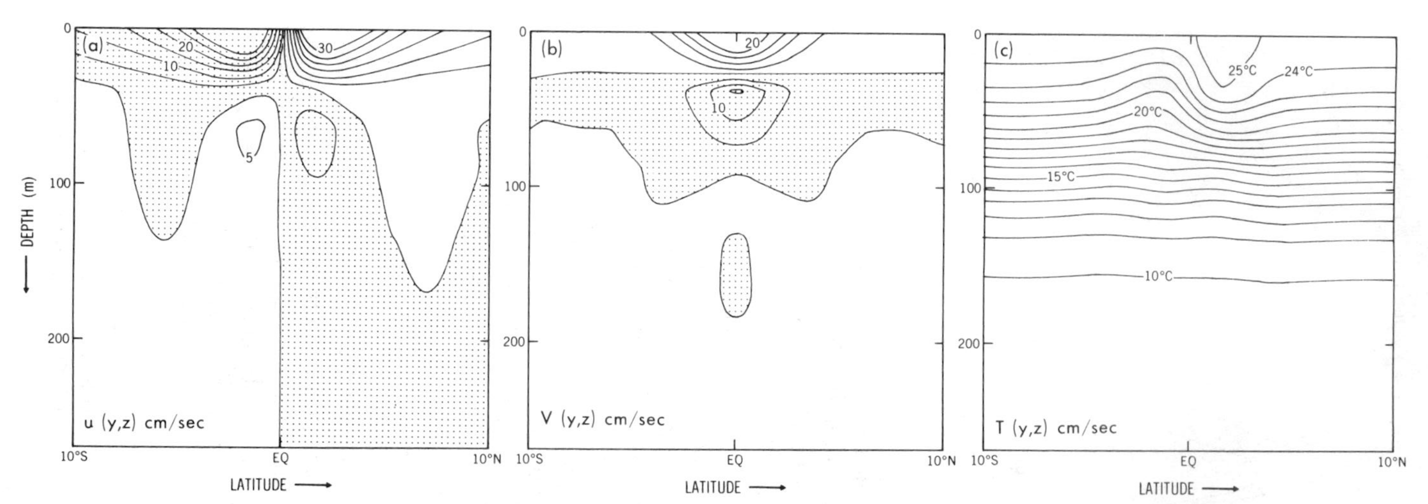

Figure 4.15. The linear equatorial currents generated by northward winds of intensity 0.5 dyne/cm^2. The meridional sections show the two horizontal velocity components (a) u eastward and (b) v northward, and (c) the temperature in degrees Celsius, in a linear, longitude-independent model forced with spatially uniform northward winds. Motion, in centimeters per second, is westward or southward in shaded areas. The basic stratification is that shown in Fig. 4.1. [From Philander and Delecluse (1983).]

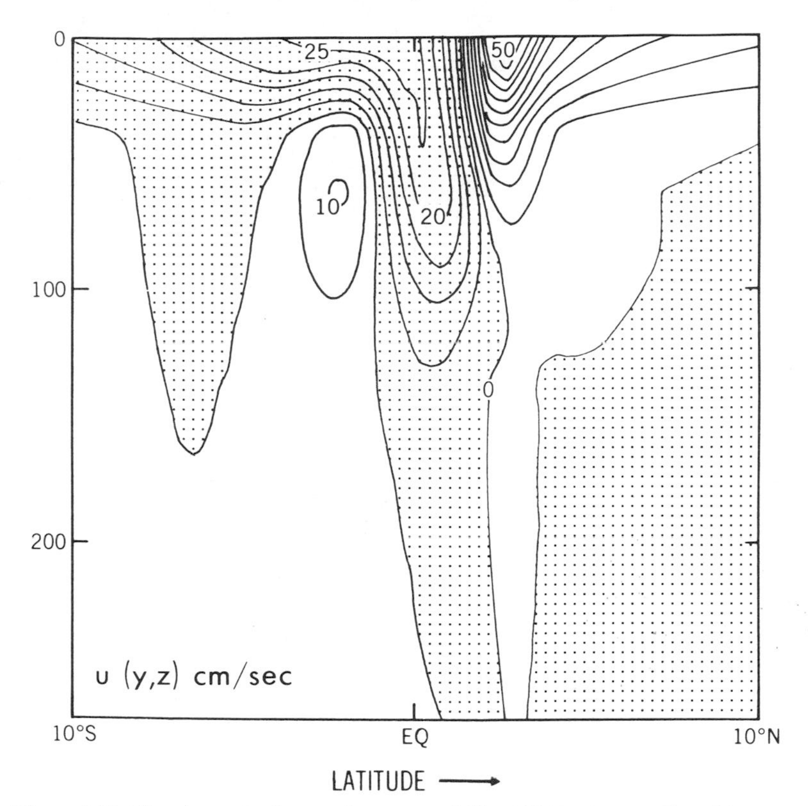

Figure 4.16. The change in the zonal currents of Fig. 4.15(a) when nonlinearities are taken into account. The westward surface current is advected across the equator and penetrates deeply because of downwelling, while the eastward surface current is intensified. [From Philander and Delecluse (1983).]

coasts. Far from the equator the alongshore winds drive an accelerating coastal jet until coastal Kelvin waves establish equilibrium conditions. Near the equator latitudinal gradients are established independently of Kelvin waves and include a southward pressure force to balance the northward winds as shown in Figs. 4.14 and 4.15. The pressure force exists across the width of the basin. It is enhanced near the eastern boundary because the onshore flow near 3°N deepens the thermocline there while the offshore flow near 3°S elevates it. Figure 4.17 illustrates the enhanced southward pressure force at a distance of 16 km from the eastern boundary. It is seen to drive a southward undercurrent below the northward surface flow. The response near the western boundary is very different. There the offshore

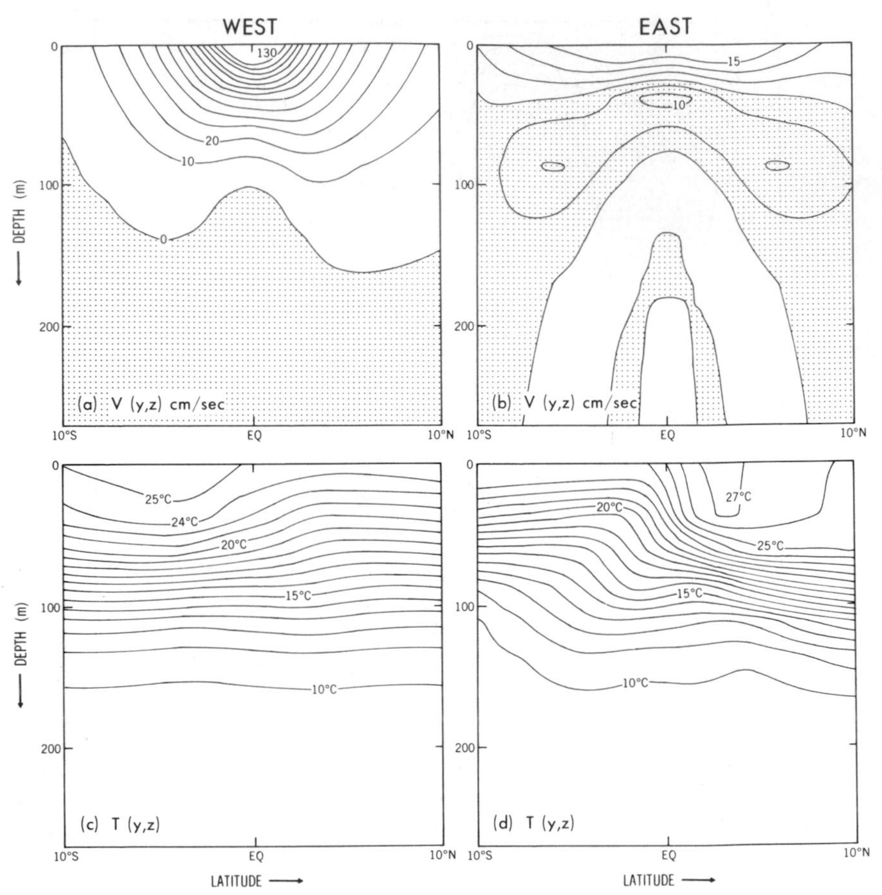

Figure 4.17. The difference in the response to alongshore winds of equatorial coastal zones along the western and eastern boundaries of ocean basins. Northward winds maintain a northward pressure force and drive an intense northward surface jet in the west. In the east these winds maintain a southward pressure force so that the surface current is weak but an undercurrent is present. The figures show sections of the alongshore current V (motion is southward in shaded areas) and temperature along a meridian 60 km from the respective coasts, 40 days after the onset of uniform northward winds of intensity 0.5 dyne/cm^2. The model is linear and the initial stratification is that of Fig. 4.1. [From Philander and Delecluse (1983).]

flow near 3°N elevates the thermocline, which is depressed by the onshore flow near 3°S. Figure 4.17 shows that the southward pressure force far from the coast not only disappears near the coast but can even be northward so that friction must be invoked to achieve a steady momentum balance near the equator along the western coast. The northward surface jet along the western boundary in Fig. 4.17 is far more intense than the one in the east and the southward undercurrent is absent since there is no southward pressure force.

Westward-propagating long Rossby waves next disperse the currents along the eastern boundary. In linear inviscid models these waves ultimately eliminate all the currents so that the equilibrium state is one of no motion in which a southward pressure force balances the uniform northward winds at all latitudes[9] (Cane and Sarachik, 1979). Dissipation attenuates the waves as they propagate westward so that the currents described earlier persist in the western and central parts of the basin, This is evident in Fig. 4.18, which shows the equilibrium response of a stratified nonlinear model to spatially uniform northward winds. The asymmetry between the eastern and western sides of the basin reflect the different properties of Rossby waves with eastward and westward group velocities. The strong resemblance between this simulated sea surface temperature pattern and the observed one in Fig. 1.10 suggests that the meridional component of the wind is principally responsible for the cold surface waters off the southwestern coasts of Africa and the Americas, and for the temperature front near 3°N. The western boundary current in Fig. 4.18 veers offshore near 5°N, where a wedge of cold surface waters appears. Offshore conditions dictate this behavior of the jet in the model; the jet has to feed the intense eastward current immediately to the north of the equator. In reality the Somali Current also veers offshore near 5°N during the early stages of the southwest monsoons but a number of factors that are absent from the model come into play. For example, the westerly component of the southwest monsoons and the inclination of the African coast to meridians inhibit the northward penetration of the coastal jet (Cox, 1979; McCreary and Kundu, 1985; Hurlburt and Thompson, 1976; Luther, 1987; Anderson and Rowlands, 1976b). Nonlinearities, on the other hand, can result in inertial overshoot (Anderson and Moore, 1979). To explain the observed northward migration of the wedge of cold surface waters towards the end of the southwest monsoon season it is necessary to take remote forcing and the curl of the wind into account. Similar factors must influence the North Brazil Current, which also veers offshore near 5°N during certain months. It does so when the eastward North Equatorial Countercurrent is intense. It can therefore be argued that the factors that control the countercurrent, the curl of the wind for example, also control the North Brazil Current.

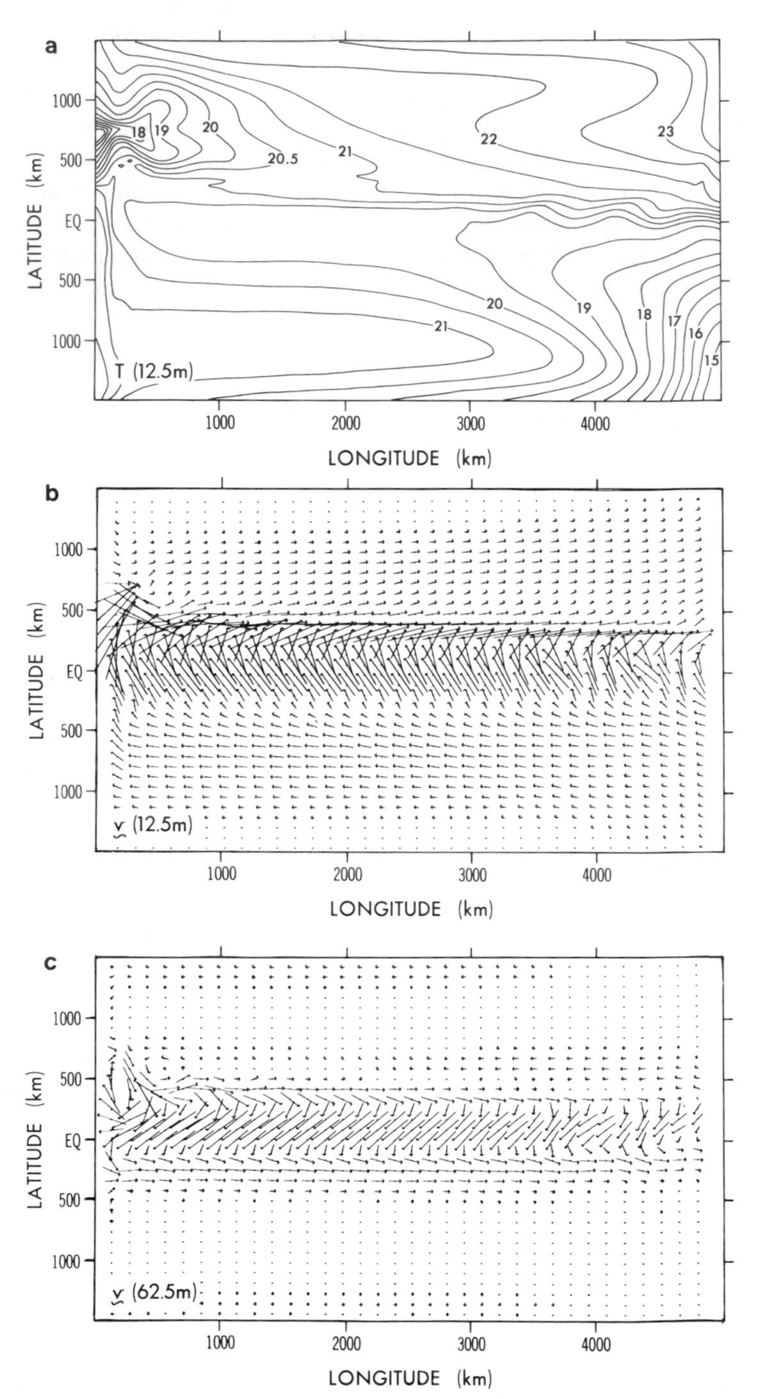

Figure 4.18. (a) Sea surface temperatures in degrees Celsius, (b) surface currents, and (c) currents at 112 m after equilibrium conditions are reached in a nonlinear model with the initial stratification of Fig. 4.1. The model is forced with spatially uniform northward winds. The undulations in the eastern equatorial region correspond to waves associated with instabilities of the currents. [From Philander and Pacanowski (1981b).]

Comparative studies of the Somali Current and North Brazil Current will be valuable.

4.9 General Circulation Models of the Ocean

Low-frequency variability of the tropical oceans is induced primarily by fluctuations in the winds that drive the oceans.[10] This is in striking contrast to the subtropical gyres of the ocean, especially the neighborhood of intense currents such as the Gulf Stream and Kuroshio, where variability is attributable primarily to instabilities of the mean currents. Attempts to simulate Gulf Stream meanders deterministically will never succeed for more than a limited period because the instabilities that amplify perturbations and thus cause meanders also amplify errors in the initial conditions. In the tropics, phenomena such as El Niño and La Niña can be simulated deterministically for an indefinite period provided the forcing function is known. Accurate initial conditions, which are of utmost importance if variability is caused by instabilities, are not too critical for the simulation of variability in low latitudes. The basic density field of the ocean must of course be specified. If it is assumed that the ocean is initially isothermal then the time it takes to establish a realistic thermal structure that includes a thermocline exceeds the time scale of the Southern Oscillation by an order of magnitude at least. The models reproduce the dynamical processes involved in the horizontal redistribution of heat that changes the topography of the thermocline on the time scale of El Niño but, at this stage, the models are less successful at reproducing the processes that maintain the thermocline itself on much longer time scales. The models can establish the correct density gradients rapidly but errors in the absolute temperature persist far longer (Philander and Hurlin, 1987). Hence deterministic simulations of El Niño and La Niña for indefinite periods require specification of the initial density field, accurate forcing functions, and a realistic model.

General Circulation Models of the ocean, which solve the nonlinear primitive equations of motion numerically (see Note 6), are at present capable of realistic simulations of the low-frequency variability of the upper tropical oceans. Simulations of the seasonal cycle of the Atlantic and Pacific Oceans, and of El Niño of 1982–1983, are sufficiently realistic for the model now to be run operationally so that it produces a monthly description of contemporary conditions in the tropical Pacific Ocean (Leetmaa and Ming Ji, 1988). This description was particularly helpful to scientists issuing monthly statements about conditions in the Pacific during 1986 and 1987, when El Niño developed in a very erratic manner.

A number of fortuitous factors contribute to the success of the General Circulation Model. The relative unimportance of instabilities of the mean flow has already been mentioned. These instabilities have spatial and temporal scales (on the order of 1000 km and 3 weeks) that are sufficiently large for the models to resolve them explicitly. In midlatitudes instabilities have small spatial scales and current computer resources barely permit a numerical grid that covers an entire ocean basin and that also resolves the eddies. In low latitudes the small latitudinal scale of the mean currents dictates the grid spacing and present computer resources readily accommodate high-resolution models of the tropical oceans. Presently the models have artificial boundaries along circles of latitude outside the tropics—40°N and 40°S typically—where the observed climatological density field is specified. These boundary conditions in effect parameterize processes in higher latitudes where deep convection, with which the models cope poorly,[11] is important. That the models correctly reproduce the difference between the meridional heat transports of the Pacific and Atlantic Oceans, a difference that is determined by conditions in high latitudes (Section 2.3), attests to the success of the boundary conditions along the northern and southern walls. These walls inhibit interactions between low and high latitudes but such interactions are important only on time scales much longer than that of the Southern Oscillation. (The situation in the atmosphere is very different and even General Circulation Models for short-term weather predictions have to be global.)

The General Circulation Models of the ocean, though reasonably realistic, are far from perfect. Flaws are attributable primarily to inaccurate forcing functions, the surface winds for example, and to inadequate parameterization of mixing processes. For vertical mixing, the Richardson number–dependent mixing of Section 2.8 is used but there are indications that it needs to be more intense near the surface. A comparison of simulated and observed surface currents in the tropical Atlantic (Richardson and Philander, 1986) indicates that the model currents are too intense and do not penetrate sufficiently deep in regions where the mixed surface layer is deep. The model is accurate in the Gulf of Guinea, where the mixed surface layer is shallow or absent, so that the observed and simulated currents in Fig. 2.21 are in excellent agreement. This, however, is not the case in the western equatorial Atlantic as is evident from Fig. 2.18. In that region, which has a deep mixed surface layer, the simulated surface currents are too intense and the model needs to be improved. The surface layers of the western equatorial Atlantic and Pacific Oceans tend to be stratified rather than mixed in the simulations. Increased mixing will improve this feature and will reduce the strength of the surface currents. It is possible, however, that the mixing is fine and that the specified heat flux across the ocean surface is too large and causes the surface layer to be stratified in the model. There is a need for further measurements and for numerical experiments with other mixing schemes.

To improve the model, quantitative comparisons between simulations and observations (rather than the qualitative comparisons that have been made thus far) are necessary. Given the uncertainty in the forcing functions, and given the error bars for the measurements, how significant are the discrepancies between the model and reality? It will be possible, once this question is answered, to decide to what extent changes in the model, improved parameterization of mixing for example, contribute to improved simulations.

A General Circulation Model is the most complex of the hierarchy of models that is available for studies of the oceanic circulation. Simpler models isolate the role of specific physical processes in certain phenomena and provide a vocabulary for the discussion of results from more complex models. For example, the concept of a Kelvin wave emerges not from a General Circulation Model but from the relatively simple analytical model of Chapter 3. Simple models are sometimes capable of simulating certain aspects of the observed motion with reasonable realism but in such cases it is imperative to check that the assumptions made in the simple models are valid. The quasi-geostrophic vorticity equation (3.8), with latitudinal gradients neglected, seems to reproduce seasonal variations in the depth of the thermocline in the region of the North Equatorial Countercurrent when forced with the observed winds, but it is not clear that the simplified version of Eq. (3.8) is valid in such low latitudes. Linear models can reproduce a realistic Equatorial Undercurrent— in response to steady winds the simulated current has a reasonable intensity—but the models are inconsistent because the neglected nonlinear terms are important. Simple models, those that attempt to reproduce sea surface temperatures for example, frequently have a number of parameters that can be adjusted to ensure realistic simulations of some fields in certain cases. [Equations (3.95) provide an example.] These models usually cannot cope with the full range of conditions that are possible in reality. Simulations with General Circulation Models are the most reliable because these models make the fewest number of assumptions and because they have the fewest number of adjustable parameters. However, results from such models are extremely difficult to interpret without the physical insight provided by simpler models. Studies with the full hierarchy of models are necessary to arrive at an ability to both explain and simulate the variability of the oceans.

In addition to being one of a hierarchy of models with which to study the dynamics of the ocean, General Circulation Models have several other uses. A potentially very important use is the generation of data sets with which to address questions that cannot be answered on the basis of measurements only, questions about the heat and mass budget of the ocean, for example. This matter is discussed in Chapter 2.

Notes

1. In a fluid contained between two rotating spherical shells, westward surface flow and eastward subsurface flow at the equator can appear for a wide range of values for the external parameters. All these models of the Equatorial Undercurrent have an eastward pressure force maintained by westward surface winds. Far from the equator the pressure force drives equatorward geostrophic flow. The physical processes that come into play near the equator to balance the pressure force—nonlinearities, vertical mixing, horizontal mixing, the horizontal component of the Coriolis force—determine the width and intensity of the Equatorial Undercurrent in a given model (Philander, 1973). Measurements that determine the parameter values that are appropriate for the observed Equatorial Undercurrent will presumably determine which model describes reality. Wacongne's (1988) analysis of data from a realistic simulation of the Equatorial Undercurrent with a General Circulation Model suggests otherwise. The dynamical balance in the Equatorial Undercurrent is different in different regions, as shown in Fig. 4.9, and in a given region the balance can change with time. In other words, the various models may all be relevant to the observed Equatorial Undercurrent, each in a different region or at different times.

2. In a nonlinear shallow-water model the eastward pressure force maintained by westward winds can be so large that motion near the equator is eastward and in effect corresponds to an Equatorial Undercurrent (Greatbatch, 1985).

3. Waves that propagate downward in the ocean can in part be reflected by the thermocline and can in part penetrate below the thermocline. There is controversy about the amount of energy that can penetrate (Gent and Luyten, 1985) but a resolution of the problem is unlikely. The reason is the following. To distinguish between upward- and downward-propagating disturbances—this needs to be done to determine the fraction of energy that penetrates—the Brunt—Väisälä frequency must be a slowly varying function of depth (so that the WKB approximation is possible in the regions above and below the thermocline). This is no problem below the thermocline but is a serious problem above the thermocline, especially if that region is mixed ($N = 0$). The assumptions about the detailed thermal structure above the thermocline and about the manner in which the windstress acts as a body force are largely arbitrary. The issue is of no great import, because it is of limited relevance to the question of how much energy penetrates into the deep ocean. In reality, the thermocline is not a fixed region through which the waves propagate but it heaves up and down seasonally and interannually and forces the deep ocean directly.

4. The eigenfunctions for which the equivalent depths are negative were first studied in connection with the atmospheric response to tidal forcing (Chapman and Lindzen, 1970). If these eigenfunctions, which have their largest amplitudes near the poles and which are best described in spherical coordinates, are important in the response of the tropics to forcing, then this particular description of the response is obviously inconvenient although it is possible to proceed with a combination of equatorial and midlatitude β-planes (Lindzen, 1967). A description in terns of vertical modes, even if none is excited, will be more efficient. Outside the tropics matters are different. The dispersion relation (3.23) can be used to calculate the negative equivalent depths when the forcing at a specified frequency and wave number does

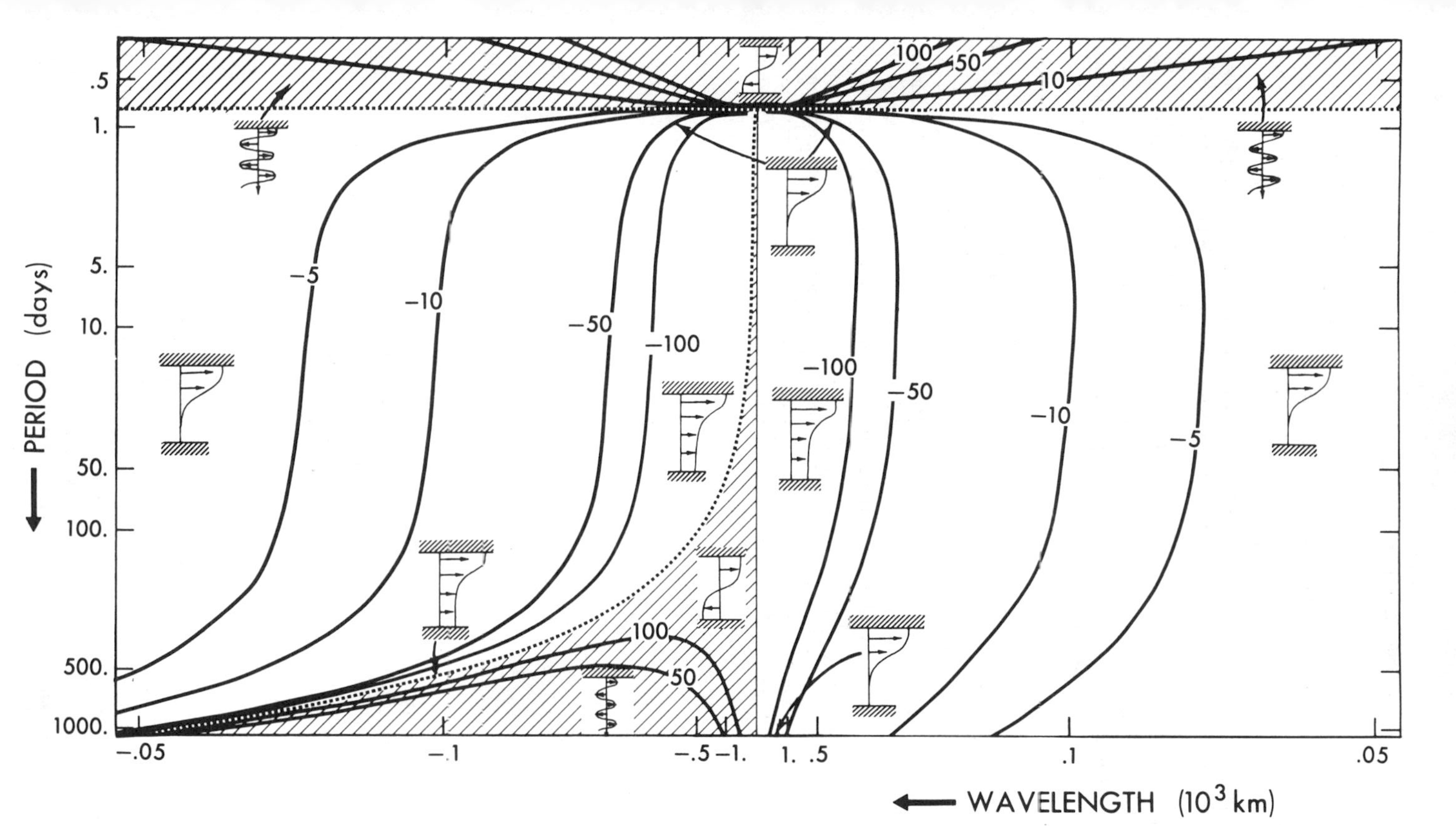

Figure 4.19. Contours of constant values of the equivalent depth h in centimeters as a function of frequency and zonal wavelength as given by Eq. (3.23) with f and β evaluated at 40°N. The insets show the vertical structure schematically. Equivalent depths h are positive in shaded areas, where inertia-gravity and Rossby waves are possible.

not excite any waves. Figure 4.19 shows these equivalent depths and shows schematically the vertical structure of the response. Large, eastward-traveling weather systems force a response that penetrates to the ocean floor, but motion forced at frequencies just below the inertial frequency is trapped near the ocean surface.

5. Spectral peaks in tide-gauge records from islands in the tropical Pacific, at periods near 3 and 5 days, correspond to equatorially trapped inertial-gravity waves (Wunsch and Gill, 1976). They appear to be resonant modes of the ocean—the vertical structure is that of the first baroclinic mode and the zonal group velocity is zero—and are not dependent on the atmospheric forcing having spectral peaks at the same periods. Similar waves observed in the tropical Atlantic, where the irregular bottom topography probably prevents the establishment of vertical modes, appear to be forced by atmospheric waves (Garzoli and Katz, 1981).

6. The General Circulation Model was developed at the Geophysical Fluid Dynamics Laboratory (Princeton, New Jersey) by K. Bryan and M. D. Cox. The model solves the primitive equations of motion, in finite difference form, using the method of Bryan (1969). In the high-resolution version the grid spacing is 1° longitude, 0.3° latitude between 10°N and 10°S, and 10 m in the vertical in the upper 100 m of the ocean. The spacing increases poleward of 10° latitude and below a depth of 100 m. In the idealized studies described in this chapter, the coefficient of vertical eddy viscosity is assigned the constant value of 10 cm^2/sec and the coefficient for vertical eddy diffusivity is 10% smaller. The heat flux across the ocean surface is specified to be zero and initially the thermal field is a function of depth only. For realistic simulations, such as those described in Chapter 2, the vertical mixing is parameterized in terms of the Richardson number of the flow. The horizontal mixing of momentum is parameterized in terms of a coefficient of eddy viscosity with a constant value of $2 \times 10^7\ cm^2/sec$. The initial conditions for the model are climatological oceanographic data (Levitus, 1982), which are also used as boundary conditions at the artificial walls along 50°N and 28°S. The heat flux across the ocean surface is the specified incoming solar radiation (500 ly per day equatorward of 20°N and 20°S), minus the long-wave back radiation (115 ly per day), minus the evaporative heat loss QE, which is calculated from the formula

$$QE = \rho C_D LV[e_S(T_0) - \gamma e_S(T_A)](0.622/p_A)$$

where the saturation vapor temperature is

$$e_S(T) = 10^{9.4-2353/T}$$

Here $\rho = 1.2 \times 10^{-3}$ g/cm^{-3}; $L = 595$ cal/g; $C_D = 1.4 \times 10^{-3}$; $p_A = 1013$ mbar; $C_P = 0.24$ cal/g/°C; T_0 is the sea surface temperature in degrees Kelvin as determined by the model; T_A, the atmospheric temperature at the surface, is specified from measurement; V is the surface windspeed; and γ, the relative humidity, is assigned the constant value 0.8. To compensate for the lack of high-frequency wind fluctuations in the climatological winds, V has a minimum value of 0.3 dyne/cm^2 in the expression for QE.

To reproduce the seasonal cycle the model is forced with the monthly mean climatological surface windstress described by Hellerman and Rosenstein (1983). After three annual cycles are simulated, the upper ocean reaches a state of equilibrium so that the fourth year is essentially identical to the third. The results appear to be relatively insensitive to changes in the parameters. Simulations of the annual cycle over a 500-year period, with an Atlantic model that has a coarser numerical resolution, that has different values for the mixing parameters, and that specifies sea surface temperatures rather than heat flux as an upper boundary condition, give very similar results. The meridional heat flux, in particular, is essentially the same as that depicted in Fig. 2.19 (Sarmiento, 1986).

There are many aspects of the model that need to be improved: the parameterization of mixing processes is inadequate; the western boundary of the Pacific, a solid wall in the model, is too idealized; and the surface boundary conditions, the fluxes of heat and momentum across the ocean surface, have errors. The forcing functions have improved considerably since the model started to be run operationally.

7. To study the generation of the Equatorial Undercurrent, Cane (1979b) cleverly exploited the dominance of a single baroclinic mode in the adjustment of the ocean by constructing a one-layer model in which the forcing does not extend throughout the depth of that layer.

8. The one-layer shallow-water model assumes that the thermocline is the interface between two fluids each of constant but different density. A multi-N-layer model approximates the continuous stratification with N discrete layers each of constant density. The density is then discontinuous at each interface that is deformed when the fluid is in motion. In level models such as the General Circulation Models the density of the ocean is assumed to vary continuously, but in the equations of motion vertical derivatives are approximated by finite differences between fixed horizontal planes (levels). If the motion is linear and inviscid then one approach has no intrinsic advantage over the other although the results may differ, especially if few levels are used, because the results depend on the details of the finite difference scheme. In the presence of mixing processes and nonlinearities, level models are preferable to layer models.

9. The linear inviscid shallow-water equations forced by steady, spatially uniform northward winds permit as solutions both

$$g'\eta_y = \tau^y/H; \qquad u = v = 0$$

and the solution of Fig. 4.14, which includes currents, if the ocean is zonally unbounded. To determine which is the relevant solution it is necessary either to solve the initial value problem (Moore and Philander, 1977) or to include dissipative processes, in which case the specified ratio of Rayleigh damping to Newtonian cooling in the limit of both approaching zero determines the appropriate solution (Yamagata and Philander, 1985).

10. It is possible for models of the tropical oceans to have a rich spectrum of variability even when forced with steady winds, provided the parameterization of mixing processes is such that these processes are almost negligible. The mean currents are then very fast and their instabilities are highly nonlinear (Semtner and Holland, 1980).

11. Calculations with a General Circulation Model that has the northernmost boundary near 45°N, so that the region of deep convection is excluded, are more realistic than calculations with the boundary farther north (Sarmiento, 1986).

Chapter 5 | Models of the Tropical Atmosphere

5.1 Introduction

The atmospheric fluctuations that constitute the Southern Oscillation are highly correlated with interannual sea surface temperature variations in the tropical Pacific Ocean. The most sophisticated models of the atmosphere, General Circulation Models, simulate the Southern Oscillation realistically and demonstrate convincingly that it is caused by these sea surface temperature variations. The models are powerful tools with a variety of uses: predictions, studies of the sensitivity of the atmosphere to external factors, and the generation of comprehensive data sets for detailed analyses of specific phenomena. These models provide a wealth of information about the atmospheric response to different sea surface temperature patterns and about the dynamical and thermodynamical balances during different phases of the Southern Oscillation. They clearly are very versatile and in effect are surrogate atmospheres. The General Circulation Models are so complex, however, that simpler models are necessary to isolate and study the role of specific physical processes in various phenomena. Another drawback of the General Circulation Models is the enormous computer resources they require. This precludes studies that explore a wide range of values for the parameters that influence interactions between the ocean and atmosphere. Relatively simple models that require modest computer resources are needed for this purpose.

The simple models that have been developed thus far fall into two groups. One group describes the response of the atmosphere to given heat sources. Changes in the intensity and location of the convective zones where the heating occurs are regarded as known so that attention focuses on various aspects of the response: the changes in the surface winds in low latitudes, the equatorially trapped waves that are excited, and the signals that propagate to high latitudes. The forcing functions for the models, the precipitation, for example, are sometimes taken from the General Circulation Models so that there is an interplay between models of different levels of complexity. The second group of simple models addresses the difficult question of how the position and intensity of the heat sources relate to a given sea surface temperature pattern. General Circulation Models have elaborate parameterizations of convection and boundary layer processes to cope with these problems but there are a number of questions to which they do not provide explicit answers. Which factors determine evaporation from the ocean surface and which factors determine where the moisture converges and condenses? What is the relative importance of sea surface temperature variations and the heating of the continents for the seasonal movements of the convergence zones? To answer these questions it is not enough to have a realistic General Circulation Model of the atmosphere; it is necessary to have a hierarchy of models of increasing complexity. The development of such a hierarchy of models is leading to rapid progress in our understanding of the tropical atmosphere.

The dynamics of the tropical atmosphere and oceans have much in common so that many of the results derived in Chapters 3 and 4 apply to the atmosphere. Indeed, the motivation for the study of equatorially trapped waves was originally provided by meteorological phenomena. On the other hand, methods developed (in Section 3.7) to determine the oceanic response to forcing can be used to calculate the atmospheric response to heat sources. It is important to keep in mind that, in addition to similarities, there are also significant differences between the ocean and atmosphere. For example, the stratification of the ocean is far weaker than that of the atmosphere, so that the equatorial radius of deformation has a value on the order of 100 km in the ocean but 1000 km in the atmosphere. Because of this difference, connections between low and high latitudes in the oceans and atmosphere are different. From the dispersion relation for Rossby waves [Eq. (3.25)] it follows that the north–south group velocity increases with increasing zonal wave number. Disturbances with east–west scales that are small relative to the radius of deformation propagate north–south efficiently; those with large zonal scales barely propagate meridionally. Meridional propagation is therefore more common in the atmosphere than in the ocean. Teleconnections from low to high latitudes form a significant aspect of the Southern Oscillation. Meteorological models, even those that predict weather on the short time scale of days, are usually global. Models of the tropical oceans, on the other hand, frequently exclude higher latitudes and often have artificial walls just outside the tropics. The link between low and high latitudes in the ocean is primarily in the form of forced motion, Ekman drift, for example. In summary, the understanding of one medium, the ocean say, is enhanced by studying the other medium (the atmosphere) too, but it is important to keep in mind that parameter values can be very different for the two media.

5.2 Waves

Satellite photographs indicate that the cloudiness of the major convective zones in the tropics varies over a wide spectrum of frequencies. The response of the atmosphere to these fluctuating heat sources can be analyzed in a manner very similar to that presented in Chapter 4. Modifications are necessary because of the compressibility of the atmosphere. The equation for the conservation of mass is

$$\frac{1}{\rho}\frac{D\rho}{Dt} + \operatorname{div} \boldsymbol{v} = 0 \tag{5.1}$$

and the thermodynamic energy equation is

$$c_v \frac{DT}{Dt} + p\frac{D\alpha}{Dt} = Q \tag{5.2}$$

Here c_v denotes the specific heat of air at constant volume, α is the specific volume, Q is the rate of heating, and the rest of the notation is the same as that used earlier. As in Section 4.2, linearization and a separation of variables lead to equations similar to (4.6) for the horizontal structure of the motion and the following equation for the vertical structure:

$$V_{zz} + \left(\frac{\gamma - 1}{\gamma h H} - \frac{1}{4H^2}\right)V = J \tag{5.3}$$

The meridional velocity component v is related to V as

$$v = \rho_0^{-1/2} V(z) \tag{5.4}$$

Here h, the equivalent depth, is the constant of separation, $\gamma = c_p/c_v = 1.4$, and J is related to the heating function. It is assumed that the atmosphere is isothermal at temperature T_0 so that the scale height $H = RT_0/g$ is a constant, where R is the universal gas constant. The basic density of the isothermal atmosphere is

$$\rho_0 = \rho_0(0)\exp(-z/H) \tag{5.5}$$

Generalization to a nonisothermal atmosphere is straightforward.

The vertical velocity component vanishes at the lower boundary. The atmosphere has no upper boundary so that the appropriate second boundary condition for Eq. (5.3) depends on the sign of $A = (\gamma - 1)/\gamma hH - 1/4H^2$. If $A < 0$ then solutions must be bounded as $z \to \infty$. If $A > 0$, so that the equation describes vertically propagating waves, then a radiation condition is necessary as $z \to \infty$. For an isothermal atmosphere with $T_0 = 256$ K and $H = 7.5$ km, vertically propagating waves are possible provided $0 < h < 8.57$ km. If $h < 0$ the motion attenuates with increasing height. If $h > 8.57$ km, the amplitude of the horizontal velocity components [which are proportional to $\rho_0^{-1/2}V(z)$] increases with height but the energy [which is proportional to $V^2(z)$] decreases with height.

In the absence of forcing, $J = 0$, Eq. (5.3) describes free oscillations of the atmosphere. For an isothermal atmosphere there is only one free mode (Lindzen, 1967), in other words, only one eigenvalue h:

$$h = \gamma H \tag{5.6}$$

This mode, known as the external or equivalent barotropic mode, is bounded as $z \to \infty$ and has phase that is independent of height. Because of the large equivalent depth, Rossby waves with this vertical structure propagate to high latitudes when excited in the tropics. If the latitudinal shear of the mean atmospheric winds is taken into account, then the Rossby wave trains are similar to that depicted in Fig. 1.26 (Hoskins and Karoly, 1981) but, as pointed out in Section 1.7, there is more to the pattern in Fig. 1.26 than Rossby waves.

Many models of the atmosphere have a rigid upper lid where the vertical velocity component is required to vanish. The situation is then analogous to that in the ocean, discussed in Section 4.3, and Eq. (5.3) yields an infinite number of free vertical modes that are artifacts of the unrealistic upper boundary condition.

Forcing in the tropics excites the poleward (horizontally) propagating equivalent barotropic mode and in addition excites vertically propagating equatorially trapped modes. If the frequency and zonal wave number of the forcing are known, then the dispersion relation (4.12) determines the equivalent depth h and Eq. (5.3) determines the vertical structure of the response. Examples of upward-propagating waves that have been analyzed in this manner include Rossby-gravity waves with a period of 4 to 5 days, zonal wave number 4 to 5, and vertical wavelength of the order of 5 km (Yanai and Maruyama, 1966) and Kelvin waves with periods between 10 and 20 days, vertical wavelengths of the order of 10 km, and zonal wave numbers 1 and 2 (Wallace and Kousky, 1968).

The periods of some of the waves observed in the atmosphere do not correspond to periods at which the atmospheric heating is particularly energetic. In other words, the atmosphere appears to respond energetically at selected frequencies even when it is forced over a continuum of scales. Analyses of the diabatic forcing indicate that the vertical wavelength of the excited waves tends to be twice the depth of the heating (Chang, 1976; Hayashi, 1976; Itoh, 1977). Theoretical studies suggest that this result is independent of the precise shape of the heating profile (Salby and Garcia, 1987; Garcia and Salby, 1987). If the vertical scale of the waves is determined by that of the forcing function, and if the zonal wave numbers of the long waves have to be small integers, then the dispersion relation for equatorially trapped waves determines a frequency for each latitudinal mode. This appears to explain the frequencies that the atmosphere selects.

Vertically propagating equatorially trapped waves play a critical role in maintaining the quasi-biennial oscillation in the stratosphere so that the effect of vertical and horizontal mean wind shear and the effects of dissipation on these waves have been studied in considerable detail. Holton (1975) describes many of these studies.

5.3 The Response to Steady Heating

In midlatitudes, where horizontal temperature gradients are large, a steady heat source can be balanced by the horizontal advection of colder air towards the source. In the tropics, horizontal gradients are small and a heat source gives rise to strong upward motion. This difference between low and high latitudes can be quantified by means of a scale analysis of the equations of motion linearized about a state in which there is a mean zonal wind U that is in geostrophic balance with a latitudinal potential temperature gradient $\bar{\theta}_y$. The time-averaged thermodynamic equation (5.2) for the perturbation fields, after the introduction of potential temperature as a dependent variable, is

$$U\theta_x + v\theta_y + w\bar{\theta}_z = \frac{\theta_0}{g}Q \tag{5.7}$$

This equation can be written

$$fUv_z - fU_z v + wN^2 = Q \tag{5.8}$$

if a thermal wind relation is assumed for the perturbation velocity v. It follows that if horizontal advection balances the heating, then

$$v \sim QH/fU \tag{5.9}$$

The scale H is the vertical scale of the heat source if zonal advection is dominant and is the vertical scale of the mean zonal wind U if meridional advection is more important. Should vertical advection be dominant—this is only possible away from the surface of the earth—then

$$w = Q/N^2 \tag{5.10}$$

The magnitude of the vertical velocity w is related to that of v by the linear vorticity equation

$$\beta v = f w_z \tag{5.11}$$

$$v \sim fQ/\beta N^2 H \tag{5.12}$$

If the mechanisms that require the smallest value of v dominate, then r is the deciding parameter, where

$$r = f^2 U/\beta N^2 H^2 \tag{5.13}$$

In midlatitudes $r \gg 1$ and heating is balanced by horizontal advection. Near the equator r is small; this is even the case at the limits of the tropics near 20° latitude provided the vertical scale of the heat source exceeds a kilometer. Equation (5.10) is therefore a good approximation to the thermodynamic equation in the tropics. For the large-scale, time-averaged flow in low latitudes the dominant balance is between diabatic (latent) heating and adiabatic cooling because of the rising motion.

In the study of the interactions between the ocean and atmosphere it is important to know the low-level winds or, even better, the surface stress exerted on the ocean by the winds in response to a given heat source. Intense vertical mixing in a boundary layer near the surface determines this stress. Let p_{T} denote the pressure at the top of the boundary layer where the stress vanishes and define the boundary layer transport as

$$U = \int_{p_{\mathrm{S}}}^{p_{\mathrm{T}}} u\,dp/g, \qquad V = \int_{p_{\mathrm{S}}}^{p_{\mathrm{T}}} v\,dp/g \tag{5.14}$$

where pressure p rather than the height z is the vertical coordinate and p_{S} is the surface pressure. (This step simplifies the equation for the conservation of mass.) The simplest parameterization of the surface stress τ is to assume that it is proportional to the mass transport of the boundary layer

$$\tau = A(U, V) \tag{5.15}$$

where A has the dimensions of inverse time. The linear momentum equations and the equation for the conservation of mass, integrated vertically as in Eq. (5.14), can then be written

$$AU - fV + \varphi_x = 0 \tag{5.16a}$$

$$AV + fU + \varphi_y = 0 \tag{5.16b}$$

$$\omega + U_x + V_y = 0 \tag{5.16c}$$

where φ is the geopotential. Equation (5.10) relates the vertical flux ω to the heat source Q, which has two main components: latent heating proportional to precipitation P and radiative cooling, which is arbitrarily assumed to be proportional to φ. These assumptions permit Eq. (5.16c) to be written

$$B\varphi + c_A^2(U_x + V_y) = -aP \tag{5.17}$$

where a is a constant, B can be interpreted as a coefficient for Newtonian cooling, and c_A is a measure of the static stability of the atmosphere. It has the dimensions of a phase speed, but given the various crude assumptions that have been made, there is no reason to expect c_A to correspond to a true phase speed (see Note 1).

If the forcing has a structure that can be expressed in terms of Hermite functions, then the analytical methods of Section 3.7 can be used to calculate the atmospheric response (Matsuno, 1966; Gill, 1980). Figure 5.1 shows the response to forcing with the latitudinal y structure

$$P = (1 - y/L)\exp(-y^2/4L^2) \tag{5.18}$$

that is confined to a band of longitude 20° wide. The length $L = 10°$ latitude. Figure 5.1 shows convergence onto the heated region, where the

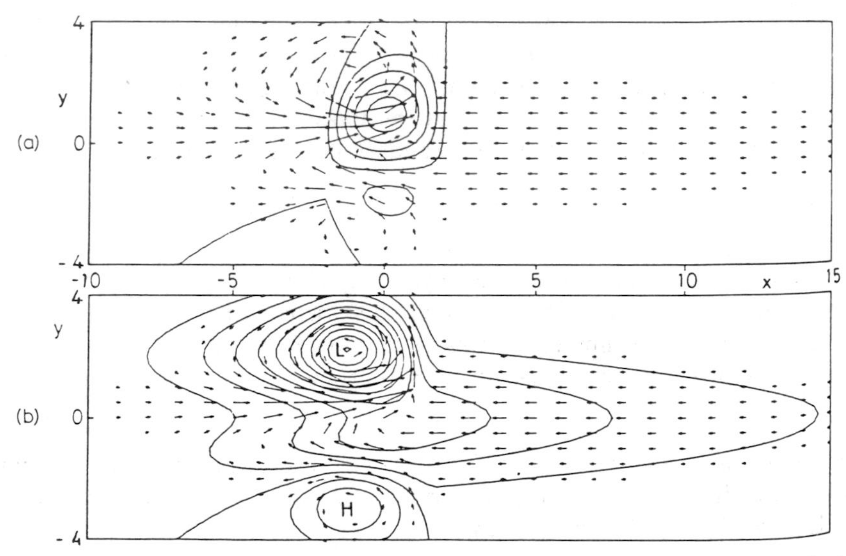

Figure 5.1. The low-level winds, shown as arrows, in response to a heat source with the meridional structure given by Eq. (5.18). The heating is confined to a band of longitude 20° wide centered on $x = 0$. (The unit of distance in the figure is the radius of deformation, which is 10°.) The contours in (a) give the vertical velocity, which has a structure close to that of the heat source. Solid lines in (b) are pressure contours. [From Gill (1982).]

rising motion causes vortex stretching and hence the acquisition of cyclonic vorticity. This qualitatively explains the meridional component of the flow near the heat source. The parameter A in Eq. (5.16) must be assigned a large value of the order of $1/(2 \text{ days})$ for the low-level winds to have a realistic amplitude. Turbulent mixing in the turbulent boundary layer can be invoked to justify such a large value. The assumptions that lead to the dissipative term B in Eq. (5.17) are more difficult to justify so that it is reassuring that certain aspects of the response are insensitive to the value assigned to B (Neelin, 1988). Specifically this is true of that part of the wind whose zonal average is zero. The meridional component of the wind, for example, can be written $V = \overline{V}(y) + v'(x, y)$, where $\overline{V}$ is the zonal average and $\overline{v'} = 0$. The v' field is insensitive to the value of B but this is not so for the zonally symmetric component $\overline{V}$, which satisfies the following equation that can be derived from (5.16) and (5.17):

$$Ac_{\mathrm{A}}\overline{V}_{yy} - B(f^2 + A^2)\overline{V} = A\left(\overline{aP}\right)_y \tag{5.19}$$

The value of B is now of considerable importance. If its value were zero, for example, then there would be no damping at all and the divergence of $\overline{V}$ would not integrate to zero over the domain unless the forcing does. Calculations to reproduce the low-level winds in a General Circulation Model by using its precipitation field as a forcing function for Eqs. (5.16) are reasonably successful for the zonally asymmetric component but not for the symmetric part (Neelin, 1988). The low-level wind fluctuations associated with the Southern Oscillation are principally in the zonally asymmetric component. Thus the simple model can be used in studies of the interaction between the ocean and the atmosphere provided it is not relied upon to simulate the climatological winds.

The low-level winds in the boundary layer tend to be independent of the motion in the upper troposphere. The flow patterns at the two levels are similar, except for opposite directions, because the divergence that drives the upper-level flow is the opposite of the convergence in low levels. The dynamical balances at the two levels are different; data from a General Circulation Model indicate that nonlinearities are important in the vorticity balance of the upper troposphere (Sardeshmukh and Hoskins, 1985).

5.4 Convection in the Tropics

The large-scale cloud systems in the atmospheric convergence zones have as their elements vigorous cumulus convective cells or "hot towers," which are relatively few in number and which occupy a small fraction of the area of the cloud system. A moist parcel of air that rises adiabatically expands, cools, and becomes saturated at a certain level, the cloud base. Some of the water vapor condenses there and becomes visible as a cloud. The accompanying release of latent heat gives extra buoyancy to the air and could help it to rise further. In so doing more vapor condenses and more latent heat is released. A convective instability that is conditional on the condensation of the water vapor in an air parcel is therefore possible. The stability of the atmosphere depends on its equivalent potential temperature θ_e, which is the potential temperature the parcel would have if all its moisture were condensed out and the latent heat thus released were used to warm the parcel. (The parcel attains its equivalent potential temperature when it is raised until all the water vapor has condensed and has fallen out and is then compressed adiabatically to a pressure of 1000 mbar. The ascent is said to be pseudoadiabatic because the liquid water that falls out carries a small among of heat with it so that the process is not truly adiabatic.) Figure 5.2, which shows the vertical profiles of θ and θ_e for a typical tropical sounding, indicates that in the lower troposphere θ_e decreases with height. However, this does not imply that convective overturning will occur spontaneously in the tropics. For that to happen it is necessary not only that θ_e decrease with height but also that the atmosphere be saturated. A parcel rising pseudoadiabatically from $z - \Delta z$ to z will conserve the value of θ_e at $z - \Delta z$. Its

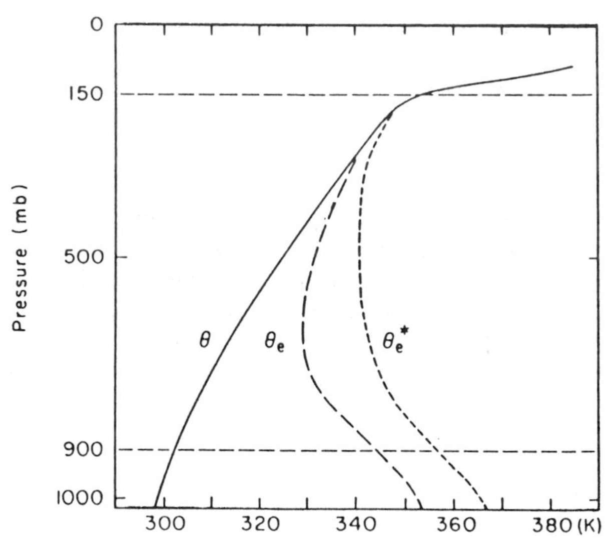

Figure 5.2. A typical sounding in the tropical atmosphere showing the vertical profiles of potential temperature θ and the equivalent potential temperature θ_e^* of a hypothetically saturated atmosphere with the same temperature. [From Ooyama (1969).]

buoyancy at z depends on the difference in density between the parcel and the environment. If the environment is unsaturated, the difference in θ_e for the parcel and the environment may be due to the difference in mixing ratios,[2] not to temperature or density differences. To estimate the buoyancy of the parcel, the value of $\theta_e(z - \Delta z)$ should instead be compared to $\theta_e^*(z)$, which is the equivalent potential temperature the environment at z would have if it were isothermally brought to saturation. The parcel could become buoyant if $\theta_e^*(z) < \theta_e(z - \Delta z)$.

In Fig. 5.2 this is the case below approximately 900 mbar but not at higher levels. This implies that in the tropics, low-level convergence is required to initiate convective overturning because only the air near the surface has a sufficiently high value of θ_e to become buoyant when it is forced upward. Forced ascent is most likely to occur in the convergence zones of the large-scale circulation. The latent heat that is released once air parcels are elevated sufficiently causes further elevation, enhanced convergence of moist air in the lower atmosphere, an increase in the release of latent heat, even more intense convergence, and so on. Interaction between the large-scale circulation and the small-scale cumulus convection can therefore result in an instability. It is known as CISK, conditional instability of the second kind (Ooyama, 1964; Charney and Eliassen, 1964). Models of this instability have to express the heating field due to the cumulus clouds in terms of the large-scale field variables, unless the detailed structure of the cumulus elements is taken into account explicitly. This is a formidable problem. Some progress has been made in explaining the development of intense tropical storms and hurricanes by invoking CISK, but many features of the tropical circulation that probably involve CISK, the ITCZ for example, are poorly understood.

The processes described so far explain how convection can be initiated over a region of high sea surface temperatures. The convection apparently alters the vertical stability of the air column so that the atmosphere becomes convectively adjusted, that is, neutrally buoyant with respect to the air in the subcloud layer. The curves in Fig. 5.2 indicate otherwise, but those are based on calculations that assume that in a tropical cloud the condensed water vapor rains out before it accumulates to the point of having a significant effect on buoyancy. Rain actually does affect the density of clouds, especially in the lower portions of deep clouds (Betts, 1982; Emanuel, 1988). When the weight of water vapor is included in calculating the density, then a typical vertical profile shows that the tropical atmosphere is convectively adjusted. Emanuel (1988) proposes that, under such conditions, the clouds are not of special importance in the circulation of the tropical atmosphere. Rather, the heat engine that drives the circulation depends critically on the thermodynamic disequilibrium between the ocean and atmosphere and operates between the high temperatures near the ocean surface where latent heat is gained and the low temperatures of the upper troposphere where sensible heat is lost to space.

5.5 The Atmospheric Response to Sea Surface Temperature Variations

Three different types of models have been developed to relate low level atmospheric motion to sea surface temperature patterns: in one type, sea surface temperature gradients drive the boundary layer flow; in a second set of models, large values of the surface heat flux are located over regions of maximum temperature and are amplified by a convergence feedback, CISK; and in a third model, sea surface temperatures affect the position of convection through the moist stability. To a first approximation the various models all give similar results, for the surface winds for example, and at this time it is unclear which model is most relevant to the atmosphere.

The Walker Circulation is most intense during the cold phase of the Southern Oscillation when longitudinal sea surface temperature gradients are at a maximum. Its intensity is minimal during El Niño when these gradients are weak. In the model of Lindzen and Nigam (1987), these gradients drive the low-level atmospheric flow. These authors argue that turbulent mixing in the atmospheric boundary layer ties the air temperature to the sea surface temperature. The horizontal pressure gradients associated with the temperature gradients appear as forcing terms in the horizontal momentum equations and drive the boundary layer flow. This flow implies a vertical velocity at the top of the boundary layer. If all this convergence is simply absorbed by the cumulus mass flux without any feedback on the boundary layer flow (which therefore is assumed to be unaffected by the stratification of the atmosphere) then the convergence near the equator is unrealistically large. The results are more reasonable if the vertical motion does work against stratification. This requires the introduction of a damping term in the shallow-water height equation. For a circulation with a reasonable magnitude, the damping time has to be very short, about 30 min, and the depth scale of the model has to correspond to the depth of the trade cumulus boundary layer, about 300 m. Neelin (1989a) describes a simple change of variables and a rescaling of the depth and of the damping time that transform the model of Lindzen and Nigam (1987) into Eqs. (5.16) and (5.17). In other words, the models of Section 5.3 are in effect models for the boundary layer flow driven by sea surface temperature gradients.

The second type of model takes into account that the winds that converge onto the convective zone carry moist air that sustains the convection, which in turn can influence the low-level winds. In some models the parameterization of CISK involves the specification of an initial heat source. It gradually amplifies as the convergent flow induced by the original heat source transports moisture towards the source (Webster, 1981). The influence of mean conditions on the feedback process can be taken into account so that a heating anomaly grows only if the total wind field (specified mean winds plus anomalies) is convergent (Zebiak, 1982). It is assumed that there is an essentially constant moisture field that can be converged by the flow. The initially specified heating Q is very important and determines the location of the convergence zone. It depends on the sea surface temperature because anomalous evaporation is assumed proportional to the anomaly in saturation vapor pressure given by the Clausius–Clapeyron equation linearized about the climatological sea surface temperature:

$$Q = aT'(b/T_{\mathrm{m}}^2)\exp(-b/T_{\mathrm{m}}) \tag{5.20}$$

Here T_{m} is the specified climatological monthly mean sea surface temperature, T' is the sea surface temperature anomaly, and a and b are constants. In this model, sea surface temperature anomalies affect the heating more in warm regions than in cold regions. This is why the convergence zones are over the warmest water. If the specified sea surface temperatures correspond to those of a "composite" El Niño (Section 1.4) or to those observed during El Niño of 1982–1983, then the surface winds as calculated from Eqs. (5.16) and (5.17) with the heat source given by (5.20) are reasonably realistic. There are, however, significant discrepancies between the simulated and observed winds at all extraequatorial latitudes and on the equator in the eastern Pacific, where the model produces easterly anomalies where none is observed (Zebiak, 1982; Gill, 1982).

A model of the second type, which is based on Eq. (5.20), cannot explain why concentrated regions of convergence occur only over part of the area of maximum sea surface temperature. The nonlinearity in the Clausius–Clapeyron equation is too weak, as noted earlier. A further problem with this model is the assumption that the moisture field is uniform.

The third type of model takes into account moisture variations which are described by the equation

$$(Lq)_t + \operatorname{div}(Lq\mathbf{v}) + \frac{\partial}{\partial p}(Lq\omega) = -QH + g\frac{\partial}{\partial p}F^{\mathrm{L}} \tag{5.21}$$

where q is the specific humidity, QH is the latent heating per unit mass, L is the latent heat of condensation, and F^{L} is the vertical latent heat flux due to the diffusion of moisture. This equation, in which pressure is the vertical coordinate and ω is the corresponding flux component, can be combined with an equation for the dry static energy s,

$$s = c_p T + \Phi \tag{5.22}$$

$$s_t + \operatorname{div}(s\mathbf{v}) + \frac{\partial}{\partial p}(s\omega) = QH + g\frac{\partial}{\partial p}F^{R} \tag{5.23}$$

to obtain an equation for the moist static energy m,

$$m = s + Lq \tag{5.24}$$

$$m_t + \operatorname{div}(m\mathbf{v}) + \frac{\partial}{\partial p}(m\omega) = g\frac{\partial}{\partial p}F \tag{5.25}$$

The geopotential is Φ, F^{R} is the upward directed vertical fluxes of energy due to radiation, and $F = F^{R} + F^{L}$ is the total energy flux. (Regard sensible heat as included in the term F^{R}.) At the lower boundary $F^{L} = LE$, where E is evaporation, and at the top of the troposphere $F = F^{R}$.

In regions of intense convection the large-scale divergence is observed to have a simple vertical structure in the tropics with one sign near the ground (where $p = p_S$), the opposite sign near the tropopause (where $p = p_T$), and small values in the middle troposphere (where $p = p_M$). This result can be exploited to simplify Eq. (5.23) as follows. Derive an equation for the lower atmosphere by integrating vertically from p_S (where $\omega = 0$) to p_M. Similarly derive an equation for the upper troposphere by integrating from p_M to p_T (where $\omega = 0$). Subtract these equations from each other and assume that the divergence in the upper layer is minus the divergence in the lower layer to obtain the approximate equation

$$\Delta s \omega_M + [\bar{\mathbf{v}} \cdot \nabla \bar{s}] + \left[\operatorname{div}\left(\overline{s'\mathbf{v}'}\right)\right] = L\bar{P} + \Delta F^{R} \tag{5.26}$$

The difference between the average values of s in the upper and lower troposphere is denoted by Δs. In this equation [ ] denotes an integral from p_S to p_T, an overbar denotes a time average, a prime is the departure from the time average, the averages of time derivatives are neglected, and P is precipitation.

In the tropics, horizontal gradients of temperature and geopotential are small so that $\bar{v} \cdot \nabla \bar{s}$ is negligible. Eddy transports of dry static energy are also small (Oort, 1983). Equation (5.26) therefore simplifies to

$$\omega_M \Delta s \sim L\bar{P} + \Delta F^{R} \tag{5.27}$$

A similar equation can be derived for the moisture

$$\omega_M \Delta q \sim -\bar{P} + \bar{E} \tag{5.28}$$

In regions of heavy time-mean rainfall the precipitation terms $\bar{P}$ dominate the right-hand side of Eqs. (5.27) and (5.28). This is possible only locally, over confined domains, because averages of ω_M over sufficiently large regions must be small. This means that the flux terms other than precipitation are important over large areas. Their role becomes clear if Eqs. (5.27) and (5.28) are summed to obtain a moist static energy equation

$$\omega_M \Delta m \sim \Delta \bar{F} \tag{5.29}$$

In the same way that the quantity Δs represents static stability in Eq. (5.27), the quantity Δm is a measure of the "gross moist stability" of the troposphere. It is assumed that Δm is positive so that the time-mean adiabatic cooling in the troposphere always exceeds the mean latent heating resulting from moisture convergence. If this were not so, then the time-mean circulation in the tropics would not be thermally direct. (In the case of a zonally symmetric climate the constraint that Δm be positive ensures that the time-mean Hadley cell transports energy in the same direction as the flow in its upper branch.)

The balance in Eq. (5.29) implies that the horizontal structure of the low-level atmospheric convergence ω_M is determined by two factors: the horizontal gradients in the flux of energy into the atmosphere through the upper and lower boundaries ΔF and the horizontal variations in the moist stability Δm. In models that do not permit moisture variations the first factor is dominant and the convergence is over the regions of high evaporation and high sea surface temperature. In reality, however, the scale of the fluxes ΔF is larger than that of the convergence. It follows that variations in Δm must be an important factor in determining the sharp horizontal structure of the tropical convection. The problem is therefore to explain the spatial variations of Δm. Suppose that Δm were to become negative in a certain region. A convective instability results so that the atmosphere adjusts (by increasing the vertical extent of the convection and hence increasing Δs) until a state of marginal gross moist stability is attained. According to this view, the static stability Δs is determined in convective areas (by moisture convergence feedbacks) and is communicated to the rest of the tropics by the large-scale atmospheric circulation so that Δs is spatially uniform (Held, 1982b). Variations in $\Delta m = \Delta s - L\Delta q$ are therefore associated with variations in the specific humidity.

To construct a model that determines the convergence ω_M it is useful to exploit an empirical relation that relates specific humidity q in the lower atmosphere to the sea surface temperature T_S:

$$q = aq_{SAT}(T_S - T') \tag{5.30}$$

The constant $a = 0.8$ is an effective relative humidity, q_{SAT} is the saturation mixing ratio at 1000 mbar, and $T' = 1$ K. If a critical temperature T_C is defined such that the uniform static stability $\Delta s = aq_{SAT}(T_C)$, then the low-level convergence is large where T_S approaches T_C. If $q_{SAT}(T)$ is linearized, then Eq. (5.29) implies that ω_M is proportional to $1/(T_C - T_S)$.

This analysis of the moisture equation by Neelin and Held (1987) attributes the presence of intense convection over regions of maximum sea surface temperature to high moisture levels in those regions. This increases the instability of the flow to moist convection and, by means of convergent moisture transport, enhances precipitation in those regions at the expense of less favored regions. Because of the various approximations made in the analysis, a simple model constructed along these lines is not likely to be accurate and will be particularly poor over deserts. The assumptions that transient moisture fluxes are negligible, that there is a simple vertical structure everywhere in the tropics, and that the static stability is spatially uniform are all questionable. Results from such a model are nonetheless encouraging and the calculated movements of the precipitation fields, given the sea surface temperatures, agree qualitatively with observed climatology and with certain El Niño episodes even though variations in the flux terms (ΔF) were ignored (Neelin and Held, 1987).

5.6 General Circulation Models of the Atmosphere

General Circulation Models of the atmosphere are capable of realistic simulations of the time-averaged circulation (the climate) of the atmosphere when forced with the observed seasonally and latitudinally varying flux of solar radiation at the top of the atmosphere. The models solve the primitive equations numerically for a grid that typically has a horizontal resolution of the order of 200 km and that, in the vertical, has between 6 and 12 levels. This means that both the external barotropic mode and the vertically propagating waves of Section 5.2 are resolved. The horizontal grid is too coarse for the cumulus clouds to be resolved so that they have to be parameterized. One possible scheme is a convective adjustment process that operates as follows: if at a grid point the vertical gradient of the temperature, the lapse rate, exceeds the moist adiabatic lapse rate, then the lapse rate is adjusted back to the moist adiabatic value in such a manner that the potential energy of the entire vertical column is preserved. In addition, the lapse rate, if it exceeds the dry adiabatic lapse rate, is similarly adjusted back to the dry adiabatic value (Manabe *et al.*, 1965). A more sophisticated parameterization that takes into account properties of the ensemble of convective cells is also available (Arakawa and Schubert, 1974). Such schemes, one of which was devised by Kuo (1965), together with the parameterization of turbulent boundary layer processes, permit the specified sea surface temperatures and the large-scale flow to interact in such a manner as to produce realistic heat sources in the tropical atmosphere. The basis for this statement is the success with which the General Circulation Models reproduce the mean atmospheric conditions, the Southern Oscillation, the associated teleconnections to high latitudes discussed in Section 1.7, the vertically propagating equatorially trapped waves mentioned in Section 5.2, and the intraseasonal 30- to 60-day oscillations described in Section 1.8. Because of the realism of the models, they can be used in experiments to determine how different external factors affect the properties of these phenomena.

One important set of experiments with the General Circulation Model developed at the Geophysical Fluid Dynamics Laboratory at Princeton University simulated the atmospheric variability over several 15-year periods (N. C. Lau, 1985). The first experiment, in which climatological sea surface temperatures were specified as a lower boundary condition, succeeded in reproducing the statistics of atmospheric variability in midlatitudes with reasonable accuracy but failed to produce any significant low-frequency variability in the tropics other than the seasonal cycle. The second experiment was a repetition of the first except that, in the tropical Pacific, the specified sea surface temperatures corresponded to those observed during the period 1962 to 1976. This calculation simulated the Southern Oscillation reasonably well as is evident from Fig. 5.3. The model is seen to reproduce both the phase and the amplitude of the interannual variations in sea level pressure, lower- and upper-level winds, and rainfall. Table 5.1 shows the correlations between certain observed atmospheric variables and the sea surface temperature in the central equatorial Pacific, and between the corresponding variables from the model and the same sea surface temperatures. The model simulates only the low-frequency variability, the smooth line in Fig. 5.3, which has time scales longer than a few months.

A third numerical experiment, identical to the second except for different initial conditions, reproduced the low-frequency fluctuations of the second experiment on time scales longer than a few months, even though there were significant differences at high frequencies. This result is confirmed by cross-spectral analysis, which shows that fluctuations in the selected indices are strongly coherent with the imposed sea surface temperature forcing only at periods longer than approximately 1 year. At this and longer periods, variations in sea surface temperature, low-level winds, and precipitation over the central equatorial Pacific are all practically in phase and are all nearly 180° out of phase with upper-level wind variations.

The degree of confidence that can be attached to the response of any one model to sea surface temperature variations, those in Fig. 5.3, for example, is considerably enhanced by the knowledge that computations with differ-

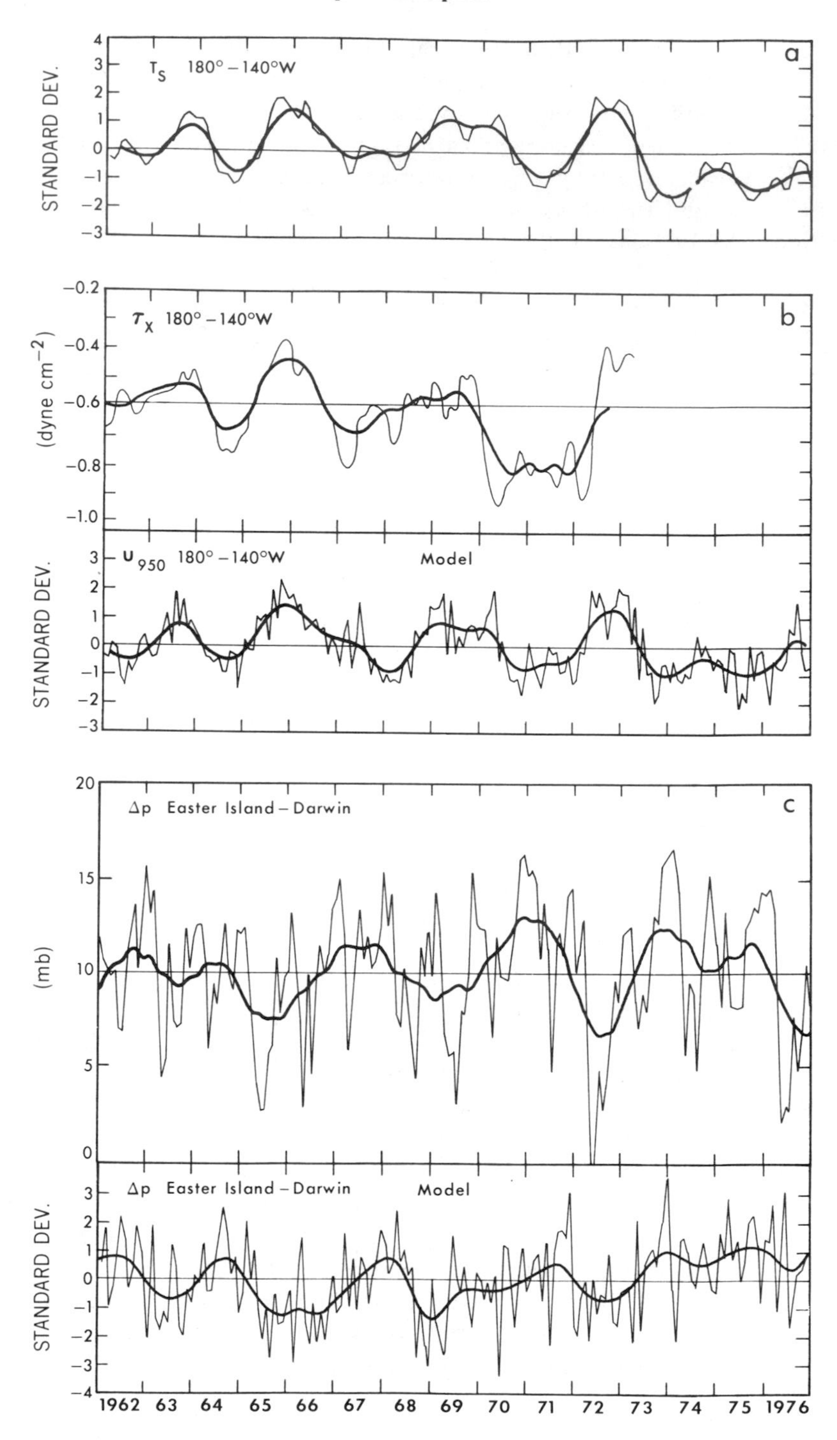
a
T_S 180° – 140°W
STANDARD DEV.
b
τ_X 180° – 140°W
(dyne cm^{-2})
u_{950} 180° – 140°W
Model
STANDARD DEV.
c
Δp Easter Island – Darwin
(mb)
Δp Easter Island – Darwin
Model
STANDARD DEV.
1962 63 64 65 66 67 68 69 70 71 72 73 74 75 1976

Table 5.1

Contemporary Correlation Coefficients between Equatorial Pacific Sea Surface Temperature Changes and Various Observed and Simulated Atmospheric Indices.[a]

	Observed[b]	Model[c]
Sea level pressure difference Tahiti–Darwin	−0.83	−0.84
200-mbar height index	0.80	0.77
700-mbar PNA[d] index	0.46	0.53
Fanning rainfall	0.79	0.71
Christmas Island rainfall	0.64	0.71
Canton rainfall	0.82	0.74

[a] Sea surface temperature changes are from the time series shown in Fig. 5.3. See Table 1.1 for a matrix of correlations between different observed variables.
[b] From Horel and Wallace (1981).
[c] From N. C. Lau (1985).
[d] Pacific North American teleconnection pattern.

ent models, which have different finite difference schemes, and different parameterizations of boundary layer processes and convection, all show a similar response (Rowntree, 1972; Julian and Chervin, 1978; Keshavamurty, 1982; Shukla and Wallace, 1983; Blackmon *et al.*, 1983; Palmer and Mansfield, 1984; Boer, 1985). Generally the models reproduce the eastward shift of the ascending branch of the Walker Circulation in the Pacific, the intensification of the meridional Hadley Circulation in the central Pacific, and the teleconnection pattern over the northern Pacific and North America in response to high sea surface temperatures in the eastern tropical Pacific. There is considerable variability from experiment to experiment but much of it appears to be high-frequency fluctuations that are not attributable to sea surface temperature variations. One variable to which

Figure 5.3. Observed and simulated time series of monthly mean anomalies during the 15-year period January 1962 through December 1976. (a) Observed sea surface temperatures (°C) averaged over the region 5°S–5°N, 180°–140°W. (b) Observed surface windstress (dynes per square centimeter) averaged over the region 4° S–4°N, 180°–140°W, and model-derived zonal winds (meters per second) at 950 mbar averaged over the area 5°S–5°N, 180°–149°W. (c) Observed and model-derived surface pressure differences between Easter Island and Darwin. The sea surface temperature anomalies in (a) and the model results in (b) and (c) are expressed in units of standard deviations, which were computed separately for each month. The smoothed curves for the observed data in (b) and (c) were obtained by applying a 12-month running mean. The smooth curves for (a) and the model data in (b) and (c) were obtained by applying a 15-point Gaussian filter that removes periods less than a few months. [From N. C. Lau (1981).]

meteorologists usually pay scant attention, and which the models simulate poorly, is the surface wind. The realization that the Southern Oscillation involves interactions between the ocean and atmosphere and that the surface wind is the meteorological variable that affects the ocean the most is leading to rapid improvements in the ability of the models to simulate this variable.

Most of the studies with General Circulation Models are for sea surface temperature variations in the tropical Pacific and focus on the interannual fluctuations associated with the east–west movements of the convergence zone over the maritime continent. In the Atlantic, a north–south movement of the ITCZ is the prominent interannual variation. Experiments with General Circulation Models indicate that sea surface temperatures that are warm south of and cold north of the equator, as in Fig. 1.23, cause a southward displacement of the ITCZ and bring heavy rains to northeastern Brazil. (Moura and Shukla, 1981). Rainfall variations in the Sahel are reproduced reasonably well if the global changes in sea surface temperature are specified (Folland *et al.*, 1986). Experiments in which the effects of sea surface temperature anomalies in the Atlantic, Pacific, and Indian Oceans are tested individually indicate that the western Sahel is affected by conditions in both the Atlantic and Pacific whereas the Indian Ocean has a strong influence on Sahel rainfall over Sudan and northern Ethiopia (Palmer, 1986).

If the dynamical response of the ocean to surface winds did not affect sea surface temperatures, then sea surface temperature gradients would be primarily meridional with a maximum near the equator. Hence experiments in which zonally uniform sea surface temperatures are specified demonstrate what the atmospheric circulation would be like in the absence of oceanic currents. It appears to resemble the circulation that is observed during intense El Niño episodes. Both the Walker Circulation along the equator and the high-pressure zone over the usually cold surface waters of the southeastern Pacific are weakened (Chervin and Druyan, 1984; Stone and Chervin, 1984; Simmons and Smith, 1986).

The principal result from these calculations with General Circulation Models is that low-frequency variations in the tropics, at periods on the order of a year and longer, are caused not by instabilities of the atmospheric circulation but primarily by variations in the boundary conditions, those at the ocean and land surfaces. The prime cause of the Southern Oscillation is the sea surface temperature variations in the tropical Pacific Ocean. Other factors, such as sea surface temperatures in higher latitudes and in other oceans, and variations in land processes, have a more modest effect on the Southern Oscillation but can at times be important because their effect can be amplified by unstable interactions between the ocean and atmosphere.

Notes

1. An alternative derivation of Eqs. (5.16a), (5.16b), and (5.17) for the response of the troposphere to heating is based on the indefensible assumption that the atmosphere has a rigid lid at the tropopause. Vertical baroclinic modes, similar to those discussed in Section 4.3, are then possible. The observed motion in the troposphere—convergence near the surface, divergence aloft—is interpreted as having the vertical structure of a first baroclinic mode. Equations (5.16) and (5.17) are viewed as describing the horizontal structure of the flow so that the forcing function is the projection of the heating function onto the first mode and c is the phase speed associated with the first mode. The dissipation parameters A and B now must have the same large value and the strong mixing is effective throughout the troposphere, not only in the boundary layer.

2. The mixing ratio r is the ratio of the mass of water vapor to the mass of dry air,

$$r = q/(1 - q)$$

where q, the specific humidity, is the mass of water vapor per unit mass of moist air.

Chapter 6 Interactions between the Ocean and Atmosphere

6.1 Introduction

In May 1982, modest westerly wind anomalies appeared to the west of the date line in the equatorial Pacific Ocean. During the subsequent months the anomalous atmospheric conditions amplified and expanded eastward until, by the end of 1982, intense westerly winds had penetrated into the central and eastern tropical Pacific. Figure 6.1 depicts these developments and also shows the concurrent eastward expansion of the warm surface waters of the western Pacific. The unusual atmospheric conditions were caused by the increase in the sea surface temperatures of the central and eastern tropical Pacific. The change in sea surface temperatures, in turn, was attributable to the eastward advection of warm water by anomalous oceanic currents driven by the westerly winds. This circular argument suggests that an explanation for what happened in the tropical Pacific Ocean in 1982 and 1983 must involve interactions between the ocean and atmosphere.

The simultaneous development of anomalous oceanic and atmospheric conditions is a feature not only of El Niño of 1982–1983, but, more generally, is a characteristic of the Southern Oscillation. Figure 6.2 illustrates how different oceanographic and meteorological variables in the tropical Pacific fluctuate practically in phase on interannual time scales. The explanation is again a circular one: the atmospheric variations cause and in turn are caused by the interannual sea surface temperature variations in the tropical Pacific. The conclusion once again is that interactions between the ocean and atmosphere are at the heart of the matter.

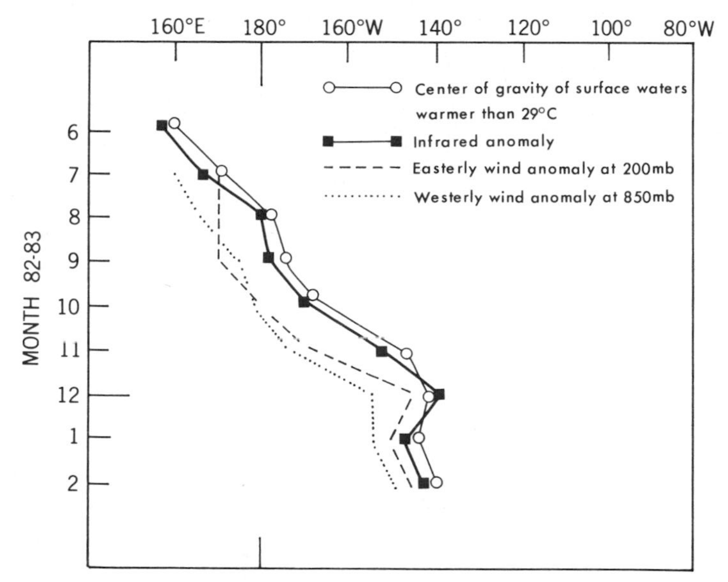

Figure 6.1. The eastward expansion of the centers of gravity of anomalous conditions in the equatorial Pacific (5°N to 5°S) during 1982–1983.

Qualitative arguments indicate that ocean–atmosphere interactions can be unstable and can amplify modest initial perturbations into El Niño or La Niña. For example, a slight relaxation of the westward trade winds, which drive the warm surface waters of the Pacific to the western side of the basin while exposing cold water to the surface in the east, can cause an eastward surge of the warm water. Such a warming of the central equatorial Pacific can induce the atmospheric convergence zone, which is over the warmest surface waters, to move or expand eastward. If this happens then there is a further relaxation of the trade winds, hence a further eastward expansion of warm surface waters, and so on. This is a plausible explanation for the development of El Niño of 1982–1983 and it also applies to other El Niño events. One of the first indications of El Niño of 1972, for example, was the appearance of unusually high sea surface temperatures off the coast of Peru early that year. If such a warm water anomaly causes a local heating of the atmosphere, then the westerly winds that converge onto that heat source drive currents that amplify the anomaly, thus initiating unstable ocean–atmosphere interactions.

The unstable interactions between the ocean and atmosphere, in mathematical terms, support interannual modes of oscillation that gradually amplify. A mode is a periodic oscillation between warm El Niño and cold

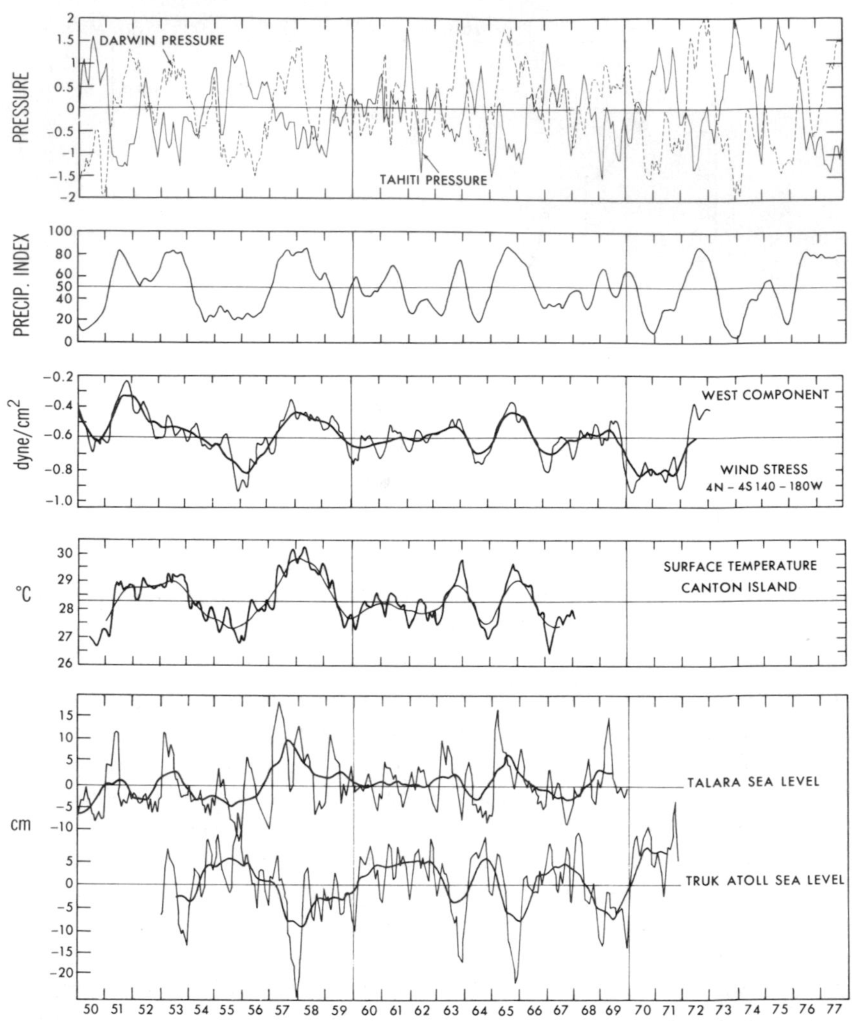

Figure 6.2. Interannual fluctuations in sea level pressure anomalies (3-month running means) for Darwin and Tahiti; in the precipitation index (6-month running means) for Ocean and Naura Islands (5°S, 167°E); in the zonal component of the windstress over the region 4°N to 4°S, 140° to 180°W (from Wyrtki, 1973a); in the sea surface temperature at Canton Island (3°S, 171°W) (from Wyrtki, 1973a); and in sea level at Talara (4°S, 81°W) and Truk Atoll (7°N, 151°E) (from Hickey, 1975). The smooth lines are 12-month running means.

La Niña states and typically has a period on the order of 3 years. Stability analyses with simple coupled ocean–atmosphere models reveal that a variety of modes exists: some have eastward phase propagation similar to that observed in 1982; some have westward phase propagation; and some are stationary. In the case of a mode with phase propagation in an eastward direction, for example, modest El Niño conditions first appear in the west and gradually expand eastward so that westerly winds start to prevail. At a certain stage in this evolution modest easterly wind anomalies appear in the west, as in Fig. 1.21c. These are the seeds that develop into the eastward-expanding La Niña that succeeds El Niño. In the case of stationary modes, oceanic waves that reflect off the western boundary of the ocean basin can play a critical role in the turnabout from El Niño to La Niña. The properties of a mode, its direction of propagation for example, depend on the processes that determine sea surface temperature variations. If changes in the rate of upwelling of cold water dominate sea surface temperature variations, then there is eastward phase propagation. If horizontal advection is dominant, then the direction of propagation depends on the sign of sea surface temperature gradients and can be westward. Sea surface temperature is the most complex of oceanic parameters and the processes that determine it vary with space and time so that each El Niño and La Niña episode tends to develop in a different manner. This means that, at different times, different coupled ocean–atmosphere modes are involved in the Southern Oscillation.

A model that captures an ocean–atmosphere mode succeeds in reproducing a Southern Oscillation with several realistic features—the time scale is correct and the oscillation is between warm El Niño and cold La Niña states—provided the model takes into account dissipative processes or nonlinearities that limit the amplitude. (For example, the rapid rate at which evaporation from the ocean surface increases as sea surface temperature increases prevents the surface waters from becoming much warmer than 30°C.) If the coupling between the ocean and atmosphere is sufficiently strong, then the oscillations attain such a large amplitude that secondary instabilities appear and the fluctuations become chaotic after a while. This parameter range is not believed to be relevant to reality because the Southern Oscillation, although it is irregular, has well-defined temporal and spatial scales. For realistic values of the parameters[1] the strength of the coupling between the ocean and atmosphere is such that the model reproduces interannual oscillations with a modest amplitude. The oscillations are unrealistically regular, however, and hence are perfectly predictable, unless the models take into account "weather," in other words random atmospheric disturbances unrelated to sea surface temperature changes. These disturbances disrupt the regular oscillations and cause them to be realistically irregular. If this were all there is to the story, then it would end on a discouraging note because the unpredictability of weather would imply that the Southern Oscillation is unpredictable. However, the effectiveness with which weather disrupts the regular oscillation varies with time because the degree to which interactions between the ocean and atmosphere are unstable varies with time. For example, there are phases of the seasonal cycle when the coupling between the ocean and atmosphere is sufficiently strong to result in energetic, self-sustaining oscillations that are not readily disturbed by random perturbations. On other occasions the coupling is so weak that only damped oscillations are possible so that perturbations have a large effect on further developments. The reasons for variations in the strength of the coupling are poorly understood, but some of the models can be used to identify those occasions when random perturbations are unlikely to disrupt an ocean–atmosphere mode and the models therefore provide reasonable predictions on those occasions.

In summary, the evidence that the Southern Oscillation involves natural modes of oscillation of the coupled ocean–atmosphere system, is persuasive. The models developed thus far capture only a few, often only one, of the many possible unstable modes so that each model reproduces only a subset of what is possible in reality. The models have several other flaws. Although they are good at reproducing the east–west movements of the convergence zone over the western tropical Pacific, they are poor at simulating the other important feature of the Southern Oscillation, the interannual movements of the ITCZ. This latter feature is the dominant aspect of interannual variability in the tropical Atlantic which, at present, is poorly understood. Perhaps the most serious flaw of the available models is their treatment of the seasonal cycle. If taken into account at all, it is specified. The models calculate only departures from the cycle which modulates the interannual modes. The assumption that the seasonal cycle plays a passive role, that of modulating the natural modes that correspond to the Southern Oscillation, is questionable. In reality, the seasonal cycle is the response of the coupled ocean–atmosphere system to forcing in the form of incoming solar radiation. The processes that determine the response to this forcing have much in common with those that determine the interannual modes of oscillation and include unstable ocean–atmosphere interactions. In the simplest possible model the response to the seasonal forcing should, at the equator, be strictly semiannual. An annual signal is nonetheless dominant over large regions along the equator and in some areas, the eastern equatorial Indian Ocean for example, there is evidence of a biennial oscillation (Meehl, 1987). This suggests that the response of the coupled system to seasonal forcing is nonlinear and is likely to affect interannual time scales. Up to now theoretical studies have focused on the unforced variability of the ocean–atmosphere system. The next step is to investigate the response to periodic, seasonal forcing and to determine to what extent the Southern Oscillation is part of that response.

6.2 Unstable Interactions

The simplest coupled model with which to explore interactions between the ocean and atmosphere has as its meteorological component the one-level model described by Eqs. (5.16) and (5.17)—it provides a reasonably accurate representation of the surface winds in response to a specified heat source—and has as its oceanographic component the shallow-water model of Eqs. (3.85), which give an acceptable description of changes in the depth of the thermocline in response to the windstress. The oceanic model also describes the surface currents but they tend to be inaccurate near the equator. These models of the ocean and atmosphere can be coupled by driving the ocean with the winds from the atmospheric model

$$(\tau^x, \tau^y) = \gamma(U, V) \tag{6.1}$$

and by relating the atmospheric heat source Q to the state of the ocean. The heat source can be assumed to depend on the sea surface temperature anomaly T:

$$Q = \alpha T \tag{6.2}$$

The coefficients α and γ are constants. It is now necessary to relate T to the oceanic variables. The simplest assumptions to explore are

$$T = s\eta \tag{6.3a}$$

$$T_t + u\overline{T}_x = -rT \tag{6.3b}$$

The first exploits the high correlation between the sea surface temperature T and the depth of the thermocline η in certain parts of the tropics. Advection is included implicitly because changes in the depth of the thermocline are attributable to a horizontal redistribution of warm surface waters. In Eq. (6.3b), advection is explicit. (The zonal gradient $\overline{T}_x$ is specified and s and r are constants.)

The equation that describes the conservation of energy in the ocean implies necessary conditions for interactions between the ocean and atmosphere:

$$\langle H(u^2 + v^2) + g\eta^2 \rangle_t = -a\langle H(u^2 + v^2)\rangle + \gamma\langle H(uU + vV)\rangle - b\langle g\eta^2 \rangle \tag{6.4}$$

In this equation, whose derivation is independent of the assumption in (6.3), the brackets $\langle\ \rangle$ denote integration in latitude y from $-\infty$ to $+\infty$ and integration over a zonal wavelength in x. The first two terms on the right-hand side of Eq. (6.4) are negative definite so that disturbances can amplify only when the atmospheric wind (U, V) and oceanic current (u, v) fluctuations are positively correlated (Yamagata, 1985). In other words, a necessary condition for unstable interactions is that convergent winds drive convergent currents.

The simplest case to analyze is a nonrotating system.[2] Assume that all disturbances have the spatial and temporal dependence $\exp i(kx + ny - \sigma t)$ and adopt relation (6.3a) rather than (6.3b). It then follows from the equations for the oceanic and atmospheric models with $f = 0$ that

$$(-i\sigma + a)^2 = c^2(K_c^2 - K^2) \tag{6.5}$$

where

$$K^2 = k^2 + n^2$$

and

$$K_c^2 = (\alpha\gamma s - A^2)/c_a^2$$

Long waves ($K < K_c$) have an imaginary frequency, are unstable, and grow exponentially in the absence of dissipation ($A = a = 0$). Short waves ($K > K_c$) are stable. The reason is the following. An initial perturbation, a depression of the thermocline for example, causes a heating of the atmosphere. The air that converges onto the depression also drives convergent motion in the ocean so that the depression grows. The gravitational restoring force owing to the stratification of the ocean, which tends to eliminate the depression, increases as the horizontal scale of the depression decreases. This is why long waves are unstable but short waves are stable in a nonrotating system.

The necessary condition for instability, that the atmospheric wind and oceanic current fluctuations be positively correlated, rules out instabilities for systems in which the Coriolis parameter has a constant value. When $f =$ constant, convergent winds drive divergent currents:

$$\operatorname{div}(u, v) = -\frac{\alpha s}{A}\operatorname{div}(U, V) \tag{6.6}$$

This means that unstable interactions between the ocean and atmosphere, of the type being discussed here, are unlikely far from the equator, where the value of f is relatively constant over large latitudinal distances. The neighborhood of the equator, however, is another matter because there it is possible for convergent surface winds to drive convergent surface currents.

A stability analysis that uses analytical methods is possible for an equatorial plane provided $c = c_a$ (provided the ocean and atmosphere have the same radius of deformation). This condition is, in general, not satisfied so that it is necessary to resort to numerical methods (Philander *et al.*, 1984; Yamagata, 1985; Hirst, 1986). Results from such stability analyses indicate that the properties of the instabilities depend critically on the processes that determine sea surface temperature variations. Consider a situation in which upwelling has a dominant influence on sea surface temperature so that Eq. (6.3a) is appropriate. For perturbations of the form $\exp i(kx - \sigma t)$, the growth rate (the imaginary part of σ) and the frequency (the real part of σ) depend on zonal wave number k as shown in Fig. 6.3. In the absence of coupling between the ocean and atmosphere ($\sigma = \gamma = 0$), the modes are the equatorially trapped waves of Section 3.4 except that they are damped. As the coupling coefficients increase—only the value of the product $\alpha\gamma$ matters—analytic continuations of the Kelvin, Rossby-gravity, and the gravest inertia-gravity modes ($n = 0$) become less damped and eventually become unstable. The Rossby waves, however, become more damped. Figure 6.3 shows that short waves are stable, that long waves are unstable, and that the most unstable wave has a wavelength near 16,000 km and an e-folding time of the order of two months.[3] The unstable waves have eastward phase propagation, essentially at the speed of the oceanic Kelvin wave, and are slightly dispersive.

The results in Fig. 6.3, which describe the growth rate of disturbances of the form $\exp(ikx)$, also give insight into the spatial amplification of disturbances. In Fig. 6.4 the initial perturbation is a Gaussian depression of the thermocline with a scale of 500 km. The resultant heating of the atmosphere, which is described by Eq. (6.3a), causes winds to converge onto the thermocline depression. To the west, the westerly winds drive convergent oceanic currents so that the thermocline deepens. To the east, matters are more complex because there are two competing mechanisms that affect the depth of the thermocline: one is equatorial upwelling induced by the local easterlies that converge onto the heat source; and the other is an eastward-traveling Kelvin wave, excited by the westerly winds to the west, which deepens the thermocline. Whether the warming caused by the Kelvin wave dominates over the local upwelling depends not only on the relative intensities of the westerlies and easterlies but also on their zonal extent. The zonal integral of the wind is the critical factor as explained in Section 3.7.

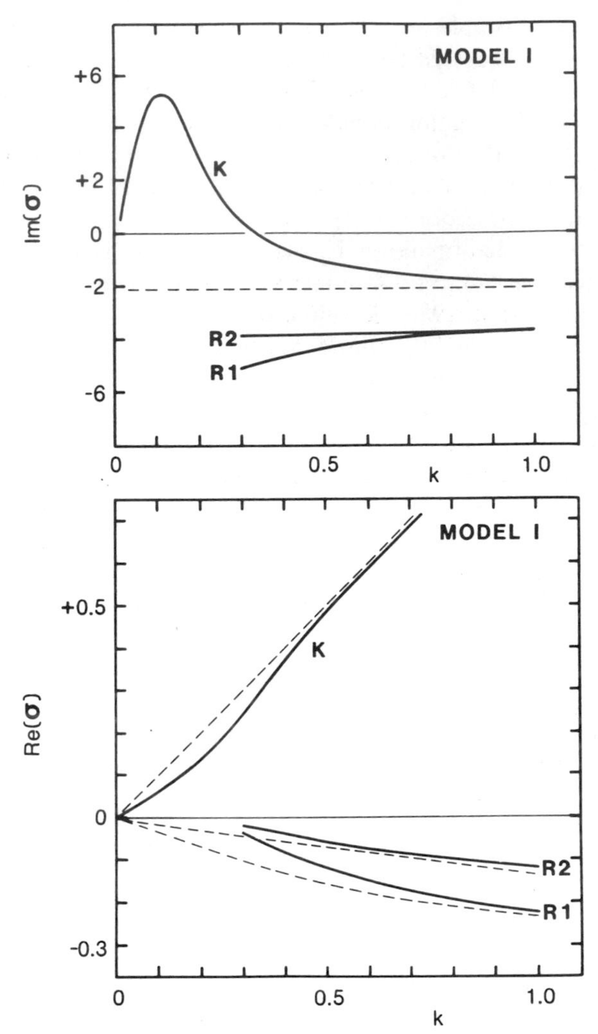

Figure 6.3. The growth rate (Im(σ)) and frequency (Re(σ)) of unstable modes, as a function of zonal wave number k, from a model in which atmospheric heating is proportional to the depth of the thermocline in accordance with Eqs. (6.2) and (6.3a). The Kelvin mode (K) is unstable but the two gravest Rossby modes (R1 and R2) are stable. The dashed lines correspond to the case in which there is no interaction between the ocean and atmosphere. The wave number $k = 0.1$ corresponds to a wavelength of 16,000 km. Im(σ) = 1 = (208 days)$^{-1}$. Note 2 gives the other parameter values. [From Hirst (1986).]

In a case where the winds are symmetrical about the heat source, the initial disturbance is stationary while the thermocline deepens steadily to the west and shoals steadily to the east (Philander *et al.*, 1984). On an equatorial β-plane the westerly winds tend to be more intense and have a larger zonal extent than the easterly winds. [This is a property of the response described by Eqs. (5.16).] It follows that the deepening of the thermocline expands to the east of the initial disturbance so that the westerly winds and the convective zone moves eastward as shown in Fig. 6.4. The speed of the eastward expansion is in accord with the stability analysis that leads to Fig. 6.3 and involves a mode that both amplifies and oscillates. For the example shown in Fig. 6.4, the growth rate is so large that huge amplitudes are attained before the next phase of the oscillation becomes evident.

The results in Fig. 6.3 are for the case in which the sea surface temperature depends on the depth of the thermocline according to Eq. (6.3a). If, instead, temperature depends on advection as in Eq. (6.3b), then the instability characteristics are those shown in Fig. 6.5. The Kelvin wave is now damped but the gravest Rossby mode ($n = 1$) is unstable. Growth rates increase with wavelength and phase speed is to the west. (There are also two modes with a structure similar to that of the $n = 2$ Rossby mode but they are essentially stable.)

The reason why unstable disturbances propagate eastward when the heating depends on the depth of the thermocline but westward when the heating depends on advection is given in Fig. 6.6. In model I, sea surface temperature is proportional to thermocline depth so that the heat source Q is positive over a depression of the thermocline. This depression tends to disperse into eastward-traveling Kelvin waves and westward-traveling Rossby waves as shown in Fig. 3.10. The Kelvin waves amplify because their structure is such that the currents are eastward over the depression of the thermocline. The winds that converge onto the heat source over the deepened thermocline accelerate the currents as shown schematically in Fig. 6.6a. The Rossby waves are damped because their structure is such that the westward currents over the heat source Q are decelerated by the winds that converge onto Q as shown in Fig. 6.6b.

To amplify the Rossby wave it is necessary to displace the heat source, as in Fig. 6.6d, so that easterly winds and easterly oceanic currents coincide. This is accomplished by assuming that heating depends on advection as in Eq. (6.3b). Such an assumption causes the Kelvin wave to be damped as shown in Fig. 6.6c provided $\overline{T}_x < 0$.

If the sea surface temperature parameterization depends on a combination of the assumptions in Eqs. (6.3a) and (6.3b), then long waves are still unstable but their properties are very sensitive to the values assigned to the

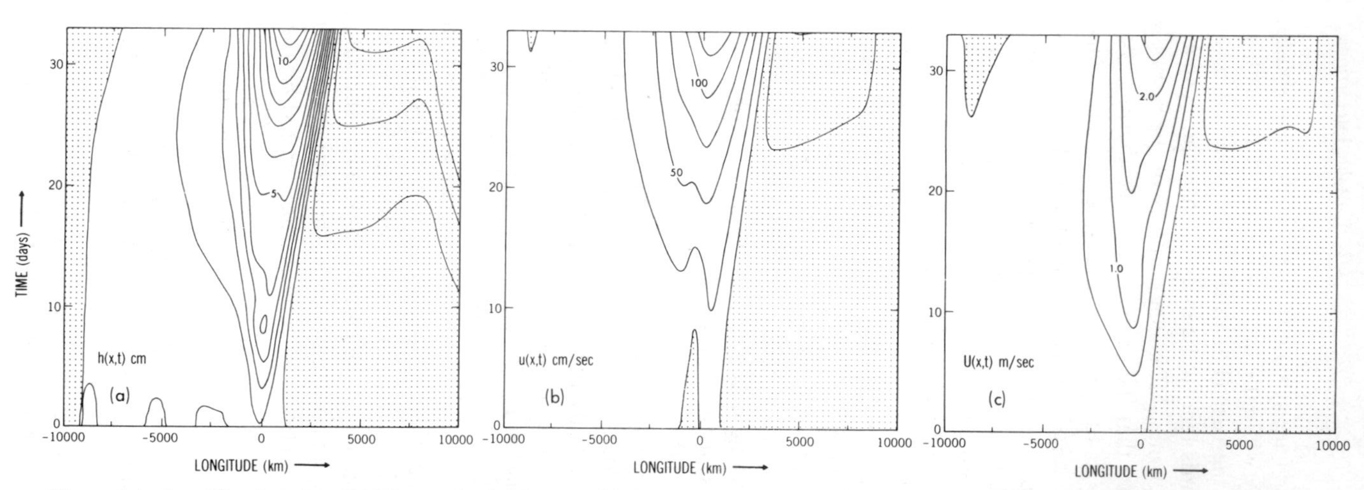

Figure 6.4. Amplification of an initial perturbation by unstable ocean–atmosphere interactions as reflected in (a) the depth of the thermocline, (b) the oceanic zonal current, and (c) low-level zonal winds in the atmosphere as a function of time along the equator. The heating of the atmosphere depends on the depth of the thermocline in accordance with Eqs. (6.2) and (6.3a). The anomalies expand in an eastward direction in agreement with the results of Fig. 6.3. (The parameter values for this case give a growth rate twice as large as that given in Fig. 6.3.) [From Philander *et al*. (1984).]

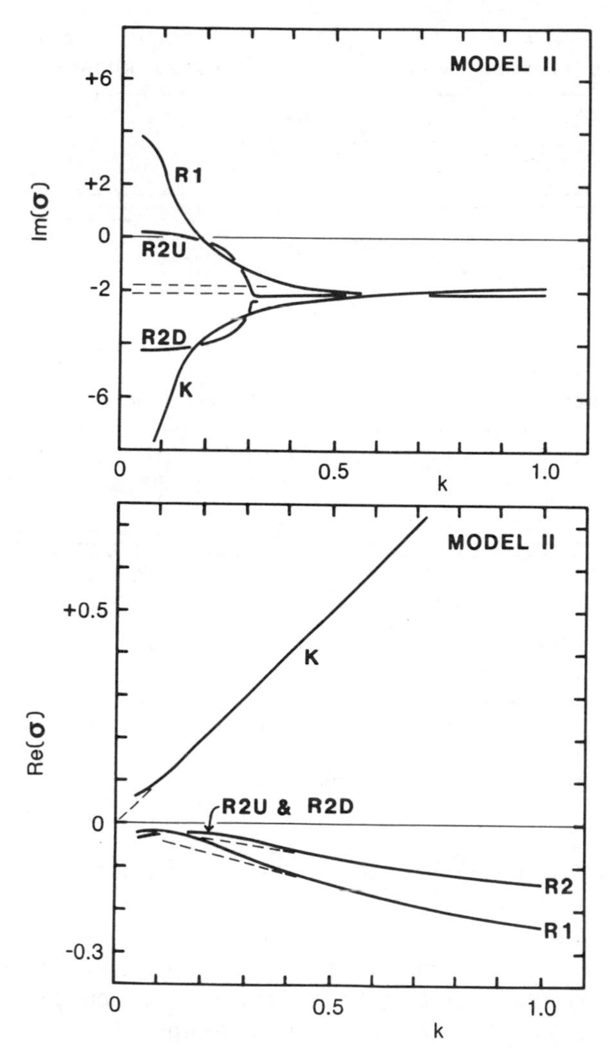

Figure 6.5. As for Fig. 6.3 but the atmospheric heating and sea surface temperature depend on advection according to Eq. (6.3b). The Kelvin mode (K) is stable but the gravest Rossby mode (R1) is unstable. The Rossby mode R2 splits into two modes, one of which (R2U) is unstable and one of which (R2D) is damped. [From Hirst (1986).]

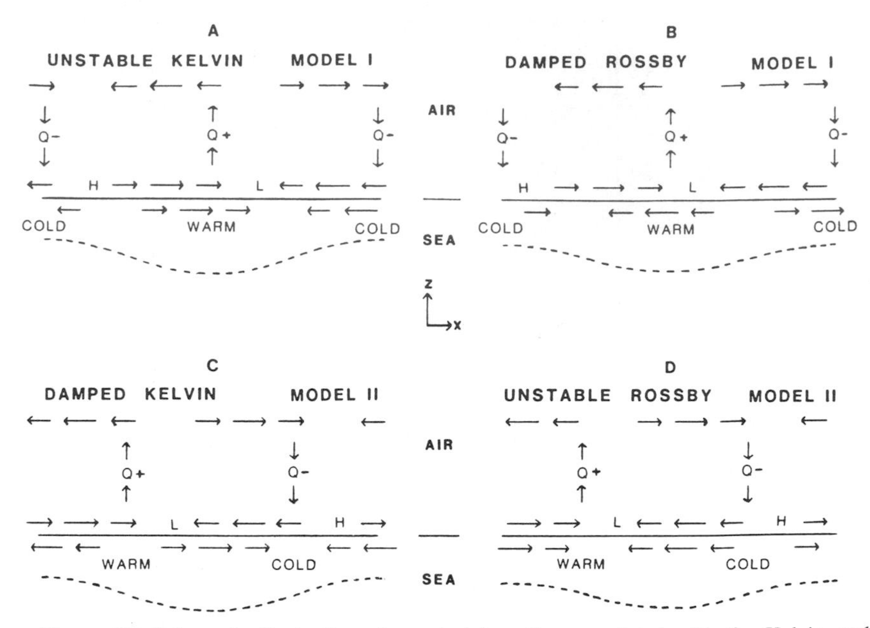

Figure 6.6. Schematic illustration of equatorial motion associated with the Kelvin and gravest Rossby modes of Fig. 6.3 (in A and B) and Fig. 6.5 (in C and D). The solid horizontal line represents the ocean surface and the dashed line the thermocline. Arrows indicate oceanic currents and atmospheric winds. Maximum perturbations of sea surface temperature are indicated by WARM and COLD, of atmospheric heating by $Q+$ and $Q-$, and of surface atmospheric pressure by H and L. [From Hirst (1986).]

coefficients r, s, and $\bar{T}_x$. It becomes possible for the unstable waves to propagate extremely slowly and even to be stationary.

If the ocean basin has a finite zonal extent L (because of north–south boundaries a distance L apart), then the unstable modes discussed above are practically unaffected except that the zonal wavelengths are quantized (Hirst, 1988). As before, only long waves are unstable so that instabilities are possible in a basin the size of the Pacific but not in basins the size of the Atlantic and Indian Oceans.

The unstable modes discussed thus far are practically unaffected by the presence or absence of north–south coasts. There also exist modes that depend critically on the presence of such coasts. They appear in models in which the sea surface temperature equation is far more complicated than Eq. (6.3) and, in the simplest case, has the form

$$T_t + u\bar{T}_x + v\bar{T}_y + \bar{u}T_x + \bar{v}T_y + w\bar{T}_z + M(\bar{w})\frac{T - s\eta}{h} + rT = 0 \quad (6.7)$$

Quantities with a bar describe a specified mean state so that both the advection of mean temperature gradients by anomalous currents and the advection of anomalous gradients by mean currents are taken into account. The function M is a Heaviside function ($M = \overline{w}$ if $\overline{w} > 0$ and $M = 0$ if $\overline{w} < 0$) and the term in which it appears represents the upwelling of a subsurface temperature that is proportional to the thermocline depth η. If the mean quantities $\bar{u}$, $\bar{v}$, and $\overline{w}$ are constants then the unstable modes are similar to the propagating unstable modes such as those in Figs. 6.3, 6.4, and 6.5. If, however, the mean quantities vary spatially in a reasonably realistic manner, then a new nonpropagating unstable mode appears (Battisti and Hirst, 1989). Its sea surface temperature has a large amplitude primarily in the eastern side of the basin and unstable ocean–atmosphere interactions are confined to that region. During the warm El Niño phase of this mode, westerly wind anomalies over the central and eastern part of the basin excite westward-traveling Rossby waves similar to those in Fig. 3.21. These waves propagate into a region where the thermocline depth is so large that they have little influence on the sea surface temperature and hence have little effect on the atmosphere. The Rossby waves excited by westerly winds elevate the thermocline. Upon reaching the western boundary of the ocean basin they reflect as eastward-traveling equatorial Kelvin waves, which also elevate the thermocline. This elevation, when it reaches the central Pacific, reduces the deepening of the thermocline caused by the local westerly winds. There is consequently a reduction in the intensity of the interactions between the ocean and atmosphere. Persistence of the westerly wind anomalies in the central Pacific implies continued excitation of Rossby and reflected Kelvin waves that further erode the development of El Niño conditions in the central Pacific until the westerly wind anomalies start to decay and La Niña starts to develop. As before, the easterly wind anomalies in the central Pacific now sow the seeds for the destruction of La Niña in the form of westward-traveling Rossby waves that deepen the thermocline. These waves reflect off the western coast and in due course El Niño replaces La Niña.[4]

Figure 6.7 depicts the structure of this "delayed oscillator" mode[5] in the coupled ocean–atmosphere model in which it was first detected, that of Schopf and Suarez (1988). The first panel (a) shows anomalies in the depth of the thermocline for the band of latitudes 5°N to 7°N, where the Rossby wave signature is more easily discernible than it is along the equator. Substantial depth perturbations are seen to originate in the central Pacific and to propagate westward. The next panel (b), which has the zonal direction reversed so that east is to the right, shows how reflection causes the westward propagation in panel (a) to continue as eastward propagation along the equator. The amplitude of the thermocline depth anomaly in-

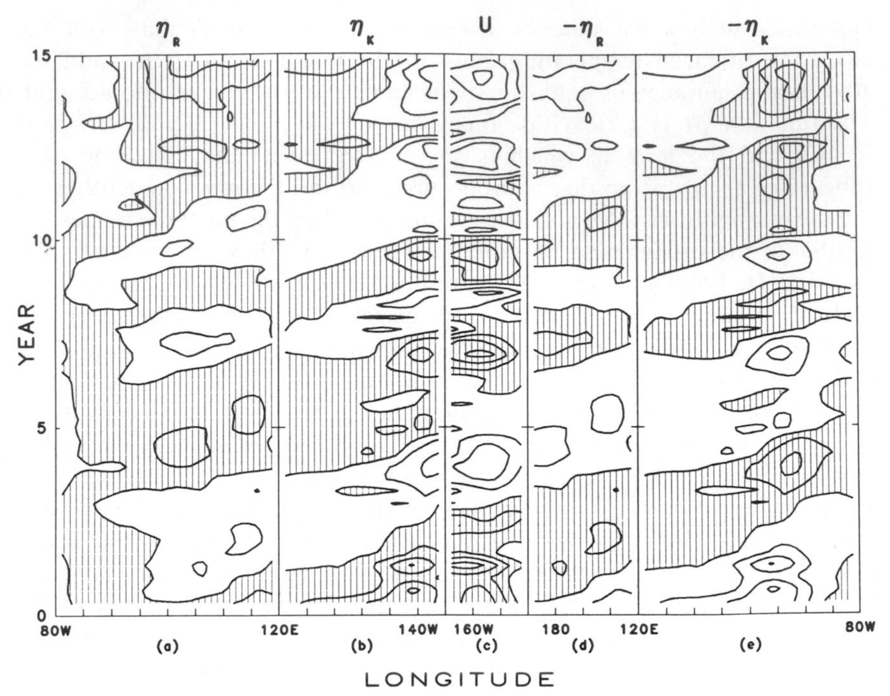

Figure 6.7. The structure of the "delayed oscillator" mode in a coupled ocean–atmosphere model. This figure is discussed in detail in the text. [From Schopf and Suarez (1988).]

creases in the central Pacific, where it is highly correlated with the zonal wind anomalies U in panel (c). Westerly wind anomalies coincide with a deepening of the thermocline and an increase in sea surface temperature and vice versa. By the arguments presented earlier, unstable interactions between the ocean and atmosphere cause the amplification of anomalies. In panel (d) the shading is reversed and west is to the right in order to underline the strong relation between positive wind anomalies and negative thermocline depth anomalies near 6°N, where the Rossby waves are prominent. The final panel (e) is a repetition of (b) but with the shading reversed to show that the Rossby waves lead to equatorial Kelvin modes that amplify primarily in the central Pacific.

The critical factor for sustained oscillations in the "delayed oscillator" models (see Note 5) is a region of unstable ocean–atmosphere interactions at a distance L from a reflecting western wall. If L is small then the reflected waves that reverse the unstable interactions in the central part of the basin get there so rapidly that initial perturbations barely amplify. A steady equilibrium state is quickly attained in such cases. It appears that the mechanism described here cannot sustain an oscillation in the tropical Atlantic Ocean because the value of L is too small (Battisti, 1988). In the Pacific, oscillations are possible and one of the factors that determine the period is the time it takes a Rossby wave to propagate from the region of ocean–atmosphere interaction to the western boundary ($3L/c$) plus the time it takes an equatorial Kelvin wave to propagate back (L/c). If this were the only factor, then the period of the oscillation would be $8L/c$, where c is the speed of long gravity waves in the ocean. For the Pacific this time is approximately 14 months if $c = 2.9$ m/sec. The models oscillate at a far longer period because the time scale associated with unstable interactions is another important factor. The period of the oscillation can readily be changed by changing the values of some of the parameters in Eq. (6.7).

In summary, the coupled ocean–atmosphere has several unstable modes whose properties depend critically on the processes that control sea surface temperature variations. Some of the modes are propagating and are essentially independent of coasts; others depend strongly on reflections off the western coast of the ocean basin. Further studies are likely to uncover additional modes, especially if models were to include ocean–atmosphere interactions over the western tropical Pacific, where small changes in sea surface temperature can have a large effect on the location of the atmospheric convergence zones.

A specific unstable mode can be isolated in an idealized model, but in reality measurements are more likely to indicate the presence, not of a specific mode, but of certain types of unstable modes. For example, if, towards the end of one phase of the Southern Oscillation, the seeds of the next phase appear in a small region and start to expand and amplify, then that would be indicative of a propagating instability. Such La Niña seeds sometimes appear during the mature phase of El Niño, in early 1973 and 1988, for example, in the form of easterly wind anomalies over the western equatorial Pacific. They appear while high sea surface temperatures and westerly wind anomalies prevail over the eastern tropical Pacific. These easterly wind anomalies contribute to the demise of El Niño and the initiation of La Niña. The degree to which factors other than the sea surface temperature pattern in the tropical Pacific, conditions over the continents and over the Indian Ocean, for example, contribute to the appearance of easterly wind anomalies is unclear. Indicators of the presence of the "delayed oscillator" mode could be in the form of westward-traveling oceanic Rossby waves. In Fig. 6.7 these waves are near 7° latitude but the latitudinal structure of the winds in that model is not particularly realistic and it is possible that, in reality, the waves are excited farther poleward. White *et al.* (1985) assert that depressions of the thermocline that are observed to drift westward in the North Equatorial Current to the north of

10°N correspond to these Rossby waves. Because of the paucity of the data the evidence is unconvincing. There is a decided possibility that, instead of moving equatorward when they reach the far west, depressions of the thermocline in the North Equatorial Current will get caught up in the northward-flowing Kuroshio Current and hence in the subtropical gyre of the northern Pacific Ocean.

6.3 The Irregularity of the Southern Oscillation

An unstable mode oscillates in a periodic manner while it amplifies. To explain the irregularity of the Southern Oscillation it is necessary either to invoke several modes that interact with each other or to introduce stochastic forcing. The simple coupled ocean–atmosphere models that have been developed to study this matter are similar to the models described in Section 6.2 and have a large number of adjustable parameters whose values determine how many modes come into play. The results described by Schopf and Suarez (1988), Cane and Zebiak (1985) and Battisti (1988) seem to involve one (or at most very few) modes—they are all of the delayed oscillator type—so that random noise is necessary to make the oscillations irregular. Neelin (1989b), whose model captures one of the propagating unstable modes of the type shown in Figs. 6.3, 6.4, and 6.5, explored how irregularities can arise when the coupled system becomes more and more unstable. He changed the stability of his model by, in effect, changing the value of the parameter γ in Eq. (6.1). The perturbation that initiates unstable ocean–atmosphere interactions in the model is a westerly wind-stress anomaly of magnitude 0.3 dyne/cm that is maintained for a month over the region 10°S to 10°N, and west of 170°W. If the ocean is uncoupled from the atmosphere ($\gamma = 0$), then the winds generate an eastward-traveling pulse of warm surface water along the equator, but after approximately 3 months sea surface temperatures have returned to their initial equilibrium values. If the changing sea surface temperatures are allowed to affect the atmosphere, then the coupled system behaves very differently from the uncoupled ocean. The initial warming now persists for several months because of anomalous westerly winds in response to the anomalous sea surface temperatures. The interactions are such that this warm phase is followed by a cold phase, whereafter a less intense warm phase occurs as shown in Fig. 6.8a where the oscillation is seen to attenuate rapidly. This damped unstable mode appears for parameter values that correspond to weak ocean–atmosphere interactions.[6] For stronger interactions the oscillation becomes very regular and has a period close to 3 years as shown in Fig. 6.8b. Conditions at the peak of the warm phase closely resemble El Niño

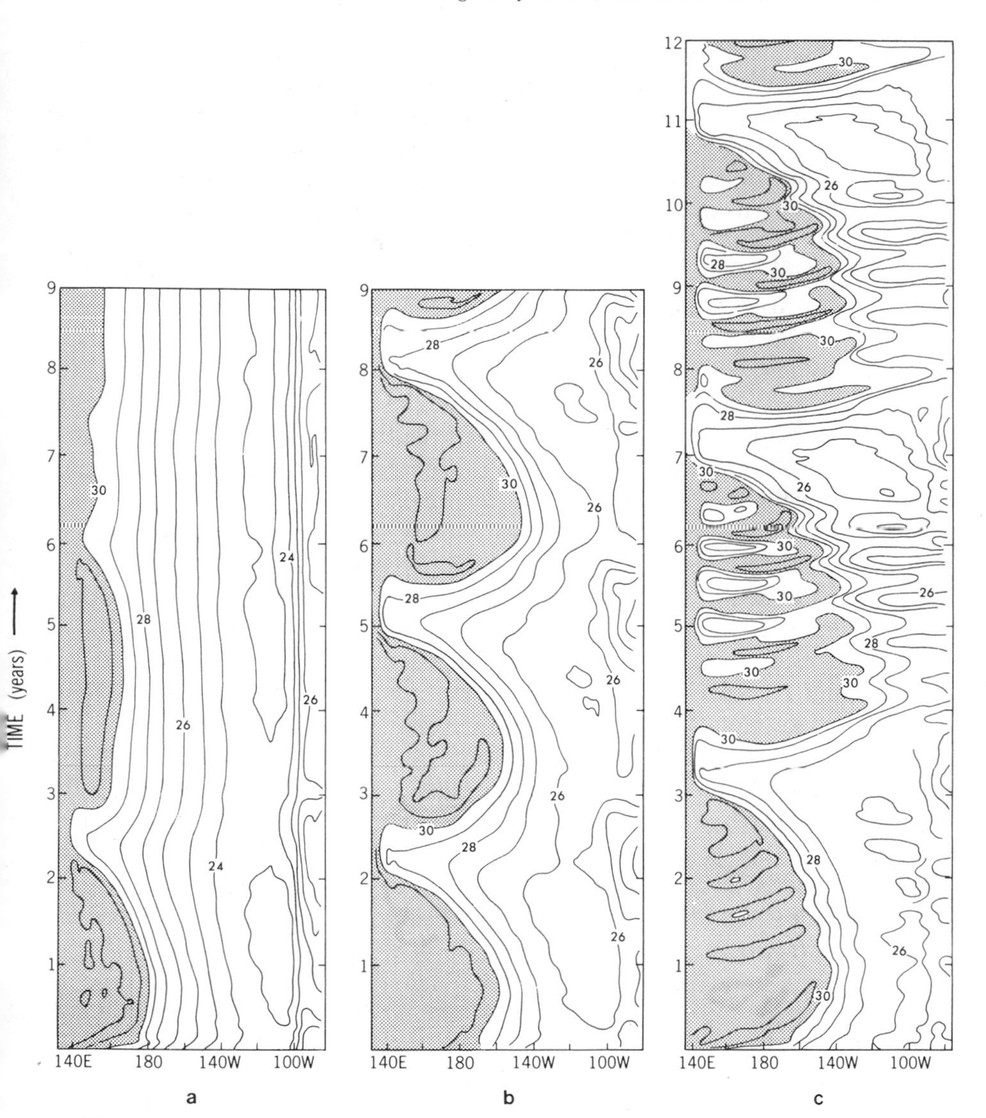

Figure 6.8. Evolution of the sea surface temperature (in °C) along the equator in the coupled model of Neelin (1989b). The strength of the ocean–atmosphere coupling increases from (a) to (b) to (c). Note 5 describes the differences between the three cases in more detail. The evolution over a 9-year period is shown in (a) and (b), and that over a 12-year period in (c). Regions warmer than 30°C are shaded.

conditions, and those at the peak of the cold phase resemble La Niña. The evolution of El Niño (and La Niña) in this model differs from that in other models because this model captures an unstable mode different from those in other models. The periodic oscillations associated with this mode become more complex as the strength of the ocean–atmosphere coupling (see note 6) increases further. The first stage is the appearance of two dominant periods. In Fig. 6.8c the one period is slightly less than 4 years and corresponds to an oscillation very similar to that in Fig. 6.8b. The other period is approximately 6 months and corresponds to an oscillation that appears during the warm El Niño phase of the long-period fluctuation. Further increases in the strength of the ocean–atmosphere coupling will presumably result in chaotic oscillations with a broad frequency spectrum. There is no evidence that the observed Southern Oscillation, which has a well-defined time scale, corresponds to the latter parameter range. The cases shown in Fig. 6.8 seem to be more relevant to reality.

The simulated Southern Oscillation of Figs. 6.7 and 6.8 are regular, not irregular as in reality. One way in which to make these oscillations irregular is by introducing random atmospheric disturbances unrelated to sea surface temperatures. In a marginally stable system, that of Fig. 6.8a for example, random perturbation can readily initiate the development of El Niño at one time and La Niña at another time. For a more unstable system, that of Fig. 6.8b for example, the regular oscillation has such a large amplitude that it becomes more difficult for noise to make it irregular. In other words, the more unstable the system the more regular its oscillations and the more predictable it becomes (within certain limits of course). Battisti and Hirst (1989) use this result to explain why, in simple models, noise is more effective at making the interannual oscillations irregular when the seasonal cycle is introduced. (The cycle is specified so that the equations that are solved describe departures from the seasonal cycle.) The stability analyses of Battisti and Hirst (1989) indicate that if March conditions were to persist, then the coupled system is marginally stable so that noise will readily lead to irregular fluctuations. If September conditions should persist, then the coupled system is sufficiently unstable for noise to be ineffective at introducing irregularities. It is not at all clear that March (or September) conditions persist sufficiently long for these arguments to be valid. Studies of the stability properties of time-dependent basic states are necessary.

The Southern Oscillation is closely related to the seasonal cycle. In the eastern tropical Pacific most El Niño episodes tend to start not at arbitrary times of the year but during the warm phase of the seasonal cycle (during the early calendar months of the year) as shown in Figs. 1.18 and 2.12. A plausible reason is the following. The initiation of unstable ocean–atmosphere interactions require that unusually warm surface waters cause a local heating of the atmosphere. Favorable large-scale meteorological conditions are necessary for this to happen. An increase in sea surface temperatures increases the latent heat released to the atmosphere locally, but condensation of that water vapor and the associated heating of the atmosphere are nonlocal if there is large-scale subsidence over the unusually warm water. Low-level atmospheric convergence and rising motion over the sea surface temperature anomaly are necessary to initiate unstable interactions between the ocean and atmosphere. In the eastern tropical Pacific this condition is satisfied during the early months of the year when the ITCZ, a band of rising air, is near the equator. During the rest of the year the eastern equatorial Pacific is a region of subsidence so that anomalously warm surface waters are unlikely to lead to El Niño. The seasonal movements of the atmospheric convergence zones effectively modulate the unstable interactions between the ocean and atmosphere. This argument suggests that the coupled system is more unstable in the northern spring than in autumn. Battisti and Hirst (1989) propose the opposite. This underlines the need for further studies of the stability properties of different states.

It is possible that the seasonal cycle does far more than merely modulate interannual variability; it may actively contribute to that variability. This possibility cannot be explored with the coupled models developed thus far because in those models the seasonal cycle and the mean conditions are simply specified. Only departures from the specified states are calculated. This approach fails to take into account that the interannual fluctuations, the seasonal cycle, and the mean conditions are all influenced by similar interactions between the ocean and atmosphere. Consider, for example, the pattern of low and high pressure zones that characterizes the mean state of the lower atmosphere in Fig. 1.7. This pattern is closely related to, and is presumably influenced by, the sea surface temperature pattern: high-pressure zones are over cold surface waters in the northeastern and southeastern Pacific and Atlantic Oceans; low-pressure zones are over the warm surface waters. The pressure gradients, through the associated low-level winds, in turn influence the sea surface temperatures. The winds around the high-pressure zones are parallel to the western coasts of Africa and the Americas and induce upwelling and divergent currents that give rise to low sea surface temperatures. It seems plausible that ocean–atmosphere interactions determine the surface pressure and sea surface temperature patterns. Similar interactions also affect the salient feature of the seasonal cycle, the north–south movements of the ITCZ. The ITCZ is close to the equator when the sea surface temperature there is at a maximum and it moves northward when the sea surface temperatures there start to fall. But from an oceanographic point of view the northward movement of the ITCZ and the associated intensification of the southeast trades near the equator cause equatorial upwelling and low sea surface temperatures. This argument suggests that not only factors such as the seasonal heating of the continents but also interactions between the ocean and atmosphere affect the movements of the ITCZ and the other atmospheric convergence zones. Small year-to-year perturbations in the heating of the continents, for example, can be amplified by unstable interactions between the ocean and atmosphere, thus causing interannual variations with a significant amplitude. Although models without a seasonal cycle are capable of interannual fluctuations, this does not exclude the possibility that part of the observed interannual variability depends on the presence of a seasonal cycle. Coupled models capable of reproducing a realistic mean state, and the seasonal cycle, are needed to explore this idea.

Perturbations that amplify because of an instability of the coupled ocean–atmosphere system modify the basic state, presumably in such a manner as to stabilize the coupled system. An important change that occurs during El Niño is a loss of heat by the equatorial Pacific Ocean (Wyrtki, 1985). As explained in Chapter 2, this happens because the northward oceanic heat transport increases and because the flux of heat into the ocean across its surface decreases as a result of higher evaporation rates. (Increased evaporation is the probable cause of the warming of the tropical troposphere during El Niño. The atmosphere loses this heat to space.) Wyrtki (1985) and Zebiak and Cane (1987) have suggested that the loss of heat during El Niño contributes to its termination and that El Niño cannot recur until the equatorial Pacific has recovered the lost heat. Although studies of the oceanic heat budget during La Niña are yet to be made, the poleward oceanic heat transport presumably decreases during La Niña while the flux of heat into the ocean across its surface increases. The heat that is stored in the tropical Pacific apparently increases during La Niña and decreases during El Niño. The relation between the heat budget of the tropical Pacific and the Southern Oscillation requires further study.

6.4 Statistical Predictions

The Southern Oscillation has a low frequency. If the superimposed high-frequency fluctuations can be eliminated, by filtering the data, then it will be possible to make predictions by extrapolating the low-frequency trend. The potential success of such a scheme can be judged by calculating the autocorrelations of various Southern Oscillation indices (sea level pressure at Darwin, sea surface temperatures in the central equatorial Pacific, etc.). For seasonal mean values of the indices, the largest correlations between adjacent seasons are found between the northern summer and autumn, and between the autumn and winter. The smallest correlations between adjacent seasons are found between the northern winter and spring, and between spring and summer (Wright, 1984a). This result was first discovered by Walker, who noted that a Southern Oscillation index for the summer quarter June, July, and August can have a correlation coefficient of 0.8 with the same index for the subsequent winter quarter, though only -0.2 with the previous winter. This means that sea surface temperature in the central equatorial Pacific during the northern summer and rainfall variations over India during those months can be used as predictors for subsequent developments.

Predictions of conditions during the northern summer, on the basis of measurements during the previous months, have relatively little skill. Pan and Parthasarathy (1981) find a low correlation between the Southern Oscillation pressure variations during the northern spring and subsequent Indian monsoon rainfall, but Shukla and Paolino (1983) find that the tendency of the pressure changes at Darwin is sometimes an indicator of Indian rainfall variations. If, during March, April, and May, the seasonal pressure anomaly at Darwin is above normal and is also increasing, then heavy rains over India in June, July, and August are unlikely. If the pressure is below normal and falling, then droughts are unlikely.

Exceptionally heavy rainfall in the central equatorial Pacific during the northern summer is often preceded by unusually warm surface waters off the coast of Peru in the early calendar months of the year (Barnett, 1981a), but this predictor failed in 1982.

Barnett (1984) and Graham *et al.* (1987) have developed statistical methods that identify the changing patterns in the surface winds that precede the onset of El Niño. The method enabled Barnett *et al.* (1988) to predict El Niño of 1986 at lead times of 3 to 9 months. The crucial feature of the wind field is the appearance of westerly anomalies over the western tropical Pacific during the early months of the year in which El Niño occurs. These anomalies persist for several months, which implies that westerly wind bursts that last a few days are of importance to El Niño only if they contribute significantly when the winds are time-averaged over a period of several months.

Long-term predictions, at lead times of more than a year, benefit from the quasi-periodicity of the Southern Oscillation. Spectra of indices of the oscillation have peaks at periods between 3 and 4 years. This means that if El Niño occurs in a certain year, then it is highly likely that it will recur 3 to 4 years later. If there is no event within that period, then the probability continues to increase with increasing time. By early 1982, 6 years had elapsed since the last event, in 1976, so that the likelihood of El Niño developing that year was extremely high.

6.5 Dynamical Predictions

The ocean, at any given time, is usually not in equilibrium with the winds at that time because it is still adjusting to earlier changes in the winds. If the winds were to remain steady, then the ocean will continue to adjust until it ultimately comes into equilibrium with the winds. This means that the ocean at present is predisposed to move in a certain manner. If this motion includes an eastward surge of warm surface waters along the equator then the ocean favors the development of El Niño. The nature of the oceanic tendencies can be determined by forcing a model of the ocean with the observed winds up to the present, whereafter the winds are held constant. The success of this method depends critically on the application of a low-pass filter to the wind data. For example, a prediction that starts at the time of a short burst of westerly winds will be unreliable unless the burst is filtered out. Inoue and O'Brien (1984) developed a suitable filter and showed, with a shallow-water model of the ocean, that in April 1982 the tropical Pacific favored the development of El Niño. The important factor during the preceding months was the westerly wind anomalies over the western equatorial Pacific. The assumption that the winds persist from the time of prediction onward severely limits the accuracy of this dynamical method of prediction. To improve prediction, it is necessary to take the interactions between the ocean and atmosphere into account.

Cane *et al.* (1986) have explored the predictability of El Niño by means of a simple coupled ocean–atmosphere model, one that captures an unstable mode very similar to that depicted in Fig. 6.7. Their calculations proceed as follows. They force the oceanic component of the model with the observed winds up to a certain time, whereafter the coupled system is used for predictions. Figure 6.9 shows different forecasts initiated at intervals of 3 months. Although some forecasts diverge significantly after a few months, the "consensus" (averaged) forecasts for 1972, 1976, 1979, and 1982 are remarkably good. The results suggest that it may be possible to anticipate some El Niño events many months in advance. The predictability of the model is at a minimum during the northern spring. Different forecasts that are started during March and April can diverge rapidly. Predictability is at a maximum during the northern summer. These results are consistent with the statistical results described in Section 6.4.

The rapid divergence of forecasts that are started in March and April could indicate that the coupled ocean–atmosphere system is unstable at that time so that perturbations, including errors of the model, amplify rapidly. This argument would be correct if the basic state of the coupled system were time-dependent. However, in the case of a time-dependent basic state, one that corresponds to the seasonal cycle for example, matters

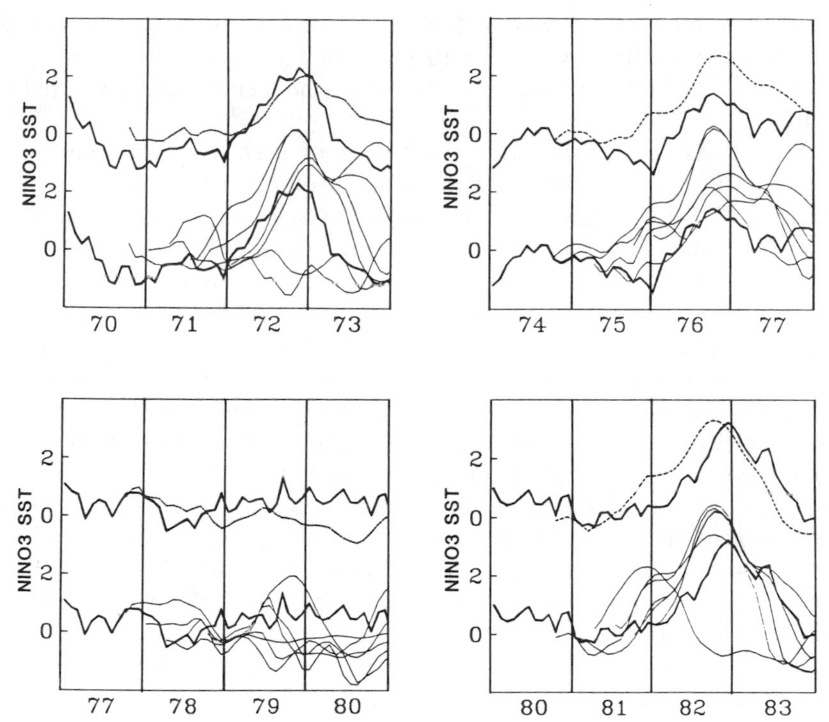

Figure 6.9. Forecasts, with the coupled ocean–atmosphere model of Cane *et al.* (1986), of sea surface temperature anomalies (°C) averaged over the eastern equatorial Pacific region (90°W to 150°W; 5°S to 5°N) for four periods centered on 1972, 1976, 1979, and 1982. The heavy curves are the observed values. The light curves in each panel are from six forecasts initiated at 3-month intervals from the October two years ahead to the January of the nominal year. The dashed curve is the average of the six forecasts (the consensus forecast). [Reprinted by permission from Cane *et al.* (1986). *Nature (London)* **321**, 827–832. Copyright ©1986 Macmillan Journals Limited.]

are different and it is possible for predictability to be associated with large growth rates for disturbances. Suppose that calculations with a coupled model indicate that, starting in September, conditions will rapidly change in such a manner that La Niña evolves. If the calculations are repeated several times, starting from slightly different initial conditions each time, and if the different predictions do not diverge, then the forecast of La Niña is likely to be an accurate one. In other words, unstable conditions with a large growth rate could imply predictability provided slightly different initial states develop in the same way.[7] In the case of a large growth rate, the coupled system is predisposed to energetic oscillations, similar to those in Fig. 6.8b. Those oscillations are stable to perturbations and hence are highly predictable. A small growth rate in March, on the other hand, could indicate that damped oscillations, similar to those in Fig. 6.8a, will ensue. Those oscillations are readily perturbed by noise and hence have low predictability.

Conditions are not the same in March of each year. In some years El Niño starts to develop in March and in other years it does not. The results in Fig. 6.9 suggest that the model of Cane *et al.* (1986) succeeds in identifying those years in which the coupled system is predisposed to the development of El Niño. The model succeeds essentially for the same reasons that the statistical methods of Section 6.4 succeed. Both depend on the information in the surface winds and in particular depend on the appearance of anomalous westerly winds over the western tropical Pacific to predict El Niño a few months later. The dynamical model uses these winds to drive its ocean, thus generating initial conditions from which to start predictions. The results of Inoue and O'Brien (1984) indicate that those initial conditions are such that the ocean is predisposed to the development of El Niño on certain occasions. That the predictions with the coupled model, and with the strictly ocean model of Inoue and O'Brien (1984), are at least as good as those using statistical methods indicates that the oceanic initial conditions generated in this manner are reasonably accurate. On the other hand, the dynamical predictions attempted thus far do not appear to be significantly better than the statistical ones, so the success of the predictions does not confirm that the oceanic processes in the model, the unstable mode that the model captures, for example, are important in reality. Barnett *et al.* (1988) point out that the various statistical and dynamical models are reasonably reliable only for predicting, 3 to 9 months in advance, whether or not a "moderate to strong" El Niño will occur. The more ambitious goal of predicting in detail how conditions (rainfall, sea level pressure, surface winds, etc.) will develop over a certain period has not been attempted yet. Coupled General Circulation Models of the ocean and atmosphere will soon be available to attempt such predictions, which, in principle, should be more accurate than statistical ones.

Notes

1. The coupled ocean–atmosphere models that have been developed thus far, and that succeed in simulating a Southern Oscillation, are highly idealized and have a large number of adjustable parameters. The results are very sensitive to the values of these parameters. For example, in the model of Zebiak and Cane (1987), an increase in the mean depth of the thermocline from 150 to 175 m decreases the amplitude of the simulated Southern Oscillation significantly, while a decrease from 150 to 125 m increases the amplitude substantially. In this

model the thermocline is the interface between two layers each of constant density, and its mean depth has a constant value. In reality the thermocline slopes considerably from east to west and, on the basis of measurements, a reasonable value for its mean depth can range from approximately 50 to 175 m. The appropriate value for the model is chosen to be that which results in oscillations with a reasonable amplitude.

2. The stability analysis of K.-M. Lau (1981) assumes that meridional velocity components are identically zero in both the ocean and atmosphere. This in effect is an assumption that sets the Coriolis accelerations equal to zero. The analysis is valid for a nonrotating system.

3. To obtain the numerical values in Figs. 6.3 and 6.5, the parameters were assigned the following values: $c_a = 30$ m/sec; $c = 1.4$ m/sec; $A = B = 50 \times 10^{-7}$ sec^{-1}; $a = b = 10^{-7}$ sec^{-1}; $\gamma = 8 \times 10^{-1}$ sec^{-1}; $\alpha = 7 \times 10^{-3}$ m^2/sec^{-3}/K.

4. McCreary (1983) pioneered studies of ocean–atmosphere interactions by coupling an oceanic model with an atmosphere that is simply in one of two possible specified states, depending on the oceanic conditions: the trade winds are either intense or relaxed. This study first suggested that the speed of off-equatorial Rossby waves could be one of the factors that determine the interval between El Niño episodes.

5. Models in which Rossby waves excited during one phase of the Southern Oscillation later reappear as Kelvin waves that terminate that phase are referred to as "delayed oscillator" models because their behavior is captured by the equation (Suarez and Schopf, 1988; Battisti and Hirst, 1989)

$$T_t = T - T^3 - rT(t - d)$$

The dependent variable T is a function of time t and could be sea surface temperature, for example. Its rate of change depends on positive feedback, the first term on the right-hand side, which represents unstable ocean–atmosphere interactions, and on dissipation, which is represented by the second term. The third term is a negative feedback that is delayed a time d. It represents the effect of the oceanic waves. In the absence of the nonlinear term (T^3) the equation has solutions that describe growing oscillations with a period long compared to $2d$ provided $r > 1$. The latter condition ensures that the effect on sea surface temperature of the reflected waves exceeds local effects. This can happen because local sea surface temperatures are affected by damping processes while the delayed waves, which propagate as thermocline displacements, are shielded from those damping processes. Nonlinearities serve only to limit the amplitude of the oscillations and do not influence the period significantly if $r > 1$. If, however, $r < 1$, a parameter range that takes into account that reflections off the western boundary are imperfect, then long-period oscillations are inherently nonlinear.

6. Neelin's (1989b) results shown in Fig. 6.8 are from an oceanic General Circulation Model (similar to that described in Chapter 4 but with a coarser resolution) coupled to a two-level, steady-state atmospheric model that relates convergence to moist static stability according to Eq. (5.29). The oceanic model is forced with the observed climatological steady winds to produce a climatological sea surface temperature pattern. The response of the atmospheric model to that pattern is defined as the atmospheric climatology. In the coupled mode the sea surface temperatures at any time drive the atmospheric motion from which is subtracted the atmospheric climatology. The anomalous windstress is then added to the observed climatological windstress to drive the ocean and to produce a new sea surface temperature. In other words, only the atmospheric component of the coupled model is run in an anomaly mode. For the results in Fig. 6.8a, the winds described by Hellerman and Rosenstein (1983) force the ocean and give rise to intense upwelling along the equator. This upwelling, which is perhaps too strong, tends to inhibit ocean–atmosphere interactions. For the results in Fig. 6.8b the climatological oceanic conditions were generated by forcing the

model with the Hellerman and Rosenstein (1983) winds reduced by a factor of 0.6. This causes less intense equatorial upwelling and results in more vigorous interaction between the ocean and atmosphere. For the results in Fig. 6.8c the strength of the interactions was enhanced further, in a more direct way, by multiplying windstress anomalies by a factor of 1.25.

7. Growth rates provide a measure of the instability of a time-dependent basic state. In the case of a time-dependent basic state it is more useful to calculate a Lyapunov exponent, which measures how rapidly the development of a basic state diverges from the development of a perturbed state. This divergence, starting at time T, could be small even though the growth rate for conditions at time T is large.

Bibliography

Adamec, D., and O'Brien, J. J. (1978). The seasonal upwelling in the Gulf of Guinea due to remote forcing. *J. Phys. Oceanogr.* **8**, 1050–1060.

Aleem, A. A. (1967). Concepts of currents, tides and winds among medieval geographers in the Indian Ocean. *Deep-Sea Res.* **14**, 459–463.

Allen, J. S. (1976). Some aspects of the forced wave response of stratified coastal regions. *J. Phys. Oceanogr.* **6**, 113–119.

Allen, J. S. (1980). Models of wind-driven currents on the continental shelf. *Annu. Rev. Fluid Mech.* **12**, 389–433.

Anderson, D. L. T. (1976). The low-level jet as a western boundary current. *Mon. Weather Rev.* **104**, 907–921.

Anderson, D. L. T., and Gill, A. E. (1975). Spin up of a stratified ocean with applications to upwelling. *Deep-Sea Res.* **22**, 583–596.

Anderson, D. L. T., and McCreary, J. P. (1985). Slowly propagating disturbances in a coupled ocean–atmosphere model. *J. Atmos. Sci.* **42**, 615–628.

Anderson, D. L. T., and Moore, D. W. (1979). Cross-equatorial inertial jets with special relevance to very remote forcing of the Somali Current. *Deep-Sea Res.* **26A**, 1–22.

Anderson, D. L. T., and Rowlands, P. B. (1976a). The role of inertia-gravity and planetary waves in the response of a tropical ocean to the incidence of an equatorial Kelvin wave on a meridional boundary. *J. Mar. Res.* **34**, 295–312.

Anderson, D. L. T., and Rowlands, P. B. (1976b). The Somali Current response to the southwest monsoon. *J. Mar. Res.* **34**, 395–417.

Angell, J. K., and Koshover, J. (1983). Global temperature variations in the troposphere and stratosphere, 1958–1982. *Mon. Weather Rev.* **111**, 901–921.

Arakawa, A., and Schubert, W. H. (1974). Interaction of a cumulus cloud ensemble with the large-scale environment. *J. Atmos. Sci.* **31**, 674–701.

Arkin, P. A. (1982). The relationship between interannual variability in the 200 mb tropical wind field and the Southern Oscillation. *Mon. Weather Rev.* **110**, 808–823.

Arkin, P. A. (1983). An examination of the Southern Oscillation in the upper tropospheric tropical and subtropical wind-field. Ph.D. Thesis, University of Maryland, College Park.

Bakun, A. (1978). Guinea Current upwelling. *Nature (London)* **271**, 147–150.

Barber, R. T., and Chavez, F. P. (1983). Biological consequences of El Niño. *Science* **222**, 1203–1210.

Barber, R. T., and Chavez, F. P. (1986). Ocean variability in relation to living resources during the 1982–83 El Niño. *Nature (London)* **319**, 279–285.

Barnett, T. P. (1977a). The principal time and space scales of the Pacific trade wind fields. *J. Atmos. Sci.* **34**, 221–236.

Barnett, T. P. (1977b). An attempt to verify some theories of El Niño. *J. Phys. Oceanogr.* **7**, 633–647.

Barnett, T. P. (1981a). Statistical relations between ocean/atmosphere fluctuations in the tropical Pacific. *J. Phys. Oceanogr.* **11**, 1043–1058.

Barnett, T. P. (1981b). Statistical prediction of North American air temperatures from Pacific predictors. *Mon. Weather Rev.* **109**, 1021–1040.

Barnett, T. P. (1983). Interaction of the Monsoon and Pacific tradewind system at interannual timescales. Part I. The Equatorial Zone. *Mon. Weather Rev.* **111**, 756–773.

Barnett, T. P. (1984). Prediction of El Niño of 1982–1983. *Mon. Weather Rev.* **112**, 1403–1407.

Barnett, T. P. (1985a). Interaction of the Monsoon and Pacific tradewind system at interannual timescales. Part II. The tropical band. *Mon. Weather Rev.* **113**, 2380–2387.

Barnett, T. P. (1985b). Interaction of the Monsoon and Pacific tradewind system at interannual timescales. Part III. The anatomy of the Southern Oscillation. *Mon. Weather Rev.* **113**, 2388–2400.

Barnett, T. P. (1985c). Variations in near-global sea level pressure. *J. Atmos. Sci.* **42**, 478–501.

Barnett, T. P. (1985d). Three dimensional structure of low frequency pressure variations in the tropical atmosphere. *J. Atmos. Sci.* **42**, 2798–2803.

Barnett, T. P., and Patzert, W. C. (1980). Scales of thermal variability in the tropical Pacific. *J. Phys. Oceanogr.* **10**, 529–540.

Barnett, T. P., Graham, N., Cane, M., Zebiak, S., Dolan, S., O'Brien, J., and Legler, D. (1988). On the prediction of El Niño of 1986–1987. *Science* **241**, 192–196.

Battisti, D. S. (1988). The dynamics and thermodynamics of a warming event in a coupled tropical atmosphere–ocean model. *J. Atmos. Sci.* **45**, 2889–2919.

Battisti, D. S., and Hirst, A. C. (1989). Interannual variability in the tropical atmosphere–ocean system: Influence of the basic state and ocean geometry. *J. Atmos. Sci.* **46** (in press).

Battisti, D. S., Hirst, A. C., and Sarachik, E. S. (1989). Instability and predictability in coupled ocean–atmosphere models. *Philos. Trans. R. Soc. London* (in press).

Berlage, H. P. (1957). Fluctuations in the general atmospheric circulation of more than one year, their nature and prognostic value. *K. Ned. Meteorol. Inst., Meded. Verh.* **69**, 1–152.

Berlage, H. P. (1966). The Southern Oscillation and world weather. *K. Ned. Meteorol. Inst., Meded. Verh.* **88**, 1–152.

Betts, A. K. (1982). Saturation point analysis of moist convective overturning. *J. Atmos. Sci.* **39**, 1484–1505.

Bigg, G. R., and Gill, A. E. (1986). The annual cycle of sealevel in the eastern tropical Pacific. *J. Phys. Oceanogr.* **16**, 1055–1061.

Bjerknes, J. (1966). A possible response of the atmospheric Hadley circulation to equatorial anomalies of ocean temperature. *Tellus* **18**, 820–829.

Bjerknes, J. (1969). Atmospheric teleconnections from the equatorial Pacific. *Mon. Weather Rev.* **97**, 163–172.

Bjerknes, J. (1972). Large-scale atmospheric response to the 1964–65 Pacific equatorial warming. *J. Phys. Oceanogr.* **2**, 212–217.

Blackmon, M. L., Madden, R. A., Wallace, J. M., and Gutzler, D. S. (1979). Geographical variations in the vertical structure of geopotential height fluctuations. *J. Atmos. Sci.* **36**, 2450–2466.

Blackmon, M. L., Geisler, J. E., and Pitcher, E. J. (1983). A General Circulation Model Study of January Climate Anomaly patterns associated with interannual variation of equatorial Pacific sea surface temperatures. *J. Atmos. Sci.* **40**, 1410–1425.

Blanc, T. V. (1985). Variation of bulk-derived surface flux, stability, and roughness results due to the use of different transfer coefficient schemes. *J. Phys. Oceanogr.* **15**, 650–669.

Blandford, R. (1966). Mixed Rossby-gravity waves in the ocean. *Deep-Sea Res.* **13**, 941–961.

Boer, G. J. (1985). Modeling of the atmospheric response to the 1982–83 El Niño. *In* "Proceedings of the 16th International Liege Colloquium on Ocean Hydrodynamics" (J. Nihoul, ed.), Elsevier Oceanogr. Ser., Elsevier, Amsterdam.

Boyd, J. P. (1980a). The nonlinear equatorial Kelvin wave. *J. Phys. Oceanogr.* **10**, 1–11.

Boyd, J. P. (1980b). Equatorial solitary waves. Part I. Rossby solitons. *J. Phys. Oceanogr.* **10**, 1699–1718.

Boyd, J. P., and Christidis, Z. D. (1982). Low wavenumber instability on the equatorial beta-plane. *Geophys. Res. Lett.* **9**, 769–772.

Brink, K. H. (1982). A comparison of long coastal trapped wave theory with observations off Peru. *J. Phys. Oceanogr.* **12**, 897–913.

Brink, K. H., Allen, J. S., and Smith, R. L. (1978). A study of low frequency fluctuations near the Peru coast. *J. Phys. Oceanogr.* **8**, 1025–1041.

Brink, K. H., Halpern, D., and Smith, R. L. (1980). Circulation in the Peruvian upwelling system near 15°S. *JGR, J. Geophys. Res.* **85**, 4036–4048.

Brockmann, C., Fahrback, E., Huyer, A., and Smith, R. L. (1980). The poleward undercurrent along the Peru coast: 5 to 15°S. *Deep-Sea Res.* **27A**, 847–856.

Broecker, W. S., Peng, T. H., and Stuiver, M. (1978). An estimate of the upwelling rate in the Equatorial Atlantic based on the distribution of bomb radiocarbon. *JGR, J. Geophys. Res.* **83**, 6179–6186.

Brown, O. B., Bruce, J. G., and Evans, R. H. (1980). Evolution of sea surface temperature in the Somali Basin during the southwest monsoon of 1979. *Science* **209**, 595–597.

Bruce, J. G. (1979). Eddies off the Somali Coast during the Southwest Monsoon. *JGR, J. Geophys. Res.* **84**, 7742–7748.

Bruce, J. G., and Kerling, J. (1984). Near equatorial eddies in the North Atlantic. *Geophys. Res. Lett.* **11**, 779–782.

Bruce, J. G., Quadfasel, D. R., and Swallow, J. C. (1980). Somali eddy formation during the commencement of the Southwest Monsoon. *JGR, J. Geophys. Res.* **85**, 6654–6660.

Bruce, J. G., Fieux, M., and Gonella, J. (1982). A note on the continuance of the Somali eddy after the cessation of the Southwest Monsoon. *Oceanol. Acta* **4**, 7–9.

Bryan, K. (1969). A numerical method for the study of the world ocean. *J. Comput. Phys.* **4**, 347–376.

Bryan, K. (1982). Poleward heat transport in the oceans: Observations and models. *Annu. Rev. Earth Planet. Sci.* **10**, 15–38.

Bryden, L. H., and Brady, E. C. (1985). Diagnostic model of the three-dimensional circulation of the upper equatorial Pacific Ocean. *J. Phys. Oceanogr.* **15**, 1255–1273.

Bryden, H. L., and Hall, M. M. (1980). Heat transport by ocean currents across 25°N latitude in the Atlantic. *Science* **207**, 884–886.

Bubnov, V. A., and Yegorikhin, V. D. (1980). Study of water circulation in the tropical Atlantic. *Deep-Sea Res.* **26**, Suppl. 2, 125–136.

Bubnov, V. A., Yegorikhin, V. D., and Osadchiy, A. S. (1982). Structure of the equatorial currents in the Central and Western Pacific. *Oceanology* **22**(2), 124–126.

Buchanan, J. Y. (1886). On similarities in the physical geography of the great oceans. *Geogr. J.* **8**, 753–770.

Buchanan, J. Y. (1888). The exploration of the Gulf of Guinea. *Scott. Geogr. Mag.* **4**, 177–200, 233–251.

Bunker, A. F. (1980). Trends of variables and energy fluxes over the Atlantic Ocean from 1948 to 1972. *Mon. Weather Rev.* **108**, 720–732.

Bunker, A. F., and Worthington, L. V. (1976). Energy exchange charts of the North Atlantic Ocean. *Bull. Am. Meteorol. Soc.* **57**, 670–678.

Burkov, V. A., and Ovchinkov, M. (1960). Structure of zonal streams and meridional circulation in the Central Pacific during the northern hemisphere winter. *Tr. Inst. Okeanol. im. P. P. Shirshova, Akad. Nauk SSSR* **40**, 93–107.

Busalacchi, A. J., and Cane, M. (1985). Hindcasts of sealevel variations during the 1982–83 El Niño. *J. Phys. Oceanogr.* **15**, 213–221.

Busalacchi, A. J., and O'Brien, J. J. (1980). The seasonal variability of the tropical Pacific. *J. Phys. Oceanogr.* **10**, 1929–1952.

Busalacchi, A. J., and O'Brien, J. J. (1981). Interannual variability of the equatorial Pacific in the 1960's. *JGR, J. Geophys. Res.* **86**, 10901–10907.

Busalacchi, A. J., and Picaut, J. (1983). Seasonal variability from a model of the tropical Atlantic Ocean. *J. Phys. Oceanogr.* **13**, 1564–1588.

Busalacchi, A. J., Takeuchi, K., and O'Brien, J. J. (1983). Interannual variability of the equatorial Pacific, revisited. *JGR, J. Geophys. Res.* **88**, 7551–7562.

Butler, J. L. (1988). La Niña or El Viego? *Science* **242**, 168.

Cane, M. A. (1979a). The response of an equatorial ocean to simple wind-stress patterns. I. Model formulation and analytic results. *J. Mar. Res.* **37**, 233–252.

Cane, M. A. (1979b). The response of an equatorial ocean to simple windstress patterns. II. Numerical results. *J. Mar. Res.* **37**, 253–299.

Cane, M. A. (1980). On the dynamics of equatorial currents with application to the Indian Ocean. *Deep-Sea Res.* **27A**, 525–544.

Cane, M. A. (1983). Oceanographic events during El Niño. *Science* **222**, 1189–1195.

Cane, M. A. (1986). El Niño. *Annu. Rev. Earth Planet. Sci.* **14**, 43–70.

Cane, M. A., and du Penhoat, Y. (1981). The effect of islands on low frequency equatorial motions. *J. Mar. Res.* **40**, 937–962.

Cane, M. A., and Gent, P. R. (1984). Reflection of low frequency equatorial waves at arbitrary western boundaries. *J. Mar. Res.* **42**, 487–502.

Cane, M. A., and Moore, D. W. (1981). A note on low-frequency equatorial basin modes. *J. Phys. Oceanogr.* **11**, 1578–1585.

Cane, M. A., and Sarachik, E. S. (1976). Forced baroclinic ocean motions. I. The linear equatorial unbounded case. *J. Mar. Res.* **34**(4), 629–665.

Cane, M. A., and Sarachik, E. S. (1977). Forced baroclinic ocean motions. II. The linear equatorial bounded case. *J. Mar. Res.* **35**(2), 395–432.

Cane, M. A., and Sarachik, E. S. (1979). Forced baroclinic ocean motions. III. The linear equatorial basin case. *J. Mar. Res.* **37**, 355–398.

Cane, M. A., and Sarachik, E. S. (1981). The response of a linear baroclinic equatorial ocean to periodic forcing. *J. Mar. Res.* **39**, 651–693.

Cane, M. A., and Sarachik, E. S. (1983a). Equatorial oceanography. *Rev. Geophys. Space Phys.* **21**, 1137–1148.

Cane, M. A., and Sarachik, E. S. (1983b). Seasonal heat transport in a forced equatorial baroclinic model. *J. Phys. Oceanogr.* **13**, 1744–1746.

Cane, M. A., and Zebiak, S. E. (1985). A theory for El Niño and the Southern Oscillation. *Science* **228**, 1085–1087.

Cane, M. A., Dolan, S. C., and Zebiak, S. E. (1986). Experimental forecasts of the 1982/83 El Niño. *Nature (London)* **321**, 827–832.

Cannon, G. A. (1966). Tropical waters in the western Pacific Ocean. *Deep-Sea Res.* **13**, 1139–1148.

Caviedes, C. N. (1973). Secas and El Niño: Two simulations of climatical hazards in South America. *Proc. Assoc. Am. Geogr.* **5**, 44–49.

Caviedes, C. N. (1984). El Niño 1982–83. *Geogr. Rev.* **74**, 267–290.

Chang, C. P. (1976). Forcing of stratospheric Kelvin waves by tropospheric heat sources. *J. Atmos. Sci.* **33**, 742–744.

Chang, C. P. (1977). Viscous internal gravity waves and low frequency oscillations in the tropics. *J. Atmos. Sci.* **34**, 901–910.

Chang, C. P., and Lau, K. M. (1982). Short-term planetary-scale interactions over the tropics and mid-latitudes during the northern winter. Part I. Contrasts between active and interactive periods. *Mon. Weather Rev.* **110**, 933–946.

Chang, P. (1988). Oceanic adjustment in the presence of mean currents. Ph.D. Dissertation, Princeton University, Princeton, New Jersey.

Chang, P., and Philander, S. G. H. (1988). Rossby wave packets in baroclinic currents. *Deep-Sea Res.* **36**, 17–37.

Chapman, S., and Lindzen, R. S. (1970). "Atmospheric Tides." 200 pp. Reidel, Hingham, Massachusetts.

Charney, J. G. (1955). The generation of oceanic currents by winds. *J. Mar. Res.* **14**, 477–498.

Charney, J. G. (1960). Non-linear theory of a wind-driven homogeneous layer near the equator. *Deep-Sea Res.* **6**, 303–310.

Charney, J. G., and Eliassen, A. (1964). On the growth of hurricane depression. *J. Atmos. Sci.* **21**, 6–8.

Charney, J. G., and Spiegel, S. L. (1971). Structure of wind-driven equatorial currents in homogeneous oceans. *J. Phys. Oceanogr.* **1**, 149–160.

Chen, W. Y. (1982). Assessment of Southern Oscillation sea level pressure indices. *Mon. Weather Rev.* **110**, 800–807.

Chen, W. Y. (1983). Fluctuations in Northern Hemisphere 700 mb height field associated with the Southern Oscillation. *Mon. Weather Rev.* **110**, 808–823.

Chervin, R. M., and Druyan, L. M. (1984). The influence of ocean surface temperature gradient and continentality on the Walker Circulation. *Mon. Weather Rev.* **112**, 1510–1523.

Chiu, W.-C., and Lo, A. (1979). A preliminary study of the possible statistical relationship between the tropical Pacific sea surface temperature and the atmospheric circulation. *Mon. Weather Rev.* **107**, 18–25.

Chiu, W.-C., Lo, A., Weidler, D. H., Jr., and Fulker, D. (1981). A study of the possible statistical relationship between the tropical Pacific sea surface temperature and atmospheric circulation. *Mon. Weather Rev.* **109**, 1013–1020.

Clarke, A. J. (1979). On the generation of seasonal coastal upwelling in the Gulf of Guinea. *JGR, J. Geophys. Res.* **84**, 3743–3751.

Clarke, A. J. (1983). The reflection of equatorial waves from oceanic boundaries. *J. Phys. Oceanogr.* **13**, 1193–1207.

Cochrane, J. D., Kelley, F. J., Jr., and Olling, C. R. (1979). Subthermocline countercurrents in the western equatorial Atlantic Ocean. *J. Phys. Oceanogr.* **9**, 724–738.

Cohen, J. M., trans. and ed. (1968). "The Discovery and Conquest of Peru." Penguin Books, Harmondsworth, U.K.

Colin, C., Henin, C., Hisard, P., and Oudot, C. (1971). Le Courant de Cromwell dans le Pacifique central en février. *Cah. ORSTOM, Ser Oceanogr.* **9**, 167–186.

Cornejo-Garrido, A. G., and Stone, P. H. (1977). On the heat balance of the Walker Circulation. *J. Atmos. Sci.* **34**, 1155–1162.

Cowles, T. J., Barber, R. T., and Guillen, O. (1979). Biological consequences of the 1975 El Niño. *Science* **195**, 285–287.

Cox, M. D. (1976). Equatorially trapped waves and the generation of the Somali Current. *Deep-Sea Res.* **23**, 1139–1152.

Cox, M. D. (1979). A numerical study of Somali Current eddies. *J. Phys. Oceanogr.* **9**, 311–326.

Cox, M. D. (1980). Generation and propagation of 30-day waves in a numerical model of the Pacific Ocean. *J. Phys. Oceanogr.* **10**, 1168–1186.

Crawford, W. R. (1982). Pacific equatorial turbulence. *J. Phys. Oceanogr.* **12**, 1137–1149.

Crawford, W. R., and Osborn, T. R. (1981). The control of equatorial ocean currents by turbulent dissipation. *Science* **212**, 539–540.

Cromwell, T., Montgomery, R. B., and Stroup, E. D. (1954). Equatorial Undercurrent in the Pacific revealed by new methods. *Science* **119**, 648–649.

De Angelis, D. (1983). Hurricane Alley. *Mar. Weather Log* **27**, 106–109.

Deser, C., and Wallace, J. M. (1987). El Niño events and their relation to the Southern Oscillation. *JGR, J. Geophys. Res.* **92**, 14189–14196.

Dickson, R. R., and Namias, J. (1976). North American influences on the circulation and climate of the North Atlantic sector. *Mon. Weather Rev.* **104**, 1255–1265.

Doberitz, R. (1968). Cross-spectrum analyses of rainfall and sea temperature at the equatorial Pacific Ocean. *Bonn. Met. Abh.* No. 8, pp. 1–61.

Donguy, J. R., and Henin, C. (1980a). Surface conditions in the equatorial Pacific related to the intertropical convergence zone of the winds. *Deep-Sea Res.* **27A**, 693–714.

Donguy, J. R., and Henin, C. (1980b). Climatic teleconnections in the western South Pacific with El Niño phenomenon. *J. Phys. Oceanogr.* **10**, 1952–1958.

Duing, W., and Leetmaa, A. (1980). Arabian Sea cooling: A preliminary heat budget. *J. Phys. Oceanogr.* **10**, 307–312.

Duing, W., Hisard, P., Katz, E., Meincke, J., Miller, L., Moroshkin, K., Philander, G., Rybnikov, A., Voigt, K., and Weisberg, R. (1975). Meanders and long waves in the equatorial Atlantic. *Nature (London)* **257**, 280–284.

Duing, W., Molinari, R. L., and Swallow, J. C. (1980). Somali Current: Evolution of the surface flow. *Science* **209**, 588–590.

du Penhoat, Y., and Triguier, A. M. (1985). The seasonal linear response of the tropical Atlantic Ocean. *J. Phys. Oceanogr.* **15**, 316–329.

Eguiguren, V. (1894). Las lluvias en Piura. *Bol. Soc. Geogr. Lima* **4**, 241–258.

Emanuel, K. A. (1987). An air–sea interaction model of interseasonal oscillations in the tropics. *J. Atmos. Sci.* **44**, 2324–2340.

Emanuel, K. A. (1988). Towards a general theory of hurricanes. *Am. Sci.* **76**, 371–379.

Enfield, D. B. (1981a). El Niño–Pacific eastern boundary response to interannual forcing. *In* "Resource Management and Environmental Uncertainty" (M. H. Glantz and J. D. Thompson, eds.), pp. 213–254, Wiley, New York.

Enfield, D. B. (1981b). Thermally driven wind variability in the planetary boundary layer above Lima, Peru. *JGR, J. Geophys. Res.* **86**, 2005–2016.

Enfield, D. B. (1981c). Annual and non-seasonal variability of monthly low-level wind fields over the southeastern tropical Pacific. *Mon. Weather Rev.* **109**, 2177–2190.

Enfield, D. B. (1986). Zonal and seasonal variability of the equatorial Pacific heat balance. *J. Phys. Oceanogr.* **16**, 1038–1054.

Enfield, D. B., and Allen, J. S. (1980). On the structure and dynamics of monthly mean sea level anomalies along the Pacific coast of North and South America. *J. Phys. Oceanogr.* **10**, 557–588.

Eriksen, C. C. (1979). An equatorial transect of the Indian Ocean. *J. Mar. Res.* **37**, 215–232.
Eriksen, C. C. (1980). Evidence for a continuous spectrum of equatorial waves in the Indian Ocean. *JGR, J. Geophys. Res.* **85**, 3285–3303.
Eriksen, C. C. (1981). Deep currents and their interpretation as equatorial waves in the western Pacific Ocean. *J. Phys. Oceanogr.* **11**, 48–70.
Eriksen, C. C. (1982). Geostrophic equatorial deep jets. *J. Mar. Res.* **40**, Suppl., 143–157.
Eriksen, C. C. (1985). Moored observations of deep low frequency motions in the central Pacific Ocean: Vertical structure and interpretation as equatorial waves. *J. Phys. Oceanogr.* **15**, 1085–1113.
Eriksen, C. C., Blumenthal, M. B., Hayes, S. P., and Ripa, P. (1983). Wind-generated equatorial Kelvin waves observed across the Pacific Ocean. *J. Phys. Oceanogr.* **13**, 1622–1640.
Esbensen, S. K., and Kushnir, V. (1981). "The Heat Budget of the Global Ocean: An Atlas Based on Estimates from Marine Surface Observations," Clim. Res. Inst., Rep. No. 29. Oregon State University, Corvallis.
Evans, R. H., and Brown, O. B. (1981). Propagation of thermal fronts in the Somali current system. *Deep-Sea Res.* **28A**, 521–527.
Fahrbach, E., Brockmann, C., and Meincke, J. (1986). Horizontal mixing in the Atlantic Equatorial Undercurrent estimated from drifting buoy clusters. *JGR, J. Geophys. Res.* **91**, 10557–10565.
Fiedler, P. C. (1984). Satellite observations of the 1982–83 El Niño along the U.S. Pacific Coast. *Science* **224**, 1251–1254.
Fieux, M. (1975). Etablissement de la Mousson de Sud-Ouest en Mer d'Arabie. *C. R. Acad. Sci., Paris, Ser. B* **281**, 563–566.
Fieux, M., and Stommel, H. (1976). Historical sea surface temperatures in the Arabian Sea. *Ann. Inst. Oceanogr. (Paris)* **52**, 5–15.
Findlater, J. (1971). Mean monthly airflow at low levels over the western Indian Ocean. *Geophys. Mem.* No. 115, pp. 1–53.
Fine, R. A. (1985). Direct evidence using tritium data for throughflow from the Pacific into the Indian Ocean. *Nature (London)* **315**, 478–480.
Fine, R. A., Peterson, W. H., Rooth, C. G. H., and Ostlund, H. G. (1983). Cross-equatorial tracer transport in the upper waters of the Pacific Ocean. *JGR, J. Geophys. Res.* **88**, 763–769.
Firing, E. (1987). Deep zonal currents in the central equatorial Pacific. *J. Mar. Res.* **45**, 791–812.
Firing, E., Lukas, R., Sadler, J., and Wyrtki, K. (1983). Equatorial Undercurrent disappears during 1982–1983 El Niño. *Science* **222**, 1121–1123.
Flagg, C. N., Gordon, R. L., and McDowell, S. (1986). Hydrographic and current observations on the continental slope and shelf of the western Equatorial Atlantic. *J. Phys. Oceanogr.* **16**, 1412–1429.
Fofonoff, N. P., and Montgomery, R. B. (1955). The Equatorial Undercurrent in the light of the vorticity equation. *Tellus* **7**, 518–521.
Folland, C. K., Palmer, T. N., and Parker, D. E. (1986). Sahel rainfall and world wide sea surface temperatures 1901–1985. *Nature (London)* **320**, 602–607.
Fu, L. (1986). Mass, heat, and freshwater fluxes in the South Indian Ocean. *J. Phys. Oceanogr.* **16**, 1683–1693.
Gadgil, S., Joseph, P. V., and Joshi, N. V. (1984). Ocean–atmosphere coupling over monsoon regions. *Nature (London)* **312**, 141–143.
Garcia, O. (1981). A comparison of two satellite rainfall estimates for GATE. *J. Appl. Meteorol.* **20**, 430–438.

Garcia, R. R., and Salby, M. L. (1987). Transient response to localized episodic heating in the tropics. Part II. Far-field behavior. *J. Atmos. Sci.* **44**, 499–530.

Gardiner-Garden, R. S. (1987). Some aspects of modeling the vertical structure of currents in wind-forced coastal upwelling systems. Ph.D. Dissertation, Princeton University, Princeton, New Jersey.

Gargett, A. E., and Osborn, T. R. (1981). Small scale shear measurements during the fine and microstructure experiment. *JGR, J. Geophys. Res.* **86**, 1929–1944.

Garrett, C. J., and Munk, W. H. (1979). Internal waves in the ocean: A progress report. *Annu. Rev. Fluid Mech.* **11**, 339–369.

Garzoli, S. L., and Katz, E. J. (1981). Observations of inertia-gravity waves in the Atlantic from inverted echo sounders during FGGE. *J. Phys. Oceanogr.* **11**, 1463–1473.

Garzoli, S. L., and Katz, E. J. (1983). The forced annual reversal of the Atlantic North Equatorial Countercurrent. *J. Phys. Oceanogr.* **13**, 2082–2090.

Gent, P. R. (1979). Standing equatorial wave modes in bounded ocean basins. *J. Phys. Oceanogr.* **9**, 653–662.

Gent, P. R., and Luyten, J. R. (1985). How much energy propagates vertically in the equatorial oceans? *J. Phys. Oceanogr.* **7**, 997–1007.

Gent, P. R., and Semtner, A. J., Jr. (1980). Energy trapping near the equator in a numerical ocean model. *J. Phys. Oceanogr.* **10**, 823–842.

Gent, P. R., O'Neill, K., and Cane, M. A. (1983). A model of the semiannual oscillation in the equatorial Indian Ocean. *J. Phys. Oceanogr.* **31**, 2148–2160.

Gibbs, R. J. (1980). Wind-controlled coastal upwelling in the western Atlantic. *Deep-Sea Res.* **27**, 857–866.

Gill, A. E. (1975). Models of Equatorial Currents. *Proc. Symp. Numer. Models Ocean Circ., 1972*, pp. 181–203.

Gill, A. E. (1980). Some simple solutions for heat-induced tropical circulation. *Q. J. R. Meteorol. Soc.* **106**, 447–462.

Gill, A. E. (1982). "Atmosphere–Ocean Dynamics." Academic Press, New York.

Gill, A. E. (1985). *In* "Coupled Ocean–Atmosphere Models" (J. Nihoul, ed.), pp. 303–327, Elsevier, Amsterdam.

Gill, A. E., and Clark, A. J. (1974). Wind-induced upwelling, coastal currents and sea-level changes. *Deep-Sea Res.* **21**, 325–345.

Gill, A. E., and Rasmusson, E. M. (1983). The 1982–1983 climate anomaly in the equatorial Pacific. *Nature (London)*, **305**, 229–234.

Godbole, R. V., and Shukla, J. (1981). Global analysis of January and July sea level pressure. *NASA Tech. Memo.* **NASA TM-X-82097**.

Godfrey, J. (1975). On ocean spin-down. I. A linear experiment. *J. Phys. Oceanogr.* **5**, 399–409.

Goldenberg, S. B., and O'Brien, J. J. (1981). Time and space variability of tropical Pacific wind stress. *Mon. Weather Rev.* **109**, 1208–1218.

Gonella, J., Fieux, M., and Philander, G. (1981). Mise en évidence d'ondes Rossby équatoriales dans l'Océan Indian au moyen de bouées dérivantes. *C. R. Seances Acad. Sci., Ser. 2* **292**, 1397–1399.

Gordon, A. L. (1986). Interocean exchange of thermocline water. *JGR, J. Geophys. Res.* **91**, 5037–5047.

Graham, N. E., and Barnett, T. P. (1987). Sea surface temperature, surface wind divergence and convection over tropical oceans. *Science* **238**, 657–659.

Graham, N. E., Michaelson, J., and Barnett, T. P. (1987). An investigation of El Niño–Southern Oscillation cycle with statistical models. *JGR, J. Geophys. Res.* **92**, 14251–14270.

Greatbatch, R. J. (1985). Kelvin wave fronts, Rossby solitary waves and the nonlinear spin-up of the equatorial oceans. *JGR, J. Geophys. Res.* **90**, 9097–9107.

Gregg, M. C. (1976). Temperature and salinity microstructure in the Pacific Equatorial Undercurrent. *JGR, J. Geophys. Res.* **81**, 1180–1196.

Gregg, M. C., and Sanford, T. B. (1980). Signatures of mixing from the Bermuda Slope, the Sargasso Sea and the Gulf Stream. *J. Phys. Oceanogr.* **10**, 105–127.

Gregg, M. C., Peters, H., Wesson, J. C., Oakley, M. S., and Shay, T. J. (1985). Intensive measurements of turbulence and shear in the equatorial undercurrent. *Nature (London)* **318**, 140–144.

Halpern, D. (1980a). A Pacific equatorial temperature section from 172°E to 110°W during winter–spring 1979. *Deep-Sea Res.* **27A**, 931–940.

Halpern, D. (1980b). Variability of near-surface currents in the Atlantic north equatorial countercurrent during GATE. *J. Phys. Oceanogr.* **10**, 1213–1220.

Halpern, D. (1987). Observations of annual and El Niño thermal and flow variations along the equator at 0° 110°W and 0° 95°W during 1980–1985. *JGR, J. Geophys. Res.* **92**, 8197–8212.

Halpern, D., and Weisberg, R. (1989). Upper ocean thermal and flow fields at 0° 28°W (Atlantic) and 0° 140°W (Pacific) during 1983–1985. *Deep-Sea Res.* **36**, 407–418.

Halpern, D., Hayes, S., Leetmaa, A., Hansen, D., and Philander, G. (1983). Oceanographic observations of the 1982 warming of the tropical Pacific. *Science* **221**, 1173–1175.

Hamilton, K., and Garcia, R. C. (1986). El Niño–Southern Oscillation events and their associated midlatitude teleconnections. *Bull. Am. Meteorol. Soc.* **67**, 1354–1361.

Haney, R. H. (1971). Surface thermal boundary conditions for ocean circulation models. *J. Phys. Oceanogr.* **1**, 241–248.

Hansen, D., and Paul, C. A. (1984). Genesis and effects of long waves in the equatorial Pacific. *JGR, J. Geophys. Res.* **89**, 10431–10440.

Hansen, D., and Paul, C. A. (1987). Vertical motion in the eastern equatorial Pacific inferred from drifting buoys. *Oceanol. Acta Spec. Vol.* **6**, 27–32.

Harrison, D. E. (1987). Monthly mean surface island winds in the central tropical Pacific and El Niño events. *Mon. Weather Rev.* **115**, 3133–3145.

Harrison, D. E., and Schopf, P. S. (1984). Kelvin wave induced anomalous advection and the onset of SST warming in El Niño events. *Mon. Weather Res.* **112**, 923–933.

Hastenrath, S. (1977). Relative role of atmosphere and ocean in the global heat budget: Tropical Atlantic and eastern Pacific. *Q. J. R. Meteorol. Soc.* **103**, 519–526.

Hastenrath, S. (1980). Heat budget of tropical ocean and atmosphere. *J. Phys. Oceanogr.* **10**, 159–170.

Hastenrath, S. (1982). On meridional heat transport in the world ocean. *J. Phys. Oceanogr.* **12**, 922–927.

Hastenrath, S., and Lamb, P. (1977). "Climatic Atlas of the Tropical Atlantic and Eastern Pacific Oceans," 97 charts. Univ. of Wisconsin Press, Madison.

Hastenrath, S., and Heller, L. (1977). Dynamics of climatic hazards in northeast Brazil. *Q. J. R. Meteorol. Soc.* **103**, 77–92.

Hastenrath, S., and Lamb, P. J. (1979a). "Climatic Atlas of the Indian Ocean. Part I. Surface Climate and Atmospheric Circulation." Univ. of Wisconsin Press, Madison.

Hastenrath, S., and Lamb, P. J. (1979b). "Climatic Atlas of the Indian Ocean. Part II. The Oceanic Heat Budget." Univ. of Wisconsin Press, Madison.

Hastenrath, S., and Lamb, P. J. (1980). On the heat budget of hydrosphere and atmosphere in the Indian Ocean. *J. Phys. Oceanogr.* **10**, 694–708.

Hayashi, Y. (1976). Non-singular resonance of equatorial waves under the radiation condition. *J. Atmos. Sci.* **33**, 183–201.

Hayashi, Y., and Golder, D. G. (1986). Tropical intraseasonal oscillations appearing in a GFDL general circulation model and FGGE data. Part I. Phase propagation. *J. Atmos. Sci.* **43**, 3058–3067.

Hayashi, Y., and Sumi, A. (1986). The 30–40 day oscillations simulated in an "Aqua Planet" model. *J. Meteorol. Soc. Jpn.* **64**, 451–467.

Hayes, S. P. (1982). A comparison of geostrophic and measured velocities in the Equatorial Undercurrent. *J. Mar. Res.* **40**, Suppl., 219–229.

Hayes, S. P., and Halpern, D. (1984). Correlation of upper ocean currents and sea level in the eastern equatorial Pacific. *J. Phys. Oceanogr.* **14**, 811–824.

Hayes, S. P., and Milburn, H. B. (1980). On the vertical structure of velocity in the eastern Equatorial Pacific. *J. Phys. Oceanogr.* **10**, 633–635.

Hayes, S. P., Toole, J. M., and Mangum, L. J. (1982). Water mass and transport variability at 100°W in the equatorial Pacific. *J. Phys. Oceanogr.* **13**, 153–168.

Hayes, S. P., Mangum, L. J., Barber, R. T., Huyer, A., and Smith, R. L. (1987). Hydrographic variability west of the Galápagos Islands during the 1982–83 El Niño. *Prog. Oceanogr.* **17**, 137–162.

Heddinghaus, T. R., and Krueger, A. F. (1981). Annual and interannual variations in outgoing longwave radiation over the tropics. *Mon. Weather Rev.* **109**, 1208–1218.

Held, I. M. (1978). The vertical scale of a baroclinically unstable wave and its importance for eddy heat flux parameterization. *J. Atmos. Sci.* **35**, 572–576.

Held, I. M. (1982a). Stationary and quasi-stationary eddies in the extratropical troposphere. *In* "The Dynamics of the Extra-Tropical Troposphere" (B. Hoskins, ed.). Academic Press, London.

Held, I. M. (1982b). On the height of the tropopause and the static stability of the troposphere. *J. Atmos. Sci.* **39**, 412–417.

Held, I. M. (1985). Pseudomomentum and the orthogonality of modes in shear flows. *J. Atmos. Sci.* **42**, 2280–2288.

Hellerman, S., and Rosenstein, M. (1983). Normal monthly windstress over the world ocean with error estimates. *J. Phys. Oceanogr.* **13**, 1093–1104.

Henin, C., and Hisard, P. (1987). The North Equatorial Countercurrent observed in the Atlantic Ocean July 1982 to August 1984. *JGR, J. Geophys. Res.* **92**, 3751–3758.

Hess, S. L. (1959). "Introduction to Theoretical Meteorology." Holt, Rinehart & Winston, New York.

Hickey, B. M. (1975). The relationship between fluctuations in sea level wind-stress and sea surface temperature in the Equatorial Pacific. *J. Phys. Oceanogr.* **5**, 460–475.

Hildebrandson, H. H. (1897). Quelque recherches sur les entres d'action de l'atmosphere. *K. Sven. Vetenskaps akad. Handl.* **29**, 1–33.

Hirst, A. C. (1986). Unstable and damped equatorial modes in simple coupled ocean–atmosphere models. *J. Atmos. Sci.* **43**, 606–630.

Hirst, A. C. (1988). Slow instabilities in tropical ocean basin–global atmosphere models. *J. Atmos. Sci.* **45**, 830–852.

Hisard, P. (1973). Variations saisonnières à l'équator dans le Golfe de Guinée. *Cah. ORSTOM, Ser. Oceanogr.* **11**, 349–358.

Hisard, P. (1980). Observation de réponses de type ≪ El Nino ≫ dans l'Atlantique tropical oriental Golfe de Guinea. *Oceanol. Acta* **3**, 69–78.

Hisard, P., and Henin, C. (1987). Response of the Equatorial Atlantic Ocean to the 1983–84 winds. *JGR, J. Geophys. Res.* **92**, 3759–3768.

Hisard, P., and Merle, J. (1980). Onset of summer surface cooling in the Gulf of Guinea during GATE. *Deep-Sea Res.* **26A**, Suppl. 2, 325–341.

Hisard, P., Merle, J., and Voituriez, B. (1970). The Equatorial Undercurrent at 170°E in March and April 1967. *J. Mar. Res.* **28**, 281–303.

Hisard, P., Citeau, J., and Morlière, A. (1976). Le système des contre-courants équatoriaux subsuperficiels permanence et extension de la branche sud dans l'océan Atlantique. *Cah. ORSTOM, Ser Oceanogr.* **14**, 209–220.

Hisard, P., Henin, C., Houghton, R., Piton, B., and Rual, P. (1986). Oceanic conditions in the tropical Atlantic during 1983 and 1984. *Nature (London)* **322**, 243–245.

Holton, J. (1975). The dynamic meteorology of the stratosphere and mesosphere. *Meteorol. Monogr.* **15**, 1–216.

Horel, J. D. (1982). The annual cycle in the tropical Pacific atmosphere and ocean. *Mon. Weather Rev.* **110**, 1863–1878.

Horel, J. D., and Wallace, J. M. (1981). Planetary scale atmospheric phenomena associated with the Southern Oscillation. *Mon. Weather Rev.* **109**, 813–829.

Horel, J. D., Kousky, V. E., and Kagano, M. T. (1986). Atmospheric conditions in the Atlantic Sector during 1983 and 1984. *Nature (London)* **322**, 248–251.

Horigan, A. M., and Weisberg, R. H. (1981). A systematic search for trapped equatorial waves in the GATE velocity data. *J. Phys. Oceanogr.* **11**, 497–509.

Hoskins, B. J., and Karoly, D. J. (1981). The steady linear response of a spherical atmosphere in thermal andorographic forcing. *J. Atmos. Sci.* **38**, 1179–1196.

Houghton, R. W. (1976). Circulation and hydrographic structure over Ghana continental shelf during the 1974 upwelling. *J. Phys. Oceanogr.* **6**, 910–924.

Houghton, R. W. (1983). Seasonal variations of the subsurface thermal structure in the Gulf of Guinea. *J. Phys. Oceanogr.* **13**, 2070–2081.

Houghton, R. W. (1984). Seasonal variation of the Gulf of Guinea thermal structure. *Geophys. Res. Lett.* **11**, 783–786.

Houghton, R. W., and Beer, T. (1976). Wave propagation during the Ghana upwelling. *JGR, J. Geophys. Res.* **81**, 4423–4429.

Hsuing, J. (1985). Estimates of global oceanic meridional heat transport. *J. Phys. Oceanogr.* **11**, 1405–1413.

Hughes, R. L. (1980). On the equatorial mixed layer. *Deep-Sea Res.* **27**, 1067–1078.

Hurlburt, H. E., and Thomson, J. D. (1976). A numerical model of the Somali Current. *J. Phys. Oceanogr.* **6**, 646–664.

Hurlburt, H. E., Kindle, J. C., and O'Brien, J. J. (1976). A numerical simulation of the onset of El Niño. *J. Phys. Oceanogr.* **6**, 621–631.

Huyer, A. (1976). A comparison of upwelling events in two locations: Oregon and Northwest Africa. *J. Mar. Res.* **34**, 531–546.

Huyer, A. (1980). The offshore structure and subsurface expression of sea level variations off Peru, 1976–1977. *J. Phys. Oceanogr.* **10**, 1755–1769.

Ichiye, T., and Petersen, J. (1963). The anomalous rainfall of the 1957–58 winter in the equatorial central Pacific arid area. *J. Meteorol. Soc. Jpn.* **41**, 172–182.

Inoue, M., and O'Brien, J. J. (1984). A forecasting model for the onset of a major El Niño. *Mon. Weather Rev.* **112**, 2326–2337.

Itoh, H. (1977). The response of equatorial waves to thermal forcing. *J. Meteorol. Soc. Jpn.* **55**, 222–239.

Jones, J. H. (1973). Vertical mixing in the Equatorial Undercurrent. *J. Phys. Oceanogr.* **3**, 286–296.

Julian, P. R., and Chervin, R. M. (1978). A study of the Southern Oscillation and Walker Circulation phenomenon. *J. Atmos. Sci.* **106**, 1433–1451.

Kang, I. N., and Lau, N. C. (1987). Principal modes of atmospheric variability in model atmospheres with and without sea surface temperature forcing in the tropical Pacific. *J. Atmos. Sci.* **43**, 2719–2735.

Katz, E. J. (1981). Dynamic topography of the sea surface in the equatorial Atlantic, 1981. *J. Mar. Res.* **39**, 53–63.

Katz, E. J. (1984). Basin wide thermocline displacements along the equator off the Atlantic in 1983. *Geophys. Res. Lett.* **11**, 729–732.

Katz, E. J. (1987a). Seasonal response of the sea surface to the wind in the equatorial Atlantic. *JGR, J. Geophys. Res.* **92**, 1885–1893.

Katz, E. J. (1987b). Equatorial Kelvin waves in the Atlantic. *JGR, J. Geophys. Res.* **92**, 1894–1898.

Katz, E. J., and Garzoli, S. K. (1982). Response of the western Equatorial Atlantic Ocean to an annual wind cycle. *J. Mar. Res.* **40**, Suppl., 307–327.

Katz, E. J., and Garzoli, S. L. (1984). Thermocline displacement across the Atlantic North Equatorial Countercurrent during 1983. *Geophys. Res. Lett.* **11**, 737–740.

Katz, E. J., and Witte, J. (1987). "Further Progress in Equatorial Oceanography," A report on the U.S. TOGA Workshop on the Dynamics of the Equatorial Oceans, Honolulu, Hawaii, August 11–15, 1986. Nova Univ. Press, Fort Lauderdale, Florida.

Katz, E. J., and collaborators (1977). Zonal pressure gradient along the equatorial Atlantic. *J. Mar. Res.* **35**(2), 293–307.

Katz, E. J., Bruce, J. G., and Petrie, B. D. (1979). Salt and mass flux in the Atlantic Equatorial Undercurrent. *Deep-Sea Res.* **26**, Suppl. 2, 137–160.

Katz, E. J., Molinari, R. L, Cartwright, D. E., Hisard, P., Lass, H. U., and deMesquita, A. (1981). The seasonal transport of the equatorial undercurrent in the western Atlantic (during the global weather experiment). *Oceanol. Acta* **4**, 445–450.

Katz, E. J., Hisard, P., Verstraete, J. M., and Garzoli, S. L. (1986). Annual change of sea surface slope along the equator of the Atlantic Ocean in 1983 and 1984. *Nature (London)* **222**, 245–247.

Keen, R. A. (1982). The role of cross-equatorial cyclone pairs in the Southern Oscillation. *Mon. Weather Rev.* **110**, 1405–1410.

Keshavamurty, R. N. (1982). Response of the atmosphere to sea surface temperature anomalies over the equatorial Pacific and the teleconnections of the Southern Oscillation. *J. Atmos. Sci.* **39**, 1241–1259.

Kessler, W. S., and Taft, B. A. (1987). Dynamic heights and zonal geostrophic transports in the central Pacific during 1979–1984. *J. Phys. Oceanogr.* **7**, 97–122.

Khalsa, S. J. S. (1983). The role of sea surface temperature in large scale air–sea interaction. *Mon. Weather Rev.* **111**, 954–966.

Kidson, J. W. (1975). Tropical eigenvector analysis and the Southern Oscillation. *Mon. Weather Rev.* **103**, 187–196.

Kilonsky, B. J., and Ramage, C. S. (1976). A technique for estimating tropical open-ocean rainfall from satellite observations. *J. Appl. Meteorol.* **15**, 972–975.

Kindle, J. C. (1979). Equatorial Pacific ocean variability—Seasonal and El Niño time scales. Ph.D. Dissertation, Dept. of Oceanography, Florida State University, Tallahassee.

Knauss, J. A. (1963). The Equatorial Current system. *In* "The Sea" (M. N. Hill, ed.), Vol. 1, pp. 235–252, Wiley (Interscience), New York.

Knauss, J. A., and King, J. E. (1958). Observations of Pacific Equatorial Undercurrent. *Nature (London)* **182**, 601–602.

Knox, R. A. (1974). Reconnaissance of the Indian Ocean equatorial undercurrent near Addu Atoll. *Deep-Sea Res.* **21**, 123–129.

Knox, R. A. (1976). On a long series of measurements of Indian Ocean equatorial currents near Addu Atoll. *Deep-Sea Res.* **23**, 211–221.

Knox, R. A. (1981). Time variability of Indian Ocean equatorial currents. *Deep-Sea Res.* **28A**, 291–295.

Knox, R. A., and Anderson, D. L. T. (1985). Recent advances in the study of the low-latitude ocean circulation. *Prog. Oceanogr.* **14**, 259–317.

Knox, R. A., and Halpern, D. (1982). Long range Kelvin wave propagation of transport variations in Pacific Ocean equatorial currents. *J. Mar. Res.* **40**, Suppl., 329–339.

Knutson, T. R., Weickman, K. M., and Kutzbach, J. E. (1986). Global scale interseasonal oscillations of outgoing longwave radiation and 250 mb zonal wind during Northern Hemisphere summer. *Mon. Weather Rev.* **114**, 605–623.

Kousky, V. E., Kagano, M. T., and Cavalcant, I. F. (1984). A review of the Southern Oscillation: Oceanic–atmospheric circulation changes and related rainfall anomalies. *Tellus* **36A**, 490–504.

Kraus, E. B., ed. (1977). "Modelling and Prediction of the Upper Layers of the Ocean." Pergamon Press, Oxford.

Kraus, E. B., and Turner, J. S. (1967). A one-dimensional model of the seasonal thermocline. *Tellus* **19**, 98–106.

Krishnamurti, T. N. (1971). Tropical east–west circulations during the northern summer. *J. Atmos. Sci.* **28**, 1342–1347.

Krishnamurti, T. N., and Subrahmanyam, D. (1982). The 30–50 day mode at 850 mb during MONEX. *J. Atmos. Sci.* **39**, 2088–2095.

Krishnamurti, T. N., Kanamitsu, M., Koss, W. J., and Lee, J. D. (1973). Tropical east–west circulations during the northern winter. *J. Atmos. Sci.* **30**, 780–787.

Krueger, A. F., and Gray, T. (1969). Long-term variations in equatorial circulation and rainfall. *Mon. Weather Rev.* **97**, 700–711.

Krueger, A.F., and Winston, J. S. (1974). A comparison of the flow over the tropics during two contracting circulation regimes. *J. Atmos. Sci.* **31**, 358–370.

Kuo, H. L. (1965). On the formation and intensification of tropical cyclones through latent heat release by cumulus convection. *J. Atmos. Sci.* **23**, 40–63.

Lamb, P. J., Peppler, R. A., and Hastenrath, S. (1986). Interannual variability in the tropical Atlantic. *Nature (London)* **322**, 238–240.

Lass, H. U., Bubnov, V., Huthnance, J. M., Katz, E. J., Meincke, J., deMesquita, A., Ostapoff, F., and Voituriez, B. (1983). Seasonal changes of the zonal pressure gradient in the equatorial Atlantic during the FGGE year. *Oceanol. Acta* **6**, 3–11.

Latif, M. (1987). Tropical ocean circulation experiments. *J. Phys. Oceanogr.* **17**, 246–263.

Latif, M., Biercamp, J., and von Storch, H. (1988). The response of a coupled Ocean–Atmosphere General Circulation Model to wind bursts. *J. Atmos. Sci.* **45**, 964–979.

Lau, K.-M. (1981). Oscillations in a simple equatorial climate system. *J. Atmos. Sci.* **38**, 248–261.

Lau, K.-M. (1985). Elements of a stochastic-dynamical theory of the long-term variability of El Niño/Southern Oscillation. *J. Atmos. Sci.* **42**, 1552–1558.

Lau, K.-M., and Chan, P. H. (1986). The 40–50 day oscillation and the El Niño Southern Oscillation. *Bull. Am. Meteorol. Soc.* **67**, 533–534.

Lau, N. C. (1981). A diagnostic study of recurrent meteorological anomalies appearing in a 15 year simulation with a GFDL general circulation model. *Mon. Weather Rev.* **109**, 2287–2311.

Lau, N. C. (1985). Modeling the seasonal dependence of the atmospheric responses to observed El Niños 1962–1976. *Mon. Weather Rev.* **113**, 1970–1996.

Lau, N.C., and Lau, K.-M. (1986). The structure and propagation of intraseasonal oscillation appearing in a GFDL general circulation model. *J. Atmos. Sci.* **43**, 2023–2047.

Leetmaa, A. (1972). The response of the Somali Current to the southwest monsoon of 1970. *Deep-Sea Res.* **19**, 319–325.

Leetmaa, A. (1983). The role of local heating in producing temperature variations in the eastern tropical Pacific. *J. Phys. Oceanogr.* **13**, 467–474.

Leetmaa, A., and Ming, J. (1988). Operational hindcasting of the tropical Pacific. *Dyn. Atmos. Oceans* (in press).

Leetmaa, A., and Spain, P. (1981). Results from a velocity transect along the equator from 125° to 159°W. *J. Phys. Oceanogr.* **11**, 1030–1032.

Leetmaa, A., and Stommel, H. (1980). Equatorial current observations in the Western Indian Ocean in 1975 and 1976. *J. Phys. Oceanogr.* **10**, 258–269.

Leetmaa, A., Rossby, H. T., Saunders, P. M., and Wilson, P. (1980). Subsurface circulation in the Somali current. *Science* **209**, 590–592.

Leetmaa, A., McCreary, J. P., Jr., and Moore, D. W. (1981). Equatorial currents; observations and theory. *In* "Evolution of Physical Oceanography," (B. A. Warren and C. Wunsch, eds.), pp. 184–196, MIT Press, Cambridge, Massachusetts.

Leetmaa, A., Quadfasel, D. R., and Wilson, D. (1982). Development of the flow field during the onset of the Somali Current, 1979. *J. Phys. Oceanogr.* **12**, 1325–1342.

Leetma, A., Behringer, D. W., Huyer, A., Smith, R. L., and Toole, J. (1987). Hydrographic conditions in the eastern Pacific before, during and after the 1982–83 El Niño. *Prog. Oceanogr.* **19**, 1–47.

Legeckis, R. (1977). Long waves in the eastern equatorial Pacific; a view from a geostationary satellite. *Science* **197**, 1177–1181.

Legeckis, R. (1986). A satellite time series of sea surface temperatures in the eastern equatorial Pacific Ocean, 1982–1986. *JGR, J. Geophys. Res.* **91**, 12879–12886.

Levitus, S. (1982). "Climatological Atlas of the World Ocean," NOAA Prof. Pap. 13. U.S. Govt. Printing Office, Washington, D.C.

Levitus, S. (1984). Annual cycle of temperature and heat storage in the world ocean. *J. Phys. Oceanogr.* **14**, 727–746.

Lighthill, M. J. (1969). Dynamic response of the Indian Ocean to the onset of the Southwest Monsoon. *Philos. Trans. R. Soc. London, Ser. A* **265**, 45–93.

Lim, H., and Chang, C. P. (1983). Dynamics of teleconnections and Walker circulations forced by equatorial heating. *J. Atmos. Sci.* **40**, 1897–1915.

Lindstrom, E., Lukas, R., Fine, R., Firing, E., Godfrey, S., Meyers, G., and Tsuchiya, M. (1987). The western equatorial Pacific Ocean Circulation Study. *Nature (London)* **330**, 533–537.

Lindzen, R. S. (1967). Planetary waves on β planes. *Mon. Weather Rev.* **95**, 441–451.

Lindzen, R. S., and Nigam, S. (1987). On the role of sea surface temperature gradients in forcing low level winds and convergence in the tropics. *J. Atmos. Sci.* **44**, 2418.

Lockyer, N., and Lockyer, W. J. S. (1902a). On some phenomena which suggest a short period of solar and meteorological changes. *Proc. R. Soc. London* **70**, 500.

Lockyer, N., and Lockyer, W. J. S. (1902b). On the similarity of the short-period pressure variation over large areas. *Proc. R. Soc. London* **71**, 134–135.

Lockyer, N., and Lockyer, W. J. S. (1904). The behavior of the short-period atmospheric pressure variation over the earth's surface. *Proc. R. Soc. London* **73**, 457–470.

Lockyer, W. J. S. (1906). Barometric variations of long duration over large areas. *Proc. R. Soc. London, Ser. A* **78**, 43–60.

Lonquet-Higgins, M. (1968). The eigenfunctions of Laplace's Tidal Equations over a sphere. *Philos. Trans. R. Soc. London, Ser. A* **262**, 511–607.

Longquet-Higgins, M., and Bond, S. (1970). The free oscillations of fluid on a hemisphere bounded by meridians of longitude. *Philos. Trans. R. Soc. London, Ser. A* **266**, 193–223.

Lorenz, E. N. (1984). Irregularity: A fundamental property of the atmosphere. *Tellus* **36A**, 98–111.

Lough, J. M. (1986). Tropical Atlantic sea surface temperatures and rainfall variations in sub saharan Africa. *Mon. Weather Rev.* **114**, 561–570.

Lough, J. M., and Fritts, H. C. (1985). The Southern Oscillation and tree rings. 1600–1961. *J. Clim. Appl. Meteorol.* **24**, 952–966.

Lukas, R. (1986). The termination of the Equatorial Undercurrent in the eastern Pacific. *Prog. Oceanogr.* **16**, 63–90.

Lukas, R. (1987). TOGA western tropical Pacific air–sea interaction studies. *In* "Further Progress in Equatorial Oceanography" (E. Katz and J. Witte, eds.). Nova Univ. Press, Fort Lauderdale, Florida.

Lukas, R., and Firing, E. (1983). The geostrophic balance of the Pacific Equatorial Undercurrent. *Deep-Sea Res.* **31**, 61–66.

Lukas, R., Hayes, S. P., and Wyrtki, K. (1984). Equatorial sea level response during the 1982–83 El Niño. *JGR, J. Geophys. Res.* **89**, 10425–10430.

Luther, D. S. (1980). Observations of long period waves in the tropical oceans and atmosphere. Ph.D. Thesis, Massachusetts Institute of Technology, Cambridge.

Luther, D.S., Harrison, D. E., and Knox, R. A. (1983). Zonal winds in the central equatorial Pacific and El Niño. *Science* **222**, 327–330.

Luther, M. (1987). Indian Ocean modeling. *In* "Further Progress in Equatorial Oceanography" (E. Katz and J. Witte, eds.). Nova Univ. Press, Fort Lauderdale, Florida.

Luyten, J. R. (1982). Equatorial current measurements. I. Moored observations. *J. Mar. Res.* **40**, 19–40.

Luyten. J. R., and Swallow, J. (1976). Equatorial undercurrents. *Deep-Sea Res.* **23**, 1005–1007.

Luyten, J. R., and Roemmich, D. H. (1982). Equatorial currents at semi-annual period in the Indian Ocean. *J. Phys. Oceanogr.* **12**, 406–413.

Luyten, J. R., Fieux, M., and Gonella, J. (1980). Equatorial currents in the western Indian Ocean. *Science* **209**, 600–602.

Madden, R. A., and Julian, P. R. (1971). Detection of a 40–50 day oscillation in the zonal wind in the tropical Pacific. *J. Atmos. Sci.* **28**, 702–708.

Madden, R. A., and Julian, P. R. (1972). Description of global scale circulation cells in the tropics with a 40–50 day period. *J. Atmos. Sci.* **29**, 1109–1123.

Manabe, S., and Hahn, D. G. (1981). Simulation of atmospheric variability. *Mon. Weather Rev.* **109**, 2260–2286.

Manabe, S., Smagorinsky, J., and Strickler, R. F. (1965). Simulated climatology of a general circulation model with a hydrological cycle. *Mon. Weather Rev.* **93**, 769–798.

Marinone, S. G., and Ripa, P. (1982). Energetics of the instability of depth-independent equatorial jet. *Geophys. Astrophys. Fluid Dyn.* **30**, 105–130.

Matsuno, T. (1966). Quasi-geostrophic motions in equatorial areas. *J. Meteorol. Soc. Jpn.* **2**, 25–43.

McBride, J. L., and Nicholls, N. (1983). Seasonal relations between Australian rainfall and the Southern Oscillation. *Mon. Weather Rev.* **111**, 1998–2004.

McCreary, J. (1976). Eastern tropical ocean response to changing wind systems with application to El Niño. *J. Phys. Oceanogr.* **6**, 632–645.

McCreary, J. (1981a). A linear stratified ocean model of the equatorial undercurrent. *Philos. Trans. R. Soc. London* **298**, 603–635.

McCreary, J. (1981b). A linear stratified ocean model of the coastal undercurrent. *Philos. Trans. R. Soc. London* **302**, 385–413.

McCreary, J. (1983). A model of tropical ocean–atmosphere interaction. *Mon. Weather Rev.* **111**, 370–389.

McCreary, J. (1984). Equatorial beams. *J. Mar. Res.* **42**, 395–430.

McCreary, J. (1985). Modeling equatorial oceanic circulation. *Annu. Rev. Fluid Mech.* **17**, 359–409.

McCreary, J., and Anderson, D. L. T. (1984). A simple model of El Niño and the Southern Oscillation. *Mon. Weather Rev.* **112**, 934–946.

McCreary, J., and Kundu, P. K. (1985). Western boundary circulation driven by an alongshore wind; with application to the Somali Current system. *J. Mar. Res.* **43**, 493–516.

McCreary, J., and Lukas, R. (1986). The response of the equatorial ocean to a moving wind field. *JGR, J. Geophys. Res.* **91**, 11691–11705.

McCreary J., Picaut, J., and Moore, D. W. (1984). Effects of remote annual forcing in the eastern tropical Atlantic Ocean. *J. Mar. Res.* **42**, 45–81.

McCreary, J. P., Kundu, P. K., and Chao, S-Y. (1987). On the dynamics of the California Current System. *J. Mar. Res.* **45**, 1–32.

McPhaden, M. J. (1981). Continuously stratified models of the steady-state equatorial ocean. *J. Phys. Oceanogr.* **11**, 337–354.

McPhaden, M. J. (1982a). Variability in the central equatorial Indian Ocean. Part I. Ocean dynamics. *J. Mar. Res.* **40**, 157–176.

McPhaden, M. J. (1982b). Variability in the central equatorial Indian Ocean. Part II. Oceanic heat and turbulent energy balances. *J. Mar. Res.* **40**, 403–419.

McPhaden, M. J. (1986). The Equatorial Undercurrent: 100 years of discovery. *EOS, Trans. Am. Geophys. Union* **67**, 762–765.

McPhaden, M. J., and Knox, R. A. (1979). Equatorial Kelvin and inertia-gravity waves in zonal shear flow. *J. Phys. Oceanogr.* **9**, 263–277.

McPhaden, M. J., Proehl, J. A., and Rothstein, L. M. (1986). The interaction of equatorial waves with realistically sheared zonal currents. *J. Phys. Oceanogr.* **16**, 1499–1515.

Meehl, G. A. (1987). The annual cycle and interannual variability in the tropical Pacific and Indian Ocean regions. *Mon. Weather Rev.* **115**, 27–50.

Meisner, B. N. (1976). "A Study of Hawaiian and Line Island Rainfall," Rep. UH-MET 76-4. Department of Meteorology, University of Hawaii, Honolulu.

Mellor, G. L., and Durbin, P. A. (1975). The structured dynamics of the ocean surface mixed layer. *J. Phys. Oceanogr.* **5**, 718–728.

Merle, J. (1980a). Seasonal heat budget in the Equatorial Atlantic Ocean. *J. Phys. Oceanogr.* **10**, 464–469.

Merle, J. (1980b). Variabilité thermique annuelle et interannuelle de l'océan Atlantique equatorial Est. L'hypothèse d'un ≪ El Niño ≫ Atlantique. *Oceanol. Acta* **3**, 209–220.

Merle, J. (1980c). Seasonal variation of heat-storage in the tropical Atlantic Ocean. *Oceanol. Acta* **3**, 455–463.

Merle, J. (1983). Seasonal variability of the subsurface thermal structure in the tropical Atlantic Ocean. *In* "Proceedings of the 14th Liege Colloquium on Ocean Hydrodynamics" (J. Nihoul, ed.), Elsevier Oceanogr. Ser., pp. 31–49, Elsevier, Amsterdam.

Merle, J., and Arnault, S. (1985). Seasonal variability of the surface dynamic topography in the tropical Atlantic Ocean. *J. Mar. Res.* **43**, 267–288.

Merle, J., and LeFloch, J. (1978). Cycle annuel moyen de la température dans les couches supérieures de l'Océan Atlantique intertropical. *Oceanol. Acta* **1**, 271–276.

Merle, J., Rotschi, H., and Voituriez, B. (1969). Zonal circulation in the tropical western South Pacific at 170°E. *Bull. Jpn. Soc. Fish. Oceanogr. Spec. No.* Prof. Uda's Commem. Pap., pp. 91–98.

Merle, J., Fieux, M., and Hisard, P. (1979). Annual signal and interannual anomalies of sea surface temperature in the eastern equatorial Atlantic Ocean. *Deep-Sea Res.* **26**, Suppl. 2, 77–102.

Metcalf, W. G., and Stalcup, M. C. (1967). Origin of the Atlantic Equatorial Undercurrent. *J. Geophys. Res.* **72**, 4959–4975.

Meyers, G. (1975). Seasonal variation in transport of the Pacific north equatorial current relative to the wind field. *J. Phys. Oceanogr.* **5**, 442–449.

Meyers, G. (1979a). On the annual Rossby wave in the tropical North Pacific Ocean. *J. Phys. Oceanogr.* **9**, 663–674.

Meyers, G. (1979b). Annual variation of the slope of the 14°C isotherm along the equator in the Pacific Ocean. *J. Phys. Oceanogr.* **9**, 885–891.

Meyers, G. (1980). Do Sverdrup transports account for the Pacific north equatorial countercurrent? *JGR, J. Geophys. Res.* **85**, 1073–1075.

Meyers, G. (1982). Interannual variation in sea level near Truk Island—A bimodal seasonal cycle. *J. Phys. Oceanogr.* **12**, 1161–1168.

Meyers, G., and Donguy, J. R. (1984). The North Equatorial Countercurrent and heat storage in the western Pacific Ocean during 1982–83. *Nature (London)* **312**, 258–260.

Meyers, G., White, W., and Hasunuma, K. (1982). Annual variation in baroclinic structure of the northwestern tropical Pacific. *Oceanogr. Trop.* **1**, 59–69.

Meyers, G., Donguy, J. R., and Reed, R. K. (1986). Evaporative cooling of the western equatorial Pacific Ocean by anomalous winds. *Nature (London)* **323**, 523–526.

Miller, L., Watts, D. R., and Wimbush, M. (1985). Oscillations of dynamic topography in the eastern equatorial Pacific. *J. Phys. Oceanogr.* **15**, 1759–1770.

Miller, L., Cheney, R., and Douglas, B. C. (1988). GEOSAT Altimeter observations of Kelvin waves and the 1986–87 El Niño. *Science* **239**, 52–54.

Mofjeld, H. O. (1981). An analytic theory on how friction affects free internal waves in the equatorial waveguide. *J. Phys. Oceanogr.* **11**, 1585–1590.

Molinari, R. L. (1982). Observations of eastward currents in the tropical South Atlantic Ocean: 1978–1979. *JGR, J. Geophys. Res.* **87**, 9707–9714.

Molinari, R. L. (1983). Observations of near-surface currents and temperature in the central and western tropical Atlantic Ocean. *JGR, J. Geophys. Res.* **88**, 4433–4438.

Molinari, R. L., and Hansen, D. V. (1987). Observational studies of near surface thermal budgets in the tropics. *In* "Further Progress in Equatorial Oceanography" (E. Katz and J. Witte, eds.). Nova Univ. Press, Fort Lauderdale, Florida.

Molinari, R. L., Voituriez, B., and Duncan, P. (1981). Observations in the subthermocline undercurrent of the equatorial south Atlantic ocean: 1978–1979. *Oceanol. Acta* **4**, 451–456.

Molinari, R. L., Festa, J. F., and Marmolejo, E. (1985). Evolution of sea-surface temperature in the tropical Atlantic Ocean during FGGE, 1979. II. Oceanographic fields and heat balance of the mixed layer. *J. Mar. Res.* **43**, 67–81.

Molinari, R. L., Garzoli, S. L., Katz, E. J., Harrison, D. E., Richardson, P. L., and Reverdin, G. (1986). A synthesis of the First GARP Global Experiment (FGGE) in the equatorial Atlantic Ocean. *Prog. Oceanogr.* **16**, 91–112.

Moore, D. W. (1968). Planetary-gravity waves in an equatorial ocean. Ph.D. Thesis, Harvard University, Cambridge, Massachusetts.

Moore, D. W., and Philander, S. G. H. (1977). Modelling of the tropical oceanic circulation. *In* "The Sea" (E. D. Goldberg, I. N. McCave, J. J. O'Brien, and J. H. Steele, eds.), Vol. 6, pp. 319–361, Wiley (Interscience), New York.

Moore, D. W., Hisard, P., McCreary, J. P., Merle, J., O'Brien, J. J., Picaut, J., Verstraete, J. M., and Wunsch, C. (1978). Equatorial adjustment in the eastern Atlantic. *Geophys. Res. Lett.* **5**, 637–640.

Moura, A. D., and Shukla, J. (1981). On the dynamics of droughts in northeast Brazil: Observations, theory and numerical experiments with a general circulation model. *J. Atmos. Sci.* **38**, 2653–2675.

Moum, J. N., and Caldwell, D. R. (1985). Local influences on shear-flow turbulence in the equatorial ocean. *Science* **230**, 315–316.

Moum, J. N., Osborn, T. R., and Crawford, W. R. (1986). Pacific equatorial turbulence: Revisited. *J. Phys. Oceanogr.* **16**, 1516–1522.

Murakami, T., Chen, L.-X., Xie, A., and Shrestha, M. L. (1986). Eastward propagation of 30–60 day perturbations as revealed from outgoing longwave radiation data. *J. Atmos. Sci.* **42**, 961–971.

Murphy, R. C. (1926). Oceanic and climatic phenomena along the west coast of South America during 1925. *Geogr. Rev.*, pp. 26–54.

Namias, J. (1969). Seasonal interaction between the North Pacific Ocean and the atmosphere during the 1960's. *Mon. Weather Rev.* **97**, 173–192.

Namias, J. (1976). Some statistical and synoptic characteristics associated with El Niño. *J. Phys. Oceanogr.* **6**, 130–138.

Neelin, D. (1987). Simple models of steady and low frequency circulations in the tropical atmosphere, with application to tropical air–sea interactions. Ph.D. Dissertation, Princeton University, Princeton, New Jersey.

Neelin, D. (1988). A simple model for surface stress and low level flow in the tropical atmosphere driven by prescribed heating. *Q. J. R. Meteorol. Soc.* **114**, 747–770.

Neelin, D. (1989a). A note on the interpretation of the Gill Model. *J. Atmos. Sci.* (in press).

Neelin, J. D. (1989b). Interannual oscillations in an ocean GCM—Simple atmospheric model. *Philos. Trans. R. Soc. London* (in press).

Neelin, D., and Held, I. M. (1987). Modeling tropical convergence based on the moist static energy budget. *J. Atmos. Sci.* **115**, 3–12.

Neelin, D., Held, I. M., and Cook, K. H. (1987). Evaporation–wind feedback and low frequency variability in the tropical atmosphere. *J. Atmos. Sci.* **44**, 2341–2346.

Newell, R. E., and Kidson, J. W. (1984). African mean wind changes between Sahelian wet and dry periods. *J. Clim.* **4**, 27–33.

Newell, R. E., and Weare, B. C. (1976). Ocean temperatures and large-scale atmospheric variations. *Nature (London)* **262**, 40–41.

Nicholls, N. (1983). Predicting Indian monsoon rainfall from sea surface temperature in the Indonesia–North Australia area. *Nature (London)* **306**, 576–577.

Nicholls, N. (1984a). The Southern Oscillation, sea surface temperature and interannual fluctuations in Australian Cyclone activity. *J. Clim.* **4**, 661–670.

Nicholls, N. (1984b). The Southern Oscillation and Indonesian sea surface temperature. *Mon. Weather Rev.* **112**, 424–432.

Niiler, P., and Stevenson, J. (1982). The heat budget of tropical ocean warm water pools. *J. Mar. Res.* **40**, 465–480.

Nitani, H. (1972). Beginning of the Kuroshio. *In* "Kuroshio, Physical Aspects of the Japan Current" (H. Stommel and K. Yoshida, eds.). Univ. of Washington Press, Seattle.

Nof, D. (1981). On the dynamics of equatorial outflows with application to the Amazon's basin. *J. Mar. Res.* **39**, 1–29.

O'Brien, J. J., and Hurlburt, H. E. (1974). Equatorial jet in the Indian Ocean. *Science* **124**, 1075–1077.

O'Brien, J. J., Busalacchi, A., and Kindle, J. (1981). Ocean models of El Niño. *In* "Resource

Management and Environmental Uncertainty" (M. H. Glantz and J. D. Thompson, eds.), pp. 159–212, Wiley, New York.

Oort, A. H. (1984). "Global Atmospheric Circulation Statistics 1958–1973," NOAA Prof. Pap. No. 14. U.S. Govt. Printing Office, Washington, D.C.

Oort, A. H., and von der Haar, T. H. (1976). On the observed annual cycle in the ocean–atmosphere heat balance over the northern hemisphere. *J. Phys. Oceanogr.* **6**, 781–800.

Ooyama, K. (1969). Numerical simulation of the life-cycle of tropical cyclones. *J. Atmos. Sci.* **26**, 3–40.

Ooyama, K. (1964). A dynamical model for the study of tropical cyclone development. *Geofis. Int.* **4**(4), 187.

Osborn, T. R., and Bilodeau, L. E. (1980). Temperature microstructure in the Atlantic Equatorial Undercurrent. *J. Phys. Oceanogr.* **10**, 66–82.

Pacanowski, R., and Philander, S. G. H. (1981). Parameterization of vertical mixing in numerical models of tropical oceans. *J. Phys. Oceanogr.* **11**, 1443–1451.

Palmer, T. N. (1986). Influence of the Atlantic, Pacific and Indian Oceans on Sahel rainfall. *Nature (London)* **322**, 251–253.

Palmer, T. N., and Mansfield, D. A. (1984). Response of two atmospheric general circulation models to sea surface temperature anomalies in the tropical east and west Pacific. *Nature (London)* **310**, 483–485.

Pan, Y.-H., and Oort, A. H. (1983). Global climate variations connected with sea surface temperature anomalies in the eastern equatorial Pacific Ocean for the 1958–73 period. *Mon. Weather Rev.* **111**, 1244–1258.

Pan, G. B., and Parthasarathy, B. (1981). Some aspects of an association between the Southern Oscillation and Indian summer monsoon. *Arch. Meteorol., Geophys. Bioklimatol., Ser. B* **29**, 245–252.

Pares-Sierra, A. F., Inoue, M., and O'Brien, J. J. (1985). Estimates of oceanic heat transport in the tropical Pacific. *JGR, J. Geophys. Res.* **90**, 3293–3303.

Pazan, S., and Meyers, G. (1982). Pacific trade wind fluctuations and the Southern Oscillation Index. *Mon. Weather Rev.* **110**, 587–600.

Pazan, S. E., White, W. B., Inoue, M., and O'Brien, J. J. (1986). Off-equatorial influence upon Pacific Equatorial Dynamic height variability during the 1982–83 El Niño/Southern Oscillation event. *JGR, J. Geophys. Res.* **91**, 8437–8449.

Pedlosky, J. (1979). "Geophysical Fluid Dynamics." Springer-Verlag, New York.

Pedlosky, J. (1987). An inertial theory of the Equatorial Undercurrent. *J. Phys. Oceanogr.* **17**, 1978–1985.

Peters, H., Gregg, M. C., Caldwell, D., and Moum, J. N. (1987). Equatorial vertical mixing. *In* "Further Progress in Equatorial Oceanography" (E. J. Katz and J. M. Witte, eds.). Nova Univ. Press, Fort Lauderdale, Florida.

Peters, H., Gregg, M., and Toole, J. M. (1988). On the parameterization of Equatorial Turbulence. *JGR, J. Geophys. Res.* **93**, 1199–1218.

Pezet, F. A. (1896). The counter-current El Niño on the coast of northern Peru. *Geogr. J.* **7**, 603–606.

Philander, S. G. H. (1973). Equatorial Undercurrent: Measurements and theories. *Rev. Geophys. Space Phys.* **2**(3), 513–570.

Philander, S. G. H. (1976). Instabilities of zonal equatorial currents. I. *JGR, J. Geophys. Res.* **81**(21), 3725–3735.

Philander, S. G. H. (1978a). Forced oceanic waves. *Rev. Geophys. Space Phys.* **16**(1), 15–46.

Philander, S. G. H. (1978b). Instabilities of zonal equatorial currents. II. *JGR, J. Geophys. Res.* **81**(21), 3725–3735.

Philander, S. G. H. (1979a). Equatorial waves in the presence of the Equatorial Undercurrent. *J. Phys. Oceanogr.* **9**, 254–262.

Philander, S. G. H. (1979b). Nonlinear equatorial and coastal jets. *J. Phys. Oceanogr.* **9**, 739–747.

Philander, S. G. H. (1981). The response of equatorial oceans to a relaxation of the trade winds. *J. Phys. Oceanogr.* **11**, 176–189.

Philander, S. G. H. (1985). El Niño and La Niña. *J. Atmos. Sci.* **42**, 2652–2662.

Philander, S. G. H. (1986). Unusual conditions in the tropical Atlantic Ocean in 1984. *Nature (London)* **222**, 236–238.

Philander, S. G. H., and Delecluse, P. (1983). Coastal currents in low latitudes (with application to the Somali and El Niño Currents). *Deep-Sea Res.* **30**, 887–902.

Philander, S. G. H., and Hurlin, W. J. (1987). Initial conditions for General Circulation Models of tropical oceans. *J. Phys. Oceanogr.* **17**, 147–157.

Philander, S. G. H., and Hurlin, W. J. (1988). The heat budget of the tropical Pacific in a simulation of the 1982–83 El Niño. *J. Phys. Oceanogr.* **18**, 926–931.

Philander, S. G. H., and Pacanowski, R. C. (1980). The generation of equatorial currents. *JGR, J. Geophys. Res.* **85**, 1123–1136.

Philander, S. G. H., and Pacanowski, R. C. (1981a). The response of equatorial oceans to periodic forcing. *JGR, J. Geophys. Res.* **86**, 1903–1916.

Philander, S. G. H., and Pacanowski, R. C. (1981b). The oceanic response to cross-equatorial winds (with application to coastal upwelling in low latitudes). *Tellus* **33**, 201–210.

Philander, S. G. H., and Pacanowski, R. C. (1986a). A model of the seasonal cycle in the tropical Atlantic Ocean. *JGR, J. Geophys. Res.* **91**, 14192–14206.

Philander, S. G. H., and Pacanowski, R. C. (1986b). The mass and heat budgets in a model of the tropical Atlantic. *JGR, J. Geophys. Res.* **91**, 14212–14220.

Philander, S. G. H., and Seigel, A. D. (1985). Simulation of El Niño of 1982–83. *In* "Coupled Ocean–Atmosphere Models" (J. Nihoul, ed.), pp. 517–541, Elsevier, Amsterdam.

Philander, S. G. H., and Yoon, J.-H. (1982). Eastern boundary currents and coastal upwelling. *J. Phys. Oceanogr.* **12**, 862–879.

Philander, S. G. H., Yamagata, T., and Pacanowski, R. C. (1984). Unstable air–sea interactions in the tropics. *J. Atmos. Sci.* **41**, 604–613.

Philander, S. G. H., Halpern, D., Hansen, D., Legeckis, R., Miller, L., Paul, C., Watts, R., Weisberg, R., and Winbush, M. (1985). Long waves in the Equatorial Pacific Ocean. *EOS, Trans. Am. Geophys. Union* **66**, 154.

Philander, S. G. H., Hurlin, W., and Pacanowski, R. C. (1986). Properties of long equatorial waves in models of the seasonal cycle in the tropical Atlantic and Pacific Oceans. *JGR, J. Geophys. Res.* **91**, 14207–14211.

Philander, S. G. H., Hurlin, W., and Seigel, A. D. (1987). A model of the seasonal cycle in the tropical Pacific Ocean. *J. Phys. Oceanogr.* **17**, 1986–2002.

Picaut, J. (1983). Propagation of the seasonal upwelling in the eastern equatorial Atlantic, *J. Phys. Oceanogr.* **13**, 18–37.

Picaut, J., and Verstraete, J. M. (1979). Propagation of a 14.7 day wave along the northern coast of the Guinea Gulf. *J. Phys. Oceanogr.* **9**, 136–149.

Pittock, A. B. (1975). Climatic change and the patterns of variation in Australian rainfall. *Search* **6**, 498–504.

Quadfasel, D. R. (1982). Low frequency variability of the 20°C isotherm topography in the western Equatorial Indian Ocean. *JGR, J. Geophys. Res.* **87**, 1990–1996.

Quinn, W. H. (1974). Monitoring and predicting El Niño invasions. *J. Appl. Meteorol.* **13**, 825–830.

Quinn, W. H., and Burt, W. V. (1972). Use of the Southern Oscillation in weather prediction. *J. Appl. Meteorol.* **11**, 616–628.

Quinn, W. H., Zopf, D. O. Short, K. S., and Kuo Yang, R. T. W. (1978). Historical trends and statistics of the Southern Oscillation, El Niño, and Indonesian droughts. *Fish. Bull.* **76**, 663–678.

Ramage, C. S. (1968). Role of a tropical maritime continent in the atmospheric circulation. *Mon. Weather Rev.* **96**, 365–370.

Ramage, C. S. (1975). Preliminary discussion of the meteorology of the 1972–73 El Niño. *Bull. Am. Meteorol. Soc.* **56**, 234–242.

Ramage, C. S. (1977). Sea surface temperature and local weather. *Mon. Weather Rev.* **105**, 540–544.

Ramage, C. S., and Hori, A. M. (1981). Meteorological aspects of El Niño. *Mon. Weather Rev.* **109**, 1827–1835.

Ramage, C. S., Adams, C. W., Hori, A. M., Kilonsky, B. J., and Sadler, J. C. (1980). "Meteorological Atlas of the 1972–73 El Niño," UHMET 80-03. Dept. of Meteorology, University of Hawaii, Honolulu.

Rasmusson, E. M., and Carpenter, T. H. (1982a). Variations in tropical sea surface temperature and surface wind fields associated with the Southern Oscillation/El Niño. *Mon. Weather Rev.* **110**, 354–384.

Rasmusson, E. M., and Carpenter, T. H. (1982b). The relationship between eastern equatorial Pacific sea surface temperatures and summer monsoon rainfall over India and Sri Lanka. *Mon. Weather Rev.* **111**, 517–528.

Rasmusson, E. M., and Wallace, J. M. (1983). Meteorological aspects of the El Niño/Southern Oscillation. *Science* **222**, 1195–1202.

Reiter, E. R. (1978). Long-term wind variability in the Tropical Pacific, its possible causes and effects. *Mon. Weather Rev.* **106**, 324–330.

Reverdin, G., and Luyten, J. (1986). Near surface meanders in the Equatorial Indian Ocean. *J. Phys. Oceanogr.* **16**, 1088–1100.

Richardson, P. L., and McKee, J. K. (1984). Average seasonal variation of the Atlantic North Equatorial Countercurrent from ship drift data. *J. Phys. Oceanogr.* **14**, 1226–1238.

Richardson, P. L., and Philander, S. G. H. (1986). Variability of surface currents in the tropical Atlantic Ocean: Ship-drift data and results from a model compared. *JGR, J. Geophys. Res.* **92**, 715–724.

Richardson, P. L., and Reverdin, G. (1987). Seasonal cycle of velocity in the Atlantic North Equatorial Countercurrent as measured by surface drifters, current meters and ship drifts. *JGR, J. Geophys. Res.* **92**, 3691–3708.

Richardson, P. L., and Walsh, D. W. (1986). Mapping climatological seasonal variations of surface currents in the tropical Atlantic using ship drifts. *JGR, J. Geophys. Res.* **91**, 10537–10550.

Ripa, P. (1982). Nonlinear wave–wave interactions in a one-layer reduced-gravity model of the equatorial β-plane. *J. Phys. Oceanogr.* **12**, 97–111.

Ripa, P. (1983). General stability conditions for zonal flows in a one-layer model on the beta-plane or the sphere. *J. Fluid Mech.* **126**, 463–487.

Ripa, P., and Hayes, S. (1981). Evidence for equatorially trapped waves at the Galápagos Islands. *JGR, J. Geophys. Res.* **86**, 6509–6516.

Ripa, P., and Marinone, S. G. (1983). The effect of zonal currents on equatorial waves. *In* "Hydrodynamics of the Equatorial Ocean" (J. C. J. Nihoul, ed.), pp. 291–317, Elsevier Oceanogr. Ser., Elsevier, Amsterdam.

Robinson, A. R. (1966). An investigation into the wind as the cause of the Equatorial Undercurrent. *J. Mar. Res.* **24**, 179–204.

Roemmich, D. (1983). The balance of geostrophic and Ekman transports in the tropical Atlantic. *J. Phys. Oceanogr.* **13**, 1534–1539.

Roemmich, D. (1987). Estimates of net transport, upwelling, and heat flux in the tropical oceans from inverse methods and related geostrophic models. *In* "Further Progress in Equatorial Oceanography" (E. Katz and J. Witte, eds.). Nova Univ. Press, Fort Lauderdale, Florida.

Ropelewski, C. F., and Halpert, M. S. (1987). Global and regional scale precipitation associated with El Niño/Southern Oscillation. *Mon. Weather Rev.* **115**, 1606–1626.

Rothstein, L. M., Moore, D. W., and McCreary, J. P. (1985). Interior reflections of a periodically forced equatorial Kelvin wave. *J. Phys. Oceanogr.* **15**, 985–996.

Rowlands, P. B. (1982). The flow of equatorial Kelvin waves and the Equatorial Undercurrent around islands. *J. Mar. Res.* **40**, 915–936.

Rowntree, P. R. (1972). The influence of tropical east Pacific Ocean temperature on the atmosphere. *Q. J. R. Meteorol. Soc.* **98**, 290–321.

Salby, M. L., and Garcia, R. R. (1987). Transient response to localized episodic heating in the tropics. Part I. Excitation and short-term near-field behavior. *J. Atmos. Sci.* **44**, 503–529.

Sarachik, E. S. (1978). Tropical sea-surface temperature: An interactive one-dimensional atmosphere–ocean model. *Dyn. Atmos. Oceans* **2**, 455–469.

Sardeshmukh, P. D., and Hoskins, B. J. (1985). Vorticity balance in the tropics during the 1982–1983 El Niño–Southern Oscillation event. *Q. J. R. Meteorol. Soc.* **111**, 261–278.

Sarmiento, J. L. (1986). On the north and tropical Atlantic heat balance. *JGR, J. Geophys. Res.* **91**, 11677–11690.

Schopf, P. S., and Cane, M. A. (1983). On equatorial dynamics, mixed layer physics and sea surface temperature. *J. Phys. Oceanogr.* **13**, 917–935.

Schopf, P. S. , and Suarez, M. J. (1988). Vacillations in a coupled ocean–atmosphere model. *J. Atmos. Sci.* **45**, 549.

Schopf, P. S., Anderson, D. L. T., and Smith, R. (1981). Beta-dispersion of low frequency Rossby waves. *Dyn. Atmos. Ocean* **5**, 187–214.

Schott, F. (1983). Monsoon response of the Somali Current and associated upwelling. *Prog. Oceanogr.* **12**, 357–381.

Schott, F. (1987). Recent studies of western Indian Ocean Circulation. *In* "Further Progress in Equatorial Oceanography" (E. J. Katz and J. M. Witte, eds.). Nova Univ. Press, Fort Lauderdale, Florida.

Schott, F., and Quadfasel, D. R. (1980). Development of the subsurface currents of the northern Somali current gyre from March to July 1979. *Science* **209**, 593–595.

Schreiber, R. W., and Schreiber, E. A. (1984). Central Pacific seabirds and the El Niño Southern Oscillation: 1982 to 1983 perspectives. *Science* **225**, 713–716.

Sears, A. F. (1895). The coast desert of Peru. *Bull. Am. Geogr. Soc.* **28**, 256–271.

Semtner, A. J., and Holland, W. R. (1980). Numerical simulation of Equatorial Ocean Circulation. Part I. A basic case in turbulent equilibrium. *J. Phys. Oceanogr.* **10**, 667–693.

Servain, J., and Legler, D. M. (1986). Empirical orthogonal function analyses of tropical Atlantic sea surface temperature and windstress: 1964–1979. *JGR, J. Geophys. Res.* **91**, 14181–14191.

Servain, J., and Seva, M. (1987). On relationships between tropical Atlantic sea surface temperature, windstress, and regional precipitation indices: 1964–1984. *Ocean-Air Interact.* **1**, 183–190.

Shannon, L. V., Boyd, A. J., Brundrit, G. B., and Taunton-Clark, J. (1986). On the existence of an El Niño type phenomenon in the Benguela system. *J. Mar. Res.* **44**, 495–520.

Shen, G. T., and Boyle, E. A. (1984). Lead and cadmium in corals: Tracers of global industrial fallout and paleo-upwelling. *EOS, Trans. Am. Geophys. Union* **65**, 964.

Skukla, J. (1975). Effect of Arabian sea surface temperature anomaly on Indian summer monsoon: A numerical experiment with the GFDL model. *J. Atmos. Sci.* **32**, 503–511.

Shukla, J., and Misra, M. (1977). Relationships between sea surface temperature over the central Arabian Sea, and monsoon rainfall over India. *Mon. Weather Rev.* **105**, 998–1002.

Shukla, J., and Charney, J. G. (1981). Predictability of the monsoons. *In* "Monsoon Dynamics" (J. Lighthill and R. Pierce, eds.), pp. 99–110, Cambridge Univ. Press, London and New York.

Shukla, J., and Mooley, D. A. (1987). Empirical prediction of the summer monsoon rainfall over India. *Mon. Weather Rev.* **115**, 695–703.

Shukla, J., and Paolino, D. A. (1983). The Southern Oscillation and long-range forecasting of the summer monsoon rainfall over India. *Mon. Weather Rev.* **111**, 1830–1837.

Shukla, J., and Wallace, J. M. (1983). Numerical simulation of the atmospheric response to equatorial Pacific sea surface temperature anomalies. *J. Atmos. Sci.* **40**, 1613–1630.

Sikka, D. R., and Gadgil, S. (1980). On the maximum cloud zone and the ITCZ over Indian longitudes during the southwest monsoon. *Mon. Weather Rev.* **108**, 1840–1853.

Simmons, A., Wallace, J. M., and Branstator, G. W. (1983). Barotropic wave propagation and instability, and atmospheric teleconnection patterns. *J. Atmos. Sci.* **40**, 1363–1392.

Simmons, I., and Smith, I. N. (1986). The effect of the prescription of zonally uniform sea surface temperatures in a General Circulation Model. *J. Climatol.* **6**, 641–659.

Simpson, J. J. (1984). A simple model of the 1982–83 Californian El Niño. *Geophys. Res. Lett.* **11**, 243–246.

Smith, R. L. (1978). Poleward propagating perturbations in current and sea levels along the Peru coast. *JGR, J. Geophys. Res.* **79**, 435–443.

Stevenson, J. W., and Niiler, P. P. (1983). Upper ocean heat budget during the Hawaii-to-Tahiti Shuttle Experiment. *J. Phys. Oceanogr.* **13**, 1894–1907.

Stommel, H. (1948). The westward intensification of wind-driven ocean currents. *Trans. Am. Geophys. Union* **29**, 202–206.

Stommel, H. (1980). Asymmetry of inter-oceanic fresh water and heat fluxes. *Proc. Natl. Acad. Sci. U.S.A.* **77**(5), 2377–2381.

Stone, P. H., and Chervin, R. M. (1984). The influences of ocean surface temperature gradient and continentality on the Walker Circulation. *Mon. Weather Rev.* **112**, 1524–1534.

Streten, N. A. (1983). Extreme distributions of Australian rainfall in relation to sea surface temperature. *J. Climatol.* **3**, 143–153.

Suarez, M. J., and Schopf, P. S. (1988). A delayed oscillator for ENSO. *J. Atmos. Sci.* **45**, 3283–3287.

Sverdrup, H. U. (1947). Winddriven currents in a baroclinic ocean with application to the equatorial currents in the eastern Pacific. *Proc. Natl. Acad. Sci., U.S.A.* **33**, 318–326.

Swallow, J. C. (1967). The Equatorial Undercurrent in the western Indian Ocean in 1964. *Stud. Trop. Oceanogr.* **5**, 15–36.

Swallow, J. C. (1981). Observations of the Somali current and its relationship to the monsoon winds. *In* "Monsoon Dynamics" (J. Lighthill and R. P. Pearce, eds.), pp. 444–452, Cambridge Univ. Press, London and New York.

Swallow, J. C., Molinari, R. L., Bruce, J. G., Brown, O. B., and Evans, R. H. (1983). Development of near surface flow pattern and water mass distribution in the Somali Basin in response to the southwest monsoon of 1979. *J. Phys. Oceanogr.* **13**, 1398–1415.

Taft, B. A. (1967). Equatorial Undercurrent in the Indian Ocean, 1963. *Stud. Trop. Oceanogr.* **5**, 3–14.

Talley, L. D. (1983). Meridional heat transport in the Pacific Ocean. *J. Phys. Oceanogr.* **14**, 231–241.

Tang, T. Y., and Weisberg, R. H. (1984). On the equatorial Pacific response to the 1982–83 El Niño–Southern Oscillation event. *J. Mar. Res.* **42**, 809–829.

Taylor, R. C. (1973). An atlas of Pacific island rainfall. *Hawaii Inst. Geophys.* [*Rep.*] *25*, **HIG-73-9**(25), 1–175 (NTIS No. AD 767073).

Thompson, L. G., Mosley-Thompson, E., and Arnao, B. M. (1984). El Niño Southern Oscillation events recorded in the stratigraphy of the tropical Quelcraya ice cap, Peru. *Science* **226**, 50–53.

Toole, J. M. (1981). Anomalous characteristics of equatorial thermohaline finestructure. *J. Phys. Oceanogr.* **11**, 871–876.

Trenberth, K. E. (1976). Spatial and temporal variations in the Southern Oscillation. *Q. J. R. Meteorol. Soc.* **102**, 639–653.

Trenberth, K. E. (1984). Signal versus noise in the Southern Oscillation. *Mon. Weather Rev.* **112**, 326–332.

Trenberth, K. E., and Shea, D. J. (1987). On the evolution of the Southern Oscillation. *Mon. Weather Rev.* **115**, 3078–3096.

Troup, A. J. (1965). The Southern Oscillation. *Q. J. R. Meteorol. Soc.* **91**, 490–506.

Tsuchiya, M. (1975). Subsurface Countercurrents in the eastern equatorial Pacific Ocean. *J. Mar. Res.* **33**, 125–175.

Tsuchiya, M. (1979). Seasonal variation of the equatorial zonal geopotential gradient in the eastern Pacific Ocean. *J. Mar. Res.* **37**, 399–407.

Tsuchiya, M. (1981). The origin of the Pacific equatorial 13°C water. *J. Phys. Oceanogr.* **11**, 794–812.

van Loon, H. (1984). The Southern Oscillation. Part III. Associations with the trades and with the trough in the westerlies of the South Pacific Ocean. *Mon. Weather Rev.* **112**, 947–954.

van Loon, H., and Madden, R. A. (1981). The Southern Oscillation. Part I. Global associations with pressure and temperature in northern winter. *Mon. Weather Rev.* **109**, 1150–1162.

van Loon, H., and Shea, D. J. (1985). The Southern Oscillation. Part IV. The precursors south of 15°S to the extremes of the oscillation. *Mon. Weather Rev.* **113**, 2063–2074.

Veronis, G., and Stommel, H. (1956). The action of variable wind-stresses on a stratified ocean. *J. Mar. Res.* **15**, 43–69.

Verstraete, H.-M., and Picaut, J. (1983). Variations duniveau de la mer de la température de surface et des hauteurs dynamiques le long de la côte nord du Golfe de Guinée. *Cah. ORSTOM, Oceanogr. Trop.* **18**, 139–162.

Verstraete, H.-M., Picaut, J., and Morliere, A. (1980). Atmospheric and tidal observations along the shelf of the Guinea gulf. *Deep-Sea Res.* **26A**, Suppl. 2, 343–356.

Voigt, K. (1961). Aquatoriale Unterstromung auch im Atlantic. *Beitr. Meereskd.* **1**, 56–60.

Voituriez, B. (1983). Les variations saisonnieres des courants equatoriaux à 4°W et l'upwelling equatorial du Golfe de Guinée. *Cah. ORSTOM, Ser. Oceanogr.* **18**, 169–199.

Wacongne, S. (1988). The dynamics of the Equatorial Undercurrent and its termination. Ph.D. Thesis, Woods Hole Oceanographic Institution, Woods Hole, Massachusetts.

Walker, G. T. (1923). Correlation in seasonal variations of weather. VIII. A preliminary study of world weather. *Mem. Indian Meteorol. Dep.* **24**(4), 75–131.

Walker, G. T. (1924). Correlation in seasonal variations of weather. IX. A further study of world weather. *Mem. Indian Meteorol. Dep.* **24**(9), 275–332.

Walker, G. T. (1928). World weather. III. *Mem. R. Meteorol. Soc.* **2**, 97–106.

Walker, G. T., and Bliss, E. W. (1930). World weather. IV. *Mem. R. Meteorol. Soc.* **3**, 81–95.

Walker, G. T., and Bliss, E. W. (1932). World weather. V. *Mem. R. Meteorol. Soc.* **4**, 53–84.

Walker, G. T., and Bliss, E. W. (1937). World weather. VI. *Mem. R. Meteorol. Soc.* **4**, 119–139.

Wallace, J. M., and Gutzler, D. S. (1981). Teleconnections in the geopotential height field during the Northern Hemisphere winter. *Mon. Weather Rev.* **109**, 784–812.

Wallace, J. M., and Kousky, V. E. (1968). Observational evidence of Kelvin waves in the tropical stratosphere. *J. Atmos. Sci.* **25**, 900–907.

Warren, B. A. (1965). Medieval Arab references to the seasonally reversing currents of the North Indian Ocean. *Deep-Sea Res.* **13**, 167–171.

Weare, B. C. (1979). A statistical study of the relationships between ocean surface temperatures and the Indian monsoon. *J. Atmos. Sci.* **36**, 2279–2292.

Weare, B. C. (1982). El Niño and tropical Pacific ocean surface temperatures. *J. Phys. Oceanogr.* **12**, 17–27.

Weare, B. C. (1983). Interannual variation in net heating at the surface of the tropical Pacific Ocean. *J. Phys. Oceanogr.* **13**, 873–885.

Weare, B. C., and Strub, P. T. (1981). Annual mean atmospheric statistics at the surface of the tropical Pacific ocean. *Mon. Weather Rev.* **109**, 1002–1012.

Weare, B. C., Navato, A. R., and Newell, R. E. (1976). Empirical orthogonal analysis of Pacific sea surface temperatures. *J. Phys. Oceanogr.* **6**, 671–678.

Weare, B. C., Strub, P. T., and Samuel, M. D. (1981). Annual mean surface heat fluxes in the tropical Pacific ocean. *J. Phys. Oceanogr.* **11**, 705–717.

Webster, P. J. (1972). Response of the tropical atmosphere to local steady forcing. *Mon. Weather Rev.* **100**, 518–541.

Webster, P. J. (1981). Mechanisms determining the atmospheric response to sea surface temperature anomalies. *J. Atmos. Sci.* **38**, 554–571.

Webster, P. J. (1983). Mechanisms of monsoon low-frequency variability: Surface hydrological effects. *J. Atmos. Sci.* **40**, 2110–2124.

Webster, P. J., and Holton, J. R. (1982). Cross-equatorial response to middle latitude forcing in a zonally varying basic state. *J. Atmos. Sci.* **39**, 722–733.

Weickmann, K. M. (1983). Intraseasonal circulation and outgoing long wave radiation during northern hemisphere winter. *Mon. Weather Rev.* **111**, 1838–1858.

Weickmann, K. M., Lussky, G. R., and Kutzbach, J. E. (1985). Intraseasonal (30–60 day) fluctuations of outgoing longwave radiation and 250 mb streamfunction during northern winter. *Mon. Weather Rev.* **112**, 941–961.

Weisberg, R. H. (1984). Instability waves observed on the equator in the Atlantic Ocean during 1983. *Geophys. Res. Lett.* **11**, 753–756.

Weisberg, R. H., and Colin, C. (1986). Equatorial Atlantic Ocean temperature and current variations during 1983 and 1984. *Nature (London)* **322**, 240–243.

Weisberg, R. H., and Horigan, A. (1981). Low-frequency variability in the equatorial Atlantic. *J. Phys. Oceanogr.* **11**, 913–920.

Weisberg, R. H., and Weingartner, T. J. (1986). On the baroclinic response of the zonal pressure gradient in the Equatorial Atlantic Ocean. *JGR, J. Geophys. Res.* **91**, 11717–11725.

Weisberg, R. H., and Tang, T. Y. (1983). Equatorial Ocean response to growing and moving wind systems with application to the Atlantic. *J. Mar. Res.* **41**, 461–486.

Weisberg, R. H., and Tang, T. Y. (1985). On the response of the equatorial thermocline in the Atlantic Ocean to the seasonally varying trade winds. *JGR, J. Geophys. Res.* **90**, 7117–7128.

Weisberg, R. H., Horigan, A., and Colin, C. (1979). Equatorially trapped Rossby-gravity wave propagation in the Gulf of Guinea. *J. Mar. Res.* **37**, 67–86.

White, W. B., and McCreary, J. P. (1974). Eastern intensification of ocean spin-down: Application to El Niño. *J. Phys. Oceanogr.* **4**, 295–303.

White, W. B., Meyers, G., Donguy, J. R., and Pazan, S. E. (1985). Short-term climatic variability in the thermal structure of the Pacific Ocean during 1979–1982. *J. Phys. Oceanogr.* **15**, 917–935.

Witte, J. (1986). "El Niño Southern Oscillation and Physical Processes of the Tropical Oceans: A Bibliography." Univ. Corp. Atmos. Res. (UCAR). Available from J. Witte, Nova University Oceanographic Center, 8000 North Ocean Drive, Dania, Florida.

Wooster, W. S., and Guillen, O. (1974). Characteristics of El Niño in 1972. *J. Mar. Res.* **32**, 378–404.

Wright, P. B. (1977). The Southern Oscillation—Patterns and mechanisms of the teleconnections and the persistence. *Hawaii Inst. Geophys.* [*Rep.*] **HIG-77-13**, 1–107.

Wright, P. B. (1984a). Relationships between indices of the Southern Oscillation. *Mon. Weather Rev.* **112**, 1913–1919.

Wright, P. B. (1984b). Possible role of cloudiness in the persistence of the Southern Oscillation. *Nature (London)* **310**, 128–130.

Wright, P. B. (1986). Precursors of the Southern Oscillation. *J. Climatol.* **6**, 17–30.

Wunsch, C. (1967). Long period tides. *Rev. Geophys. Space Phys.* **5**, 447–475.

Wunsch, C. (1977). Response of an equatorial ocean to a periodic monsoon. *J. Phys. Oceanogr.* **7**, 497–511.

Wunsch, C. (1980). Meridional heat flux of the North Atlantic Ocean. *Proc. Natl. Acad. Sci. U.S.A.* **77**(9), 5043–5047.

Wunsch, C. (1984). An estimate of the upwelling rate in the equatorial Atlantic based on the distribution of bomb radiocarbon and quasi-geostrophic dynamics. *JGR, J. Geophys. Res.* **89**, 7971–7978.

Wunsch, C., and Gill, A. E. (1976). Observations of equatorially trapped waves in Pacific sea level variations. *Deep-Sea Res.* **23**, 371–390.

Wyrtki, K. (1965). The annual and semiannual variation of sea surface temperature in the North Pacific Ocean. *Limnol. Oceanogr.* **10**, 307–313.

Wyrtki, K. (1973a). Teleconnections in the Equatorial Pacific Ocean. *Science* **180**, 66–68.

Wyrtki, K. (1973b). An equatorial jet in the Indian Ocean. *Science* **181**, 262–264.

Wyrtki, K. (1974). Sea level and seasonal fluctuations of the equatorial currents in the western Pacific Ocean. *J. Phys. Oceanogr.* **4**, 91–103.

Wyrtki, K. (1975). El Niño—The dynamic response of the equatorial Pacific Ocean to atmospheric forcing. *J. Phys. Oceanogr.* **5**, 572–584.

Wyrtki, K. (1977a). Sea level during the 1972 El Niño. *J. Phys. Oceanogr.* **7**, 779–787.

Wyrtki, K. (1977b). Advection in the Peru current as observed by satellite. *JGR, J. Geophys. Res.* **82**, 3939–3943.

Wyrtki, K. (1978a). Monitoring the strength of equatorial currents from XBT sections and sea level. *JGR, J. Geophys. Res.* **83**, 1935–1940.

Wyrtki, K. (1978b). Lateral oscillations of the Pacific Equatorial Countercurrent. *J. Phys. Oceanogr.* **8**, 530–532.

Wyrtki, K. (1979). The response of sea surface topography to the 1976 El Niño. *J. Phys. Oceanogr.* **9**, 1223–1231.

Wyrtki, K. (1981). An estimate of equatorial upwelling in the Pacific. *J. Phys. Oceanogr.* **11**, 1205–1214.

Wyrtki, K. (1982a). Eddies in the Pacific North Equatorial Current. *J. Phys. Oceanogr.* **12**, 746–749.

Wyrtki, K. (1982b). The Southern Oscillation, ocean–atmosphere interaction and El Niño. *Mar. Tech. Soc. J.* **16**, 3–10.

Wyrtki, K. (1984a). Monthly maps of sea level in the Pacific during El Niño of 1982 and 1983. *In* "Time Series of Ocean Measurements," Vol. 2, I.O.C. Tech. Ser., 25. UNESCO, Paris.

Wyrtki, K. (1984b). The slope of sea level along the equator during the 1982/1983 El Niño. *JGR, J. Geophys. Res.* **89**, 10419–10424.

Wyrtki, K. (1985). Water displacements in the Pacific and the genesis of El Niño cycles. *JGR, J. Geophys. Res.* **90**, 7129–7132.

Wyrtki, K., and Elden, G. (1982). Equatorial upwelling events in the central Pacific. *J. Phys. Oceanogr.* **12**, 984–988.

Wyrtki, K., and Kilonsky, B. (1984). Mean water and current structure during the Hawaii to Tahiti shuttle experiment. *J. Phys. Oceanogr.* **14**, 242–254.

Wyrtki, K., and Meyers, G. (1976). The trade wind field over the Pacific Ocean *J. Appl. Meteorol.* **15**, 698–704.

Wyrtki, K., Stroup, E., Patzert, W., Williams, R., and Quinn, W. (1976). Predicting and observing El Niño. *Science* **191**, 343–346.

Wyrtki, K., Firing, E., Halpern, D., Knox, R., McNally, G. J., Patzert, W. C., Stroup, E. D., Taft, B. A., and Williams, R. (1981). The Hawaii to Tahiti shuttle experiment. *Science* **211**, 22–28.

Yamagata, T. (1985). Stability of a simple air–sea coupled model in the tropics. *In* "Coupled Ocean–Atmosphere Models" (J. Nihoul, ed.), pp. 637–658, Elsevier, Amsterdam.

Yamagata, T., and Hayashi, Y. (1984). A simple diagnostic model for the 30–50 day oscillation in the tropics. *J. Meteorol. Soc. Jpn.* **62**, 709–717.

Yamagata, T., and Philander, S. G. H. (1985). The role of damped equatorial waves in the oceanic response to winds. *J. Oceanogr. Soc. Jpn.* **41**, 345–357.

Yanai, M., and Marayama, T. (1966). Stratospheric wave disturbances propagating over the equatorial Pacific. *J. Meteorol. Soc. Jpn.* **44**, 291–294.

Yasunari, T. (1980). A quasi-stationary appearance of 30 to 40 day period in the cloudiness fluctuations during the summer monsoon over India. *J. Meteorol. Soc. Jpn.* **58**, 225–229.

Yoon, H.-G. (1981). Effects of islands on equatorial waves. *JGR, J. Geophys. Res.* **86**, 10913–10920.

Yoon, J.-H., and Philander, S. G. H. (1982). The generation of coastal undercurrents. *J. Oceanogr. Soc. Jpn.* **38**, 215–224.

Yoshida, K. (1959). A theory of the Cromwell Current and equatorial upwelling. *J. Oceanogr. Soc. Jpn.* **15**, 154–170.

Zebiak, S. E. (1982a). A simple atmospheric model of relevance to El Niño. *J. Atmos. Sci.* **39**, 2017–2027.

Zebiak, S. E. (1982b). Atmospheric convergence feedback in a simple model for El Niño. *Mon. Weather Rev.* **114**, 1263–1271.

Zebiak, S. E., and Cane, M. A. (1987). A model ENSO. *Mon. Weather Rev.* **115**, 2262–2278.

Index

International Geophysics Series

Volume 1 BENO GUTENBERG. Physics of the Earth's Interior. 1959*

Volume 2 JOSEPH W. CHAMBERLAIN. Physics of the Aurora and Airglow. 1961*

Volume 3 S. K. RUNCORN (ed.). Continental Drift. 1962*

Volume 4 C. E. JUNGE. Air Chemistry and Radioactivity. 1963*

Volume 5 ROBERT G. FLEAGLE AND JOOST A. BUSINGER. An Introduction to Atmospheric Physics. 1963*

Volume 6 L. DUFOUR AND R. DEFAY. Thermodynamics of Clouds. 1963*

Volume 7 H. U. ROLL. Physics of the Marine Atmosphere. 1965*

Volume 8 RICHARD A. CRAIG. The Upper Atmosphere: Meteorology and Physics. 1965

*Out of print.

Volume 9 WILLIS L. WEBB. Structure of the Stratosphere and Mesosphere. 1966*

Volume 10 MICHELE CAPUTO. The Gravity Field of the Earth from Classical and Modern Methods. 1967

Volume 11 S. MATSUSHITA AND WALLACE H. CAMPBELL (eds.). Physics of Geomagnetic Phenomena. (In two volumes.) 1967

Volume 12 K. YA. KONDRATYEV. Radiation in the Atmosphere. 1969

Volume 13 E. PALMÉN AND C. W. NEWTON. Atmospheric Circulation Systems: Their Structure and Physical Interpretation. 1969

Volume 14 HENRY RISHBETH AND OWEN K. GARRIOTT. Introduction to Ionospheric Physics. 1969*

Volume 15 C. S. RAMAGE. Monsoon Meteorology. 1971*

Volume 16 JAMES R. HOLTON. An Introduction to Dynamic Meteorology. 1972*

Volume 17 K. C. YEH AND C. H. LIU. Theory of Ionospheric Waves. 1972

Volume 18 M. I. BUDYKO. Climate and Life. 1974

Volume 19 MELVIN E. STERN. Ocean Circulation Physics. 1975

Volume 20 J. A. JACOBS. The Earth's Core. 1975*

Volume 21 DAVID H. MILLER. Water at the Surface of the Earth: An Introduction to Ecosystem Hydrodynamics. 1977

Volume 22 JOSEPH W. CHAMBERLAIN. Theory of Planetary Atmospheres: An Introduction to Their Physics and Chemistry. 1978*

Volume 23 JAMES R. HOLTON. Introduction to Dynamic Meteorology, Second Edition. 1979

Volume 24 ARNETT S. DENNIS. Weather Modification by Cloud Seeding. 1980

Volume 25 ROBERT G. FLEAGLE AND JOOST A. BUSINGER. An Introduction to Atmospheric Physics, Second Edition. 1980

Volume 26 KUO-NAN LIOU. An Introduction to Atmospheric Radiation. 1980

Volume 27 DAVID H. MILLER. Energy at the Surface of the Earth: An Introduction to the Energetics of Ecosystems. 1981

Volume 28 HELMUT E. LANDSBERG. The Urban Climate. 1981

Volume 29 M. I. BUDYKO. The Earth's Climate: Past and Future. 1982

Volume 30 ADRAIN E. GILL. Atmosphere–Ocean Dynamics. 1982

Volume 31 PAOLO LANZANO. Deformations of an Elastic Earth. 1982

*Out of print.

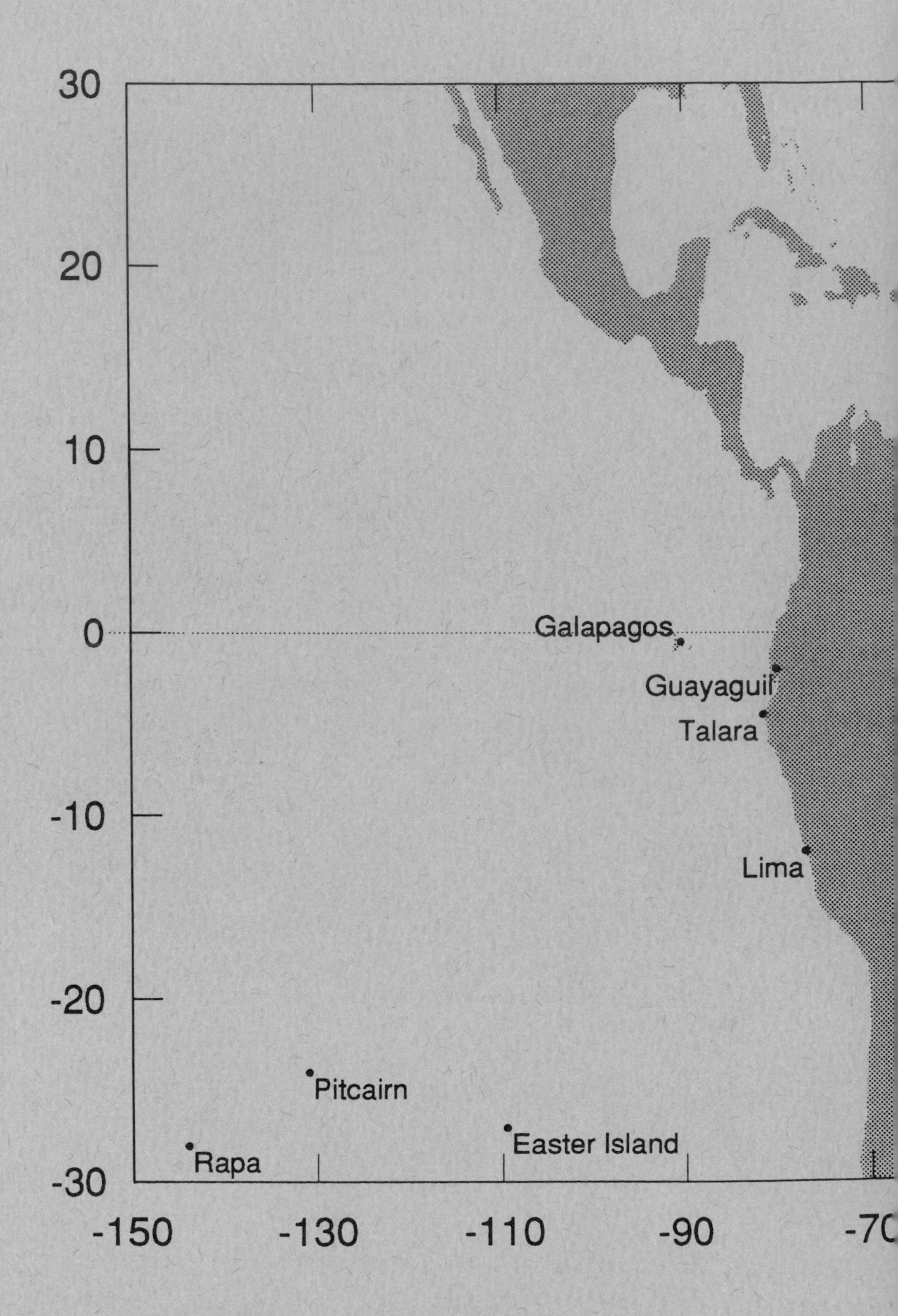

30
20
10
0
-10
-20
-30
-150
-130
-110
-90
-70
Galapagos
Guayaguil
Talara
Lima
Pitcairn
Easter Island
Rapa